BIOCHEMISTRY RESEARCH TRENDS

Additional books in this series can be found on Nova's website under the Series tab.

Additional E-books in this series can be found on Nova's website under the E-book tab.

BIOCHEMISTRY RESEARCH TRENDS

OLIGOSACCHARIDES: SOURCES, PROPERTIES AND APPLICATIONS

BIOCHEMISTRY RESEARCH TRENDS

OLIGOSACCHARIDES: SOURCES, PROPERTIES AND APPLICATIONS

NICOLE S. GORDON
EDITOR

Nova Science Publishers, Inc.
New York

For permission to use material from this book please contact us:
Telephone 631-231-7269; Fax 631-231-8175
Web Site: http://www.novapublishers.com

Additional color graphics may be available in the e-book version of this book.

LIBRARY OF CONGRESS CATALOGING-IN-PUBLICATION DATA

Oligosaccharides : sources, properties, and applications / editor, Nicole S. Gordon.
p. ; cm.
Includes bibliographical references and index.
ISBN 978-1-61122-193-0 (hardcover)
1. Oligosaccharides. I. Gordon, Nicole S.
[DNLM: 1. Oligosaccharides--physiology. 2. Oligosaccharides--biosynthesis. QU 83]
QP702.O44O455 2010
572'.565--dc22

2010038641

Published by Nova Science Publishers, Inc. † New York

CONTENTS

PREFACE

An oligosaccharide is a saccharide polymer containing a small number of component sugars, also known as simple sugars (monosaccharides). Oligosaccharides can have many functions; for example, they are commonly found on the plasma membrane of animal cells where they can play a role in cell-cell recognition. This book presents topical research data in the study of oligosaccharides, including the study of human and bovine milk oligosaccharides; the health benefits and commercial applications of prebiotic oligosaccharides; classification of oligosaccharides synthesized by enzymes; oligosaccharides from sucrose via glycansucrases; and the Mannan oligosaccharides in aquaculture.

Chapter 1- Mammalian milk/colostrum contains trace ~ 10% carbohydrate, of which lactose (Gal(β1-4)Glc) usually constitutes the largest amount [1, 2]. Exceptions to this are the milks of monotremes, marsupials and most Canoidea species of eutherians in which lactose is less prominent than a variety of oligosaccharides [1,2]. Human mature milk and colostrum contain 12 ~ 13 g/L and 22 ~ 24 g/L of oligosaccharides, respectively [3-5]. In contrast, bovine colostrum contains little more than 1 g/L of oligosaccharides and this concentration rapidly decreases after 48 hr post partum [6, 7]. Most human milk oligosaccharides (HMO) are resistant to digestion and absorption within the small intestine [8] and reach the infant colon, where they can act as prebiotics that stimulate the growth of beneficial microorganisms such as various species of Bifidobacterium. They can also act as receptor analogues that inhibit the attachment of harmful microorganisms to the infant's colonic mucosa. A small part of the milk oligosaccharides are absorbed intact into the circulation [9] and it has been hypothesized that these may act as immunomodulators. It is generally believed that bovine milk oligosaccharides are not absorbed by human adults or infants, thus making them available to be utilized as prebiotics or anti-infection materials. The colostrum of cows and other domestic farm animals is a potential source of free oligosaccharides, and oligosaccharides isolated from this natural source can be utilized as functional foods or animal feed stuffs on the industrial scale.

Studies on human and bovine milk oligosaccharides have been performed for over 50 years, especially in the field of structural analysis. At present studies on their biological functions are very active areas. In this chapter, the authors will introduce the history of structural studies on human milk oligosaccharides, on their fate within the gastrointestinal tract as well as on their possible biological functions as prebiotics, anti infection agents and immunomodulation factors. In addition, they will discuss the possibility of the commercial utilization of bovine milk oligosaccharides and those of other domestic farm animals as well as the future industrial utilization of milk-derived glycoproteins, glycopeptides and glycolipids as materials based on their biofunctional significance.

Chapter 2- In recent years, oligosaccharides and their derivatives have become useful for health applications in various fields because of their specific biological activities. In the food industries, several oligosaccharides have received increasing attention as key components for functional foods and nutraceutical products. Prebiotics are non-digestible oligosaccharides which have been shown to have properties that can modulate gastrointestinal problems and improve general health and well being. The benefits of prebiotic include relief of constipation, reduced risk of colon cancer, inhibition of pathogens in gastrointestinal tract, increased minerals absorption, immune modulation, short chain fatty acid and vitamin production, reduced blood cholesterol and lipids and improved microbial balance in the gut. Prebiotics are distinguished from other dietary fibers due to its ability to selectively promote fermentation by bifidobacteria and/or lactobacilli within the gut. The "prebiotics" was named by Gibson and Roberfroid in 1995 as "a non digestible food ingredient that beneficially affects the host by selectively stimulating the growth and/or activity of one or a limited number of bacteria in the colon, that can improve the host health". It was redefined several times until 2007 when FAO defined as a "non-viable food component that confers a health benefit on the host associated with modulation of the microflora.

Currently, commercial prebiotics are non-digestible but fermentable oligosaccharides which are consisted of sugar moieties between 2 and 10 saccharide units, with an exception of inulin and beta-glucans which are polysaccharides. Prebiotic oligosaccharides can be obtained from natural sources such as chicory and artichoke and produced through chemical synthesis (lactulose) or by enzymatic synthesis (hydrolysis or transfer reaction). Currently commercialized prebiotic oligosaccharides for food use include fructo-oligosaccharides, galacto-oligosaccharides, lactosucrose, isomalto-oligosaccharides, gentio-oligosaccharides, xylo-oligosaccharides and soybean oligosaccharides (whereas lactulose has been used as a laxative). Recommendation on prebiotics consumption is 3-10 g per day. Effectiveness on each type of prebiotic varies; for example, lactulose, lactosucrose and xylo-oligosaccharides have been reported more effective at low doses. Prebiotics are popular food ingredients used in functional foods and nutraceutical products such as low calories and mind sweetness for weight control or diabetes and for well being of gut health. Combination of prebiotics and probiotics (synbiotic products) offer synergistic benefits to the GI tract, such as in infant powder formula, beverage, yoghurt and dairy products. Functional claims of prebiotic depend on local law of each country, however, most country accepted as a source of dietary fiber.

Chapter 3- Hyperthermophilic microorganisms thrives at temperatures higher than 80°C and proteins and enzymes extracted from these sources are optimally stable and active in the presence of temperatures close to the boiling point of water and of other denaturants, i.e. chaotropic agents, pH, organic solvents, detergents, etc. Therefore, hyperstable enzymes are considered attractive alternatives in biocatalysis and in chemo-enzymatic synthesis. In addition, the molecular bases of the extreme stability to heat and to the ability to work optimally at high temperatures are not completely understood and intrigued biochemists, enzymologists, and biophysics in the last twenty years. In particular, hyperstable glycosidases, enzymes catalysing the hydrolysis of *O*- and *N*-glycosidic bonds, have been studied in detail as they are simple model systems promoting single-substrate reactions, and, more importantly, can be exploited for the enzymatic synthesis of oligosaccharides. The importance of these molecules increased enormously in recent years for their potential application in biomedicine. Hyperstable glycosidases, working in transglycosylation mode, can be excellent alternatives to the classical chemical methods helping in the control of regio- and stereoselectivity as conventional enzymes, but also resisting to the organics used in chemical synthesis. We will review here recent advances in the isolation and characterization of glycosidases from hyperthermophilic microorganisms and the methods used for their application in oligosaccharide synthesis.

Chapter 4- For oligosaccharide synthesis, an enzymatic process is preferred over the complicated multi-step reactions of chemical synthesis. The advantages of the enzymatic process are the regio- and stereo-selectivity, potential for large-scale synthesis, and also the environmental-friendly synthesis conditions. Two main classes of enzymes, the glycosidases and the glycosyltransferases, are exploited in the common approach for oligosaccharide synthesis via reverse hydrolysis and transglycosylation reactions. The glycosynthase class has recently been developed from specific mutations of glycosidases through site-directed mutagenesis to minimize the natural hydrolysis activity and thereby boost up the synthetic activity of the enzyme. In this chapter, oligosaccharides synthesized by enzymes are classified into three main types, the linear, cyclic and complex oligosaccharides. Our work on three major enzymes, comprised of two glycosyltransferases and one glycosidase, that are capable of synthesizing oligosaccharides is outlined. The use of bacterial cyclodextrin glycosyltransferase and amylomaltase in intermolecular transglucosylation to produce functional linear oligosaccharides is presented. In addition, these two enzymes, through the intramolecular transglucosylation reaction, are used for the synthesis of small and large cycloamyloses (cyclodextrins) that are widely used as stabilizers and solubilizers. For the complex-type oligosaccharides, the transglycosylation activity of endo-β-*N*-acetylglucosaminidase to yield the high mannose-GlcNAc-Glc oligosaccharides with potential applications in biological functions, such as therapeutic agents, is discussed. The chapter ends with concluding remarks on the commercially available oligosaccharides, with the emphasis on the significance of oligosaccharides and oligosaccharide-producing enzymes and their promising future.

Chapter 5- The majority of eukaryotes share a common N-linked oligosaccharide precursor, $Glc_3Man_9GlcNAc_2$, which is transferred to proteins during glycoprotein biosynthesis. This precursor is then sequentially processed in the endoplasmic reticulum (ER), creating a series of oligosaccharide structures that are recognized as codes by specific lectins and that inform of the folding status of the glycoprotein. For example, the chaperones/ lectins calnexin and calreticulin bind to monoglucosylated oligosaccharides after the excision of the two terminal glucose residues by glucosidases I and II. The last glucose and sometimes one or even two mannose residues are excised, and the resulting structures $Man_{7-9}GlcNAc_2$ are recognized by the lectins ERGIC53, VIP36 and others that are involved in transport of glycoproteins to the Golgi. These lectins associate with properly folded glycoproteins that can exit the ER. In contrast, N-linked oligosaccharides on misfolded glycoproteins are more extensively trimmed to $Man_{5-6}GlcNAc_2$. A high local concentration of ER mannosidase I in an ER-derived quality control compartment is mainly responsible for this trimming, with the possible participation of other mannosidases. $Man_{5-6}GlcNAc_2$ oligosaccharides are then recognized by the lectins OS9 and XTP3-B that target the misfolded glycoprotein for ER-associated degradation.

Chapter 6- Glycansucrases are a class of microbial enzymes that polymerize either the fructosyl or the glucosyl moiety of sucrose to give β-D-fructans or α-D-glucans. They are also capable of transferring fructosyl or glucosyl units to acceptor molecules to yield oligosaccharides. Although the glycosyl donor specificity is limited to sucrose and related sugars, the acceptor specificity is very broad, and includes numerous carbohydrates as well as non-carbohydrate molecules. This chapter describes the enzymes and the variety of oligosaccharide structures that result from their acceptor reactions. Applications can range from modified drugs to food ingredients. Although very few of the products have thus far been commercialized, the potential is being actively studied and shows great promise.

Chapter 7- Carbohydrates are widely distributed in nature, being responsible for different functions in almost all living organisms. They are present in the living systems as polysaccharides, glycoproteins, and glycolipids. Their properties and applications are dependent on their detailed structure, namely, their composition in monosaccharides, type of linkages, branching, and anomeric configurations. Mass spectrometry (MS) is a useful tool for the study of carbohydrates, especially when soft ionization methods are used. For the analysis of oligosaccharides, electrospray mass spectrometry (ESI-MS) and tandem mass spectrometry (ESI-MS^n) has been elected in the last recent years as the method of choice, having the advantage of using underivatised oligomers, even when present in mixtures and with low abundance.

The analysis of polysaccharides in their native form is difficult due to their high molecular weight. In order to overcome this, polysaccharides are usually cleaved into oligosaccharides.

In the present chapter the structural features of glucuronoxylans, galactomannans, and type-II arabinogalactans achieved by the analysis by ESI-MS of their oligosaccharides are discussed. In addition, it is shown how can the anomeric configurations of reducing disaccharides of glucose using quadrupole time-of-flight (Q-TOF2), linear ion trap (LIT), and triple quadrupole (QqQ) mass spectrometers be obtained.

Chapter 8- Biologically active oligosaccharides, referred to as oligosaccharins, are the specific group of complex carbohydrates that function in plant cells as molecular signals. Indeed, oligosaccharins participate in the regulation of growth, development, and survival in different environmental conditions. Significant research has helped to shed light on the oligosaccharin concept and to make progress in this field. For instance, various oligosaccharins have been structurally characterized, and for some of them, specific receptors have been discovered. Separate signaling events have also been elucidated. Currently, a number of scientific articles and comprehensive reviews present a large body of evidence supporting the idea of oligosaccharin existence and function. This review provides a brief overview of most recent information obtained about oligosaccharin functions, signal perception, their possible origins, and movement within the plant. The discussion is focused only on oligosaccharins that function in plant cells, either originating from the plant cell wall or from the fungal wall. Please note that there are a number of other biologically active carbohydrate-containing molecules, such as fragments of glycoproteins or lipo-oligosaccharides, which are not discussed in this review. Despite gaps in our knowledge of oligosaccharin function, there is no doubt that this group of signaling molecules plays distinct roles in plant cells. The identification of new structures and activities of oligosaccharins in a broader range of organisms, together with the elucidation of their signal transduction pathways, will broaden our understanding of the important roles of these molecules in different aspects of plant life and will yield valuable insight into how cells progress through developmental programs and respond to environmental changes.

Chapter 9- Increased concern over effects of antibiotics usage in aquaculture, on environment and human health, has prompted the search for alternative products. Recently, immunostimulants such as probiotics and prebiotics have shown promising results as preventive and environmentally friendly alternatives to the antibiotics. Among common immunostimulants used, mannan oligosaccharides (MOS) have received heightened attention in aquaculture. Since the first use of MOS in aquaculture, there has been increased number of studies demonstrating their ability to increase the survival, growth performance and control of the potential pathogens in fish and crustacean including, marron (*Cherax tenuimanus*), yabbies (*Cherax destructor*) and tropical rock lobsters (*Panulirus ornatus*). The chapter reviews the role of MOS on the culture of crayfish and fishes while detailing the effects of MOS on the growth performance, physiology and immune response of these aquatic animals. Suggestions for further research on the application of MOS in crayfish aquaculture are also included in the chapter.

Chapter 10- Unnatural oligosaccharides can be expected to exhibit new functions and applications in glycoscience such as potential as drug candidates. Enzymatic glycosylation is a useful tool for the preparation of oligosaccharides with well-defined structure. α-1,4-Glucosidic linkages can be prepared by an enzymatic polymerization through the successive phosphorylase-catalyzed glucosylation using α-D-glucose 1-phosphate as a glycosyl donor. Since enzymes often express loose specificity for recognition of substrate structures, extension of the phosphorylase-catalyzed glycosylation using different substrates is useful to obtain new unnatural oligosaccharides. In this chapter, on the basis of above viewpoints, the facile synthesis of unnatural oligosaccharides by the phosphorylase-catalyzed enzymatic glycosylations using new glycosyl donors of glycose 1-phosphates is described. It has been found that α-D-xylose, α-D-mannose, 2-deoxy-α-D-glucopyranose, 3- or 4-deoxy-α-D-glucose, and α-(*N*-formyl)-D-glucosamine 1-phosphates are recognized by phosphorylase as the glycosyl donor, to occur the transfer reaction of the corresponding sugar residues to maltooligosaccharides, giving the unnatural oligosaccharides. Consequently, construction of the new oligosaccharide chains containing the different units by the phosphorylase-catalysis probably leads to development on new applications of unnatural substrates, such as the drug candidate.

Chapter 11- It has been demonstrated that oligosaccharides have diverse bioactivities and functions. In plants, oligosaccharides act as early signal molecules and play an important role in the plant growth, development and morphogenesis as well as plant defense responses. Oligosaccharides are characterized as elicitors to trigger plant systemic acquired resistance against a variety of pathogens. Furthermore, oligosaccharides have been used to restore and maintain the intestinal balance by stimulating the growth and reproduction of probiotic strains such as *Lactobacillus* and *Bifidobacterium* and restraining the infection of pathogenic bacteria. In murine and human, the immunemodulatory activities of oligosaccharides which can stimulate the proliferation of splenocytes and the activation of macrophages are confirmed by many studies. Meanwhile, oligosaccharides significantly induce the accumulation of immune regulation factors and inhibit the growth of cancer cell, which act as assistant agents in treating immune system diseases such as hepatitis and cancer. Oligosaccharides also can regulate blood lipid, blood sugar and blood pressure. Therefore, these biological properties of oligosaccharides have made great contribution in the application of these oligosaccharides in agriculture, functional food and pharmaceutical products.ter

In: Oligosaccharides: Sources, Properties and Applications
Editor: Nicole S. Gordon
ISBN 978-1-61122-193-0

Chapter 1

Milk Oligosaccharides

Tadasu Urashima[1], Motomitsu Kitaoka[2], Takashi Terabayashi[3], Kenji Fukuda[1], Masao Ohnishi[4] and Akira Kobata[5]

[1]Graduate School of Animal Hygiene, Obihiro University of Agriculture and Veterinary Medicine, Obihiro, Hokkaido, 080-8555, Japan

[2]National Food Research Institute, National Agriculture and Food Research Organization, Tsukuba, Ibaraki, 305-8642, Japan

[3]Department of Chemistry, Faculty of Science, Kitasato University, Sagamihara, Kanagawa, 042-778-9484, Japan

[4]Department of Food Science, Obihiro University of Agriculture & Veterinary Medicine, Obihiro, Hokkaido, 080-8555, Japan

[5]The Noguchi Institute, Itabashi, Tokyo, 173-0003, Japan

1. Introduction

Mammalian milk/colostrum contains trace ~ 10% carbohydrate, of which lactose (Gal(β1-4)Glc) usually constitutes the largest amount [1, 2]. Exceptions to this are the milks of monotremes, marsupials and most Canoidea species of eutherians in which lactose is less prominent than a variety of oligosaccharides [1,2]. Human mature milk and colostrum contain 12 ~ 13 g/L and 22 ~ 24 g/L of oligosaccharides, respectively [3-5]. In contrast, bovine colostrum contains little more than 1 g/L of oligosaccharides and this concentration rapidly decreases after 48 hr post partum [6, 7]. Most human milk oligosaccharides (HMO) are resistant to digestion and absorption within the small intestine [8] and reach the infant colon, where they can act as prebiotics that stimulate the growth of beneficial microorganisms such as various species of Bifidobacterium. They can also act as receptor analogues that inhibit the attachment of harmful microorganisms to the infant's colonic mucosa. A small part of the milk oligosaccharides are absorbed intact into the circulation [9] and it has been hypothesized that these may act as immunomodulators. It is generally believed that bovine milk oligosaccharides are not absorbed by human adults or infants, thus making them available to be utilized as prebiotics or anti-infection materials. The colostrum of cows and other domestic farm animals is a potential source of free oligosaccharides, and oligosaccharides isolated from this natural source can be utilized as functional foods or animal feed stuffs on the industrial scale.

Studies on human and bovine milk oligosaccharides have been performed for over 50 years, especially in the field of structural analysis. At present studies on their biological functions are very active areas. In this chapter, we will introduce the history of structural studies on human milk oligosaccharides, on their fate within the gastrointestinal tract as well as on their possible biological functions as prebiotics, anti infection agents and immunomodulation factors. In addition, we will discuss the possibility of the commercial utilization of bovine milk oligosaccharides and those of other domestic farm animals as well as the future industrial utilization of milk-derived glycoproteins, glycopeptides and glycolipids as materials based on their biofunctional significance.

2. History of Structural Analysis of Human Milk Oligosaccharides

In 1933 Polonovski and Lespagnol observed the presence of nitrogen -containing oligosaccharides in human milk and designated them as 'gynolactose' [10]. At that time they attempted to produce crystals from these oligosaccharides, but did not succeed. In 1954 Polonovski and Montreuil found that 'gynolactose' was a mixture containing more than ten oligosaccharides and that only some of them contained nitrogen, as clarified by observations using paper chromatography [11]. This finding was supported by the studies of Richard Kuhn et al. that were aimed at clarifying the structure of the Bifidobacterium growth stimulating factor in milk.

This growth factor was relevant to the field of paediatric nutrition because of the finding by Tissier and Moro that the feces of breast-fed infants were more acidic than those of bottle-fed infants [12, 13]. Thereafter Grulee et al. reported that diarrhea, otitis and respiratory diseases less frequently affected breast-fed infants than bottle-fed infants [14], upon which the finding by Tissier and Moro received more attention. Paul György, who was a professor of paediatrics at Pennsylvania University, believed that the high acidity of the feces of breast-fed infants was due to the action of lactobacilli present in their colonic microflora. It was suggested that these lactobacilli digest milk carbohydrates, producing large amounts of lactic and acetic acids; the resulting acidity would inhibit the growth of harmful bacteria within the infant colon. Schönfeld found that the growth stimulating factor for *Lactobacillus bifidus* var. pennsylvanicus, one of the lactic acid producing bacteria, occurs in the human milk whey fraction and termed it bifidus factor [15]. Richard Kuhn, who was in the Max Plank Institute in Heidelberg, Germany, collaborated with Paul György in order to clarify the structures of the actual components of the bifidus factor in human milk and found that they consist of a variety of oligosaccharides [16-18]. By 1965 they had characterized 14 oligosaccharides structures as shown in Table 1 [19-31].

All these oligosaccharides were found to contain a lactose unit at their reducing ends. It was then shown that some of them had the haptenic activities of H antigen, Le^a antigen or Le^b antigen [27, 32], which are related to human blood groups, and they were utilized to characterize the epitope structures of these blood group antigens [33].

Table 1. Human milk oligosaccharides characterized until 1965

Name	Structures	References
2'-Fucosyllactose (2'-FL)	Fuc(α1-2)Gal(β1-4)Glc	19
3-Fucosyllactose (3-FL)	Gal(β1-4)Glc Fuc(α1-3)	20
Lactodifucotetraose(LD)	Fuc(α1-2)Gal(β1-4)Glc Fuc(α1-3)	21
Lacto-N-tetraose(LNT)	Gal(β1-3)GlcNAc(β1-3)Gal(β1-4)Glc	22
Lacto-N-neotetraose(LNnT)	Gal(β1-4)GlcNAc(β1-3)Gal(β1-4)Glc	23
Lacto-N-fucopentaoseI (LNFP I)	Fuc(α1-2)Gal(β1-3)GlcNAc(β1-3)Gal(β1-4)Glc	24
Lacto-N-fucopentaoseII (LNFP II)	Gal(β1-3)GlcNAc(β1-3)Gal(β1-4)Glc Fuc(α1-4)	25
Lacto-N-difucohexaoseI (LNDFH I)	Fuc(α1-2)Gal(β1-3)GlcNAc(β1-3)Gal(β1-4)Glc Fuc(α1-4)	26
Lacto-N-difucohexaoseII (LNDFH II)	Gal(β1-3)GlcNAc(β1-3)Gal(β1-4)Glc Fuc(α1-4) Fuc(α1-3)	27
3'-Sialyllactose(3'SL)	Neu5Ac(α2-3)Gal(β1-4)Glc	28
6'-Sialyllactose(6'SL)	Neu5Ac(α2-6)Gal(β1-4)Glc	29
LST a	Neu5Ac(α2-3)Gal(β1-3)GlcNAc(β1-3)Gal(β1-4)Glc	30
LST b	Neu5Ac(α2-6) Gal(β1-3)Gal(β1-3)GlcNAc(β1-3)Gal(β1-4)Glc	30
LST c	Neu5Ac(α2-6)Gal(β1-4)GlcNAc(β1-3)Gal(β1-4)Glc	31

The finding that the presence or absence of fucosyl oligosaccharides is related to the donor's blood group status assisted in the clarification of the biochemical basis of non-secretor and Lewis negative phenomena

In 1967 Grollman and Ginsburg made the interesting finding that milk samples collected from non-secretors do not contain Fuc(α1-2)Gal(β1-4)Glc (2'-fucosyllactose) [34]. A non-secretor is a donor whose ABO gene expresses the ABO blood group antigens on her erythrocyte cell membranes, but does not express these antigens in soluble glycoproteins which are biosynthesized by mucus epithelial cells and secreted into body fluids. To extend this notable finding, Kobata et al. [35] developed a method for profiling the oligosaccharides present in each milk sample. Using this technique they studied 50 milk samples and found that there were three different oligosaccharide patterns. All the oligosaccharides listed in Table 1 were found in the milk of around 80% donors, as shown in Figure 1, left, whereas four oligosaccharides were absent from the milk of 15% donors as in Figure 1, middle. On the other hand, three oligosaccharides were not detected in the milk of 5% of donors (Figure 1, right).

A notable finding was that the pattern as shown in Figure 1, middle, was always observed in milk samples from non-secretors. The fucosyl oligosaccharides Fuc(α1-2)Gal(β1-4)[Fuc(α1-3)]Glc (LDFT), Fuc(α1-2)Gal(β1-3)GlcNAc(β1-3)Gal(β1-4)Glc (LNFP-I) and Fuc(α1-2)Gal(β1-3)[Fuc(α1-4)]GlcNAc(β1-3)Gal(β1-4)Glc (LNDFH-I) as well as 2'-FL were always absent from the non-secretor milk. Since these four oligosaccharides contain the Fuc(α1-2)Gal unit at their non-reducing termini, this suggested that the lactating mammary glands of non-secretors do not have the activity of the fucosyltransferase that catalyzes the synthesis of this unit. This suggestion was supported by the enzymological study of Shen et al. [36].

The pattern as in Figure 1, right was always observed in milk collected from Lewis negative donors that do not express Le^a and Le^b antigens on their erythrocyte cell membrane or in soluble secreted glycoproteins. The oligosaccharides that were not detected were Gal(β1-3)[Fuc(α1-4)]GlcNAc(β1-3)Gal(β1-4)Glc (LNFP-II), Fuc(α1-2)Gal(β1-3)[Fuc(α1-4)]GlcNAc(β1-3)Gal(β1-4)Glc (LNDFH-I) and Gal(β1-3)[Fuc(α1-4)]GlcNAc(β1-3)Gal(β1-4)[Fuc(α1-3)]Glc (LNDFH-II), suggesting that the activity of another fucosyltransferase, which catalyzes synthesis of the Fuc(α1-4)GlcNAc unit, is absent from the mammary glands. This suggestion was also confirmed using enzymatic methods [37].

Kobata et al. succeeded in synthesizing oligosaccharides containing the blood group A antigenic determinant (GalNAc(α1-3)[Fuc(α1-2)]Gal) or group B antigenic determinant (Gal(α1-3)[Fuc(α1-2)]Gal) using 2'-FL or LNFP-I as an acceptor and UDP-GalNAc or UDP-Gal as a donor, and using the milk from blood groups A and AB individuals or from blood groups B and AB individuals as the enzyme source, respectively [38-40]. Interestingly, the enzymatic activities were detected irrespective of the secretor status of the donors. These results indicated that the absence of A or B antigen in the body fluids of non-secretors was due to the absence of the H antigen (Fuc(α1-2)Gal), which is a precursor to the biosynthesis of A or B antigen.

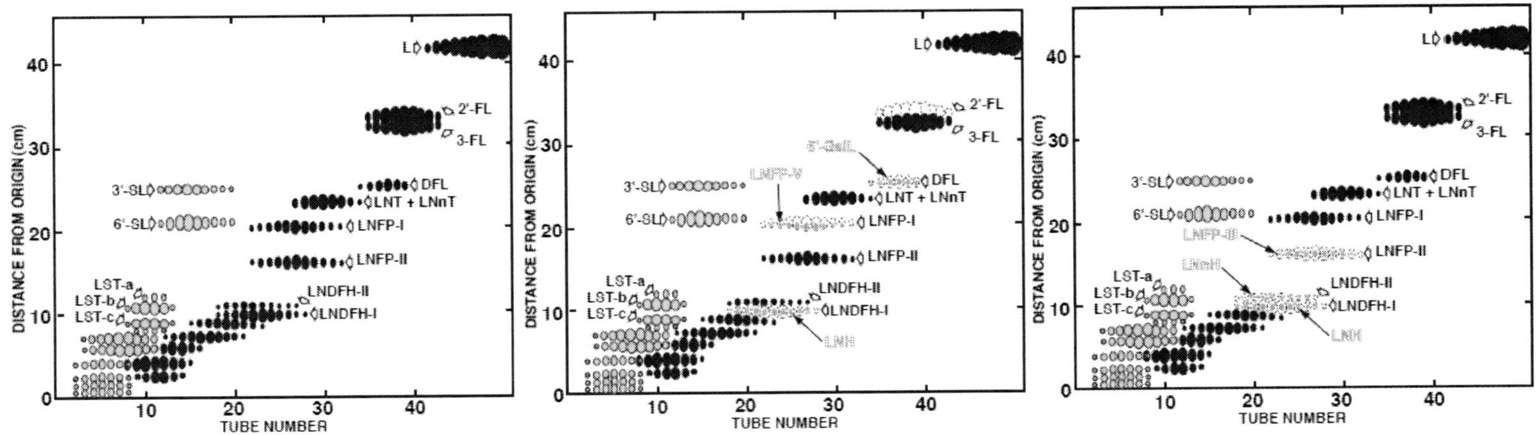

Revised from the figure in *Glycoconj. J.* 17, 443-464 (2000) with permission from Springer Science+Business Media.

Figure 1. Fingerprintings of human milk oligosaccharides. Fraction numbers as indicated by "tube number" in abscissa were obtained by Sephadex G-25 column chromatography of human milk oligosaccharide fraction of an individual donor. Aliquots of the fractions were spotted at the origin of a sheet of filter paper, and subjected to chromatography using ethyl acetate/pyridine/acetic acid/water (5:5:1:3) as solvent. Black spots represent oligosaccharides visualized by alkaline-$AgNO_3$ reagent, and hatched ones encircled by black line represent those detected by both alkaline-$AgNO_3$ reagent and thiobarbituric acid reagent. Three typical patterns are shown in the figures, left, middle and right. Names and structures of the oligosaccharides indicated by abbreviations with white arrows are listed in Table 1 ~ Table 3. Spots indicated by dotted lines in the figures of middle and right are those missing in the fingerprints.

Structural Characterization of Novel Minor Oligosaccharides in the Milk of Non-Secretor or Lewis Negative Donor Milk

The structures of novel minor oligosaccharides, indicated by the arrows in Figs. 1, middle and right, were subsequently elucidated as shown in Table 2 [41-45]. Among them, Gal(β1-4)[Fuc(α1-3)]GlcNAc(β1-3)Gal(β1-4)Glc (LNFP-III) has been utilized to clarify the role of Le^{x}, which is a selectin ligand, in studies on the homing of lymphocyte and migration of cancer cells through blood vessels [46], while Gal(β1-3)GlcNAc(β1-3)[Gal(β1-4)GlcNAc(β1-6)]Gal(β1-4)Glc (LNH) and Gal(β1-4)GlcNAc(β1-3)[Gal(β1-4)GlcNAc(β1-6)]Gal(β1-4)Glc (LN*n*H) have been useful in structural studies of branched type ABO antigens [47, 48].

Table 2. Human milk oligosaccharides separated from non-secretor or Lewis negative donors

Name	Structures	References
6'-Galactosyllactose (6'-GalL)	Gal(b1-6)Gal(β1-4)Glc	41
Lacto-*N*-fucopentaose V (LNFP V)	Gal(b1-3)GlcNAc(β1-3)Gal(β1-4)Glc Fuc(β1-3)	42
Lacto-*N*-fucopentaose III (LNFP III)	Gal(b1-4)GlcNAc(b1-3)Gal(β1-4)Glc Fuc(β1-3)	43
Lacto-*N*-hexaose (LNH)	Gal(β1-4)GlcNAc(β1-6) \ Gal(β1-4)Glc Gal(β1-3)GlcNAc(β1-3) /	44
Lacto-*N*-neohexaose (LN*n*H)	Gal(β1-4)GlcNAc(β1-6) \ Gal(β1-4)Glc Gal(β1-4)GlcNAc(β1-3) /	45

Before the structures of these oligosaccharides had been characterized, it was thought that lactose, Gal(β1-3)GlcNAc(β1-3)Gal(β1-4)Glc (LNT) and Gal(β1-4)GlcNAc(β1-3)Gal(β1-4)Glc (LN*n*T) were core units of HMO. At that stage, LNH and LN*n*H were then added as other core units, and their fucosyl and sialyl derivatives were characterized as well.

For structural studies of these HMO, sequence analysis with tritium labeling [49] and successive glycosidase digestions [50, 51] and highly sensitive methylation analysis for oligosaccharides, containing amino sugars [52], were developed; these analytical methods proved to be powerful new techniques for subsequent studies on the structural characterization of carbohydrate moieties of glycoproteins. It has also been shown that physicochemical methods, such as mass spectrometric analyses with several ionization methods, and NMR, are valuable for these structural studies. In addition, recycling chromatography as developed by Donald has been found to be useful for the isolation of specific oligosaccharides [53]. It is also worth noting that high performance liquid chromatography (HPLC) using triethyl amine as an ion pair reagent [54-56], or affinity chromatography with a monoclonal antibody against a specific epitope such as sialyl Le^{a} [57-62] are now available for the separation or purification of oligosaccharides.

Additional Core Units O\of Human Milk Oligosaccharides

Having been stimulated by the fact that several HMO have been utilized to clarify the relationship between function and structure of carbohydrate moieties of cell surface glycoconjugates, structural studies of higher molecular HMO were performed.

Structures of the fucosyl and sialyl derivatives of *para*-lacto-*N*-hexaose [63], *para*-lacto-*N*-*neo*hexaose [63], lacto-*N*-octaose [64] and lacto-*N*-*neo*octaose [64] were characterized by Kobata et al. In addition, structures of oligosaccharides, with *iso*-lacto-*N*-octaose [65], *para*-lacto-*N*-octaose [66], lacto-*N*-decaose [67] and lacto-*N*-*neo*decaose [68] (which will be provisionally named lacto-*N*-*neodecaose* in this review) as core units were characterized by other groups; 13 core units in total have been proposed as shown in Table 3 until now.

Since all of these core structures have lactose units at their reducing ends, it is believed that HMOs are synthesized by the collaborative actions of several glycosyltransferases. The biosynthesis of the sugar moieties of glycoproteins and glycolipids of epithelial cells, including mammary cells, is thought to proceed in a similar manner. As shown in Figure 2, it is considered that iGnT [69] activity produces GlcNAc(β1-3)Gal(β1-4)Glc by transfer of GlcNAc to lactose, the latter being synthesized by lactose synthase, a complex of α-lactalbumin and β4GalT. As shown in the left hand part of Figure 3, it is assumed that the linear core units of HMO are synthesized by the successive actions of β4GalT [70] and iGnT on the above trisaccharide unit; the products are *neo* type saccharides which contain the repeating unit of Gal(β1-4)GlcNAc(β1-3). If, however, a non reducing Gal(β1-3)GlcNAc unit is formed by the action of iGnT in combination with β3GalT [71] instead of β4GalT, further elongation at the non reducing end will stop. Further, it is considered that branch type core units are synthesized by the action of IGnT [72] (see Figure 2 and the right hand part of Figure 3); this enzyme is known to synthesize branch structures in glycoconjugates.

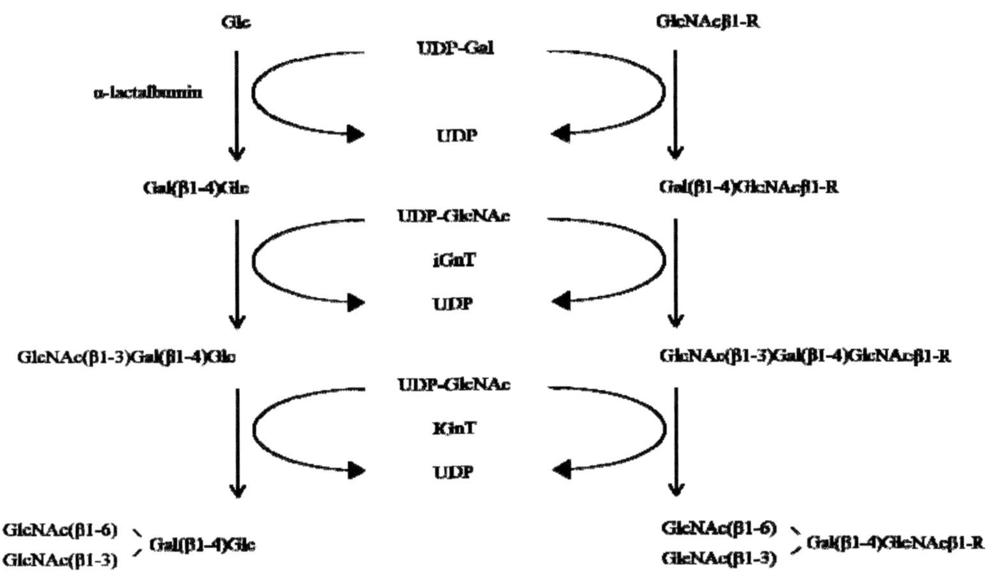

Revised from the figure in *Chang Gung Medical J.* 26, 620-636 (2003).

Figure 2. The elongation of sugar chains by the action of iGnT or IGnT starting from lactose or the non-reducing *N*-acetyllactosamine group of the glycoconjugates. R represents aglycons, including the glycoconjugates at the reducing end of the *N*-acetyllactosamine group.

Addition of fucose and/or sialic acid to these 13 core units can produce a great variety of oligosaccharides. To date, 115 structures have been characterized (see Table 4), and around 200 oligosaccharides have been detected using microfluidic HPLC – chip mass spectrometry as described by Ninonuevo et al. [73]. Recently Amano et al. [68] purified HMO using anion exchange chromatography, *Aleurai aurantia* lectin affinity chromatography and reverse phase HPLC after derivatization of the fractions; they characterized 22 oligosaccharides, whose core units are lacto-*N*-decaose or lacto-*N*-*neo*decaose, with Matrix-assisted laser desorption/ionization quadrupole ion trap time-of-flight mass sepectrometry (MALDI-QIT-TOFMS) in negative ion mode. It can be expected that further novel oligosaccharides will be characterized using advances in mass spectrometry. Although only one tetrafucosyl derivative [66] of *para*-lacto-*N*-octaose has been characterized so far, additional fucosyl or sialyl derivatives of this core oligosaccharide will presumably be characterized in the near future.

Table 3. The 13 core structures of human milk oligosaccharides

Structures	*Name*
Gal(β1-4)Glc	Lactose
Gal(β1-3)GlcNAc(β1-3)Gal(β1-4)Glc	Lacto-*N*-tetraose
Gal(β1-4)GlcNAc(β1-3)Gal(β1-4)Glc	Lacto-*N*-neotetraose
Gal(β1-4)GlcNAc(β1-6)↘ Gal(β1-4)Glc Gal(β1-3)GlcNAc(β1-3)↗	Lacto-*N*-hexaose
Gal(β1-4)GlcNAc(β1-6)↘ Gal(β1-4)Glc Gal(β1-4)GlcNAc(β1-3)↗	Lacto-*N*-neohexaose
Gal(β1-3)GlcNAc(β1-3)Gal(β1-4)GlcNAc(β1-3)Gal(β1-4)Glc	*para*-Lacto-*N*-hexaose
Gal(β1-4)GlcNAc(β1-3)Gal(β1-4)GlcNAc(β1-3)Gal(β1-4)Glc	*para*-Lacto-*N*-neohexaose
Gal(β1-4)GlcNAc(β1-3)Gal(β1-4)GlcNAc(β1-6)↘ Gal(β1-4)Glc Gal(β1-3)GlcNAc(β1-3)↗	Lacto-*N*-octaose
Gal(β1-3)GlcNAc(β1-3)Gal(β1-4)GlcNAc(β1-6)↘ Gal(β1-4)Glc Gal(β1-4)GlcNAc(β1-3)↗	Lacto-*N*-neooctaose
Gal(β1-3)GlcNAc(β1-3)Gal(β1-4)GlcNAc(β1-6)↘ Gal(β1-4)Glc Gal(β1-3)GlcNAc(β1-3)↗	*iso*-Lacto-*N*-octaose
Gal(β1-3)GlcNAc(β1-3)Gal(β1-4)GlcNAc(β1-3)Gal(β1-4)GlcNAc(β1-3)Gal(β1-4)Glc	*para*-Lacto-*N*-octaose
Gal(β1-4)GlcNAc(β1-6)↘ Gal(β1-4)GlcNAc(β1-6)↘ Gal(β1-3)GlcNAc(β1-3)↗ Gal(β1-4)Glc Gal(β1-3)GlcNAc(β1-3)↗	Lacto-*N*-decaose
Gal(β1-4)GlcNAc(β1-6)↘ Gal(β1-4)GlcNAc(β1-6)↘ Gal(β1-4)GlcNAc(β1-3)↗ Gal(β1-4)Glc Gal(β1-3)GlcNAc(β1-3)↗	Lacto-*N*-neodecaose

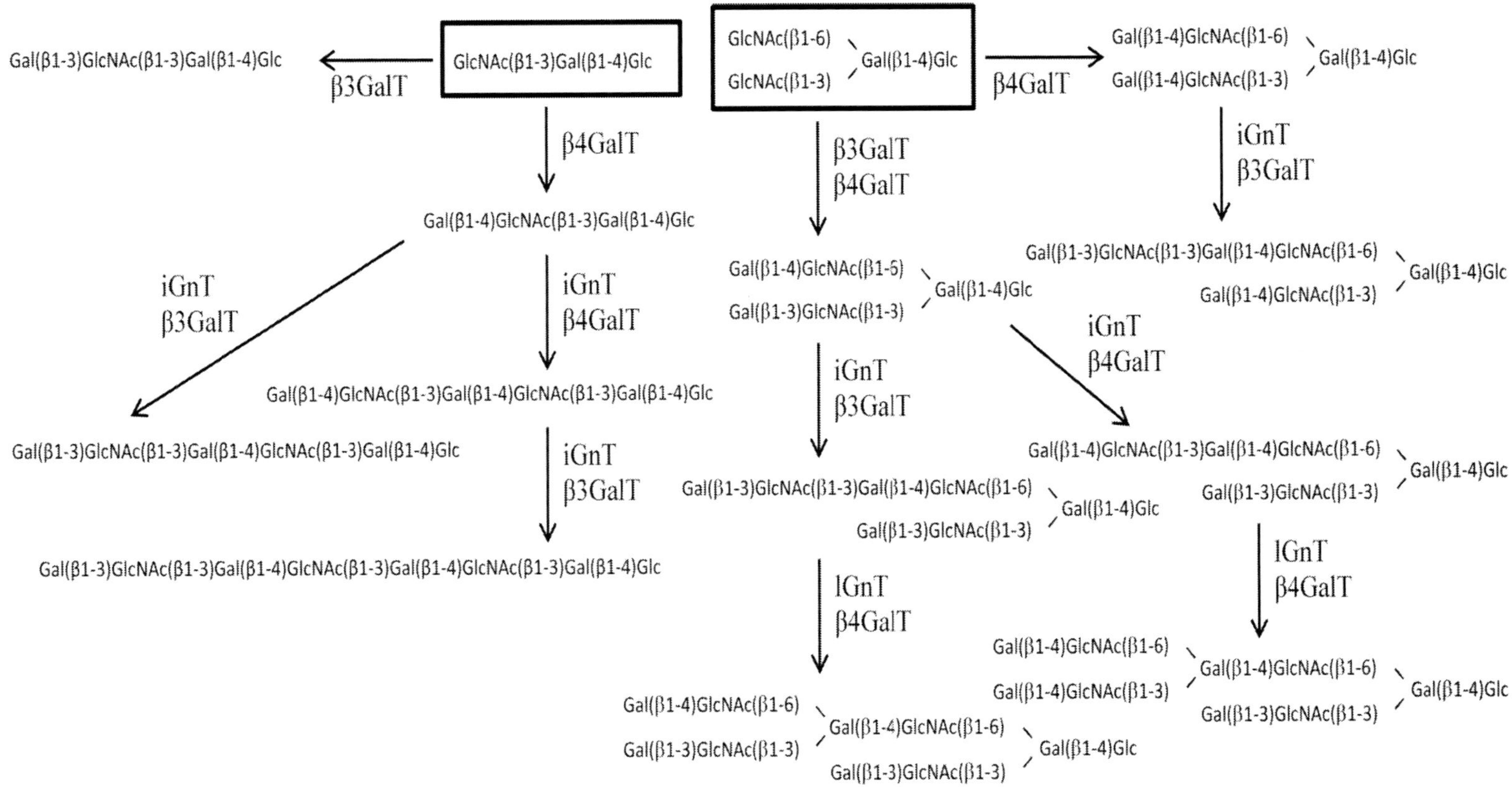

Figure 3. The biosynthetic pathway of several human milk oligosaccharides, which starts from GlcNAc(β1-3)Gal(β1-4)Glc or GlcNAc(β1-3)[GlcNAc(β1-6)]Gal(β1-4)Glc.

Table 4. Human milk oligosaccharides

No.	*Oligosaccharide*	*Structure*	*References*
Neutral oligosaccharides			
1	2'-FL	Fuc(α1-2)Gal(β1-4)Glc	19
2	3-FL	Gal(β1-4)Glc | Fuc(α1-3)	20
3	β3'-GL	Gal(β1-3)Gal(β1-4)Glc	53
4	β4'-GL	Gal(β1-4)Gal(β1-4)Glc	240
5	β6'-GL	Gal(β1-6)Gal(β1-4)Glc	41
6	LNTri II	GlcNAc(β1-3)Gal(β1-4)Glc	241
7	DF-L(LDFT)	Fuc(α1-2)Gal(β1-4)Glc | Fuc(α1-3)	21
8	LNT	Gal(β1-3)GlcNAc(β1-3)Gal(β1-4)Glc	22
9	LN*n*T	Gal(β1-4)GlcNAc(β1-3)Gal(β1-4)Glc	23
10	LNFP I	Fuc(α1-2)Gal(β1-3)GlcNAc(β1-3)Gal(β1-4)Glc	24
11	LNFP II	Gal(β1-3)GlcNAc(β1-3)Gal(β1-4)Glc | Fuc(α1-4)	25
12	LNFP III	Gal(β1-4)GlcNAc(β1-3)Gal(β1-4)Glc | Fuc(α1-3)	43
13	LNFP V	Gal(β1-3)GlcNAc(β1-3)Gal(β1-4)Glc | Fuc(α1-3)	42
14	LNDFH I	Fuc(α1-2)Gal(β1-3)GlcNAc(β1-3)Gal(β1-4)Glc | Fuc(α1-4)	26
15	LNDFH II	Gal(β1-3)GlcNAc(β1-3)Gal(β1-4)Glc | | Fuc(α1-4) Fuc(α1-3)	27
16	LNDFH III	Gal(β1-4)GlcNAc(β1-3)Gal(β1-4)Glc | | Fuc(α1-3) Fuc(α1-3)	53
17	LNH	Gal(β1-4)GlcNAc(β1-6) \\ Gal(β1-4)Glc Gal(β1-3)GlcNAc(β1-3) /	44
18	LN*n*H	Gal(β1-4)GlcNAc(β1-6) \\ Gal(β1-4)Glc Gal(β1-4)GlcNAc(β1-3) /	45

No.	*Oligosaccharide*	*Structure*	*References*
19	F-LNH I	Gal(β1-4)GlcNAc(β1-6) \\ Gal(β1-4)Glc Fuc(α1-2)Gal(β1-3)GlcNAc(β1-3) /	48
20	F-LNH II	Fuc(α1-3) Gal(β1-4)GlcNAc(β1-6) \\ Gal(β1-4)Glc Gal(β1-3)GlcNAc(β1-3) /	242
21	F-LN*n*H	Fuc(α1- { Gal(β1-4)GlcNAc(β1-6) \\ Gal(β1-4)Glc Gal(β1-4)GlcNAc(β1-3) /	45
22	F-*para*-LNH I	Gal(β1-3)GlcNAc(β1-3)Gal(β1-4)GlcNAc(β1-3)Gal(β1-4)Glc Fuc(α1-3)	243
23	F-*para*-LNH II	Gal(β1-3)GlcNAc(β1-3)Gal(β1-4)GlcNAc(β1-3)Gal(β1-4)Glc Fuc(α1-4)	244
24	DF-LNH II	Fuc(α1-3) Gal(β1-4)GlcNAc(β1-6) \\ Gal(β1-4)Glc Gal(β1-3)GlcNAc(β1-3) / Fuc(α1-4)	242
25	DF-LNH I	Fuc(α1-3) Gal(β1-4)GlcNAc(β1-6) \\ Gal(β1-4)Glc Fuc(α1-2)Gal(β1-3)GlcNAc(β1-3) /	40
26	DF-LN*n*H	Fuc(α1-3) Gal(β1-4)GlcNAc(β1-6) \\ Gal(β1-4)Glc Gal(β1-4)GlcNAc(β1-3) / Fuc(α1-3)	66
27	DF-*para*-LNH	Gal(β1-3)GlcNAc(β1-3)Gal(β1-4)GlcNAc(β1-3)Gal(β1-4)Glc Fuc(α1-4) Fuc(α1-3)	63
28	DF-*para*-LN*n*H	Gal(β1-4)GlcNAc(β1-3)Gal(β1-4)GlcNAc(β1-3)Gal(β1-4)Glc Fuc(α1-3) Fuc(α1-3)	63
29	TF-LNH	Fuc(α1-3) Gal(β1-4)GlcNAc(β1-6) \\ Gal(β1-4)Glc Fuc(α1-2)Gal(β1-3)GlcNAc(β1-3) / Fuc(α1-4)	245

Table 4 (continued)

No.	*Oligosaccharide*	*Structure*	*References*
30	TF-*para*-LNH I	Fuc(α1-2)Gal(β1-3)GlcNAc(β1-3)Gal(β1-4)GlcNAc(β1-3)Gal(β1-4)Glc Fuc(α1-4) Fuc(α1-3)	246
31	TF-*para*-LNH II	Gal(β1-3)GlcNAc(β1-3)Gal(β1-4)GlcNAc(β1-3)Gal(β1-4)Glc Fuc(α1-4) Fuc(α1-3) Fuc(α1-3)	244
32	TF-*para*-LN*n*H	Gal(β1-4)GlcNAc(β1-3)Gal(β1-4)GlcNAc(β1-3)Gal(β1-4)Glc Fuc(α1-3) Fuc(α1-3) Fuc(α1-3)	244
33	F-LNO	Fuc(α1-3) Gal(β1-4)GlcNAc(β1-3)Gal(β1-4)GlcNAc(β1-6) \\ Gal(β1-4)Glc Gal(β1-3)GlcNAc(β1-3) /	247
34	F-LN*n*O	Fuc(α1-3) Gal(β1-3)GlcNAc(β1-3)Gal(β1-4)GlcNAc(β1-6) \\ Gal(β1-4)Glc Gal(β1-4)GlcNAc(β1-3) /	247
35	F-*iso*-LNO	Fuc(α1-3) Gal(β1-3)GlcNAc(β1-3)Gal(β1-4)GlcNAc(β1-6) \\ Gal(β1-4)Glc Gal(β1-3)GlcNAc(β1-3) /	248
36	DF-LNO I	Fuc(α1-3) Fuc(α1-3) Gal(β1-4)GlcNAc(β1-3)Gal(β1-4)GlcNAc(β1-6) \\ Gal(β1-4)Glc Gal(β1-3)GlcNAc(β1-3) /	64
37	DF-LNO II	Fuc(α1-3) Gal(β1-4)GlcNAc(β1-3)Gal(β1-4)GlcNAc(β1-6) \\ Gal(β1-4)Glc Gal(β1-3)GlcNAc(β1-3) / Fuc(α1-4)	64
38	DF-LN*n*O I	Fuc(α1-4) Fuc(α1-3) Gal(β1-3)GlcNAc(β1-3)Gal(β1-4)GlcNAc(β1-6) \\ Gal(β1-4)Glc Gal(β1-4)GlcNAc(β1-3) /	64
39	DF-LN*n*O II	Fuc(α1-3) Gal(β1-3)GlcNAc(β1-3)Gal(β1-4)GlcNAc(β1-6) \\ Gal(β1-4)Glc Gal(β1-4)GlcNAc(β1-3) / Fuc(α1-3)	64

Table 4 (continued)

No.	*Oligosaccharide*	*Structure*	*References*
40	DF-*iso*-LNO I	Fuc(α1-4) Fuc(α1-3) Gal(β1-3)GlcNAc(β1-3)Gal(β1-4)GlcNAc(β1-6) \\ Gal(β1-4)Glc Gal(β1-3)GlcNAc(β1-3) /	249
41	DF-*iso*-LNO II	Fuc(α1-3) Gal(β1-3)GlcNAc(β1-3)Gal(β1-4)GlcNAc(β1-6) \\ Gal(β1-4)Glc Fuc(α1-2)Gal(β1-3)GlcNAc(β1-3) /	249
42	TF-LNO	Fuc(α1-3) Fuc(α1-3) Gal(β1-4)GlcNAc(β1-3)Gal(β1-4)GlcNAc(β1-6) \\ Gal(β1-4)Glc Gal(β1-3)GlcNAc(β1-3) / Fuc(α1-4)	64
43	TF-LN*n*O	Fuc(α1-4) Fuc(α1-3) Gal(β1-3)GlcNAc(β1-3)Gal(β1-4)GlcNAc(β1-6) \\ Gal(β1-4)Glc Gal(β1-4)GlcNAc(β1-3) / Fuc(α1-3)	64
44	TF-*iso*-LNO I	Fuc(α1-3) Fuc(α1-2)Gal(β1-3)GlcNAc(β1-3)Gal(β1-4)GlcNAc(β1-6) \\ Gal(β1-4)Glc Fuc(α1-2)Gal(β1-3)GlcNAc(β1-3) /	65
45	TF-*iso*-LNO II	Fuc(α1-4) Fuc(α1-3) Gal(β1-3)GlcNAc(β1-3)Gal(β1-4)GlcNAc(β1-6) \\ Gal(β1-4)Glc Gal(β1-3)GlcNAc(β1-3) / Fuc(α1-4)	248
46	Tetra-F-*iso*-LNO	Fuc(α1-3) Fuc(α1-2)Gal(β1-3)GlcNAc(β1-3)Gal(β1-4)GlcNAc(β1-6) \\ Gal(β1-4)Glc Fuc(α1-2)Gal(β1-3)GlcNAc(β1-3) / Fuc(α1-4)	66
47	Tetra-F-*para*-LNO	Fuc(α1-2)Gal(β1-3)GlcNAc(β1-3)Gal(β1-4)GlcNAc(β1-3)Gal(β1-4)GlcNAc(β1-3)Gal(β1-4)Glc Fuc(α1-4) Fuc(α1-3) Fuc(α1-3)	66

Table 4 (continued)

No.	*Oligosaccharide*	*Structure*	*References*
48	Penta-F-*iso*-LNO	Fuc(α1-4) Fuc(α1-3) Fuc(α1-2)Gal(β1-3)GlcNAc(β1-3)Gal(β1-4)GlcNAc(β1-6) \\ Gal(β1-4)Glc Fuc(α1-2)Gal(β1-3)GlcNAc(β1-3) / Fuc(α1-4)	66
49	F-LND I	Fuc(α1-3) Gal(β1-4)GlcNAc(β1-6) \\ Gal(β1-4)GlcNAc(β1-6) \\ Gal(β1-3)GlcNAc(β1-3) / Gal(β1-4)Glc Gal(β1-3)GlcNAc(β1- 3) /	67
50	DF-LND I	Fuc(α1-3) Gal(β1-4)GlcNAc(β1-6) \\ Gal(β1-4)GlcNAc(β1-6) \\ Gal(β1-3)GlcNAc(β1-3) / Gal(β1-4)Glc Fuc(α1-2)Gal(β1-3)GlcNAc(β1-3) /	68
51	DF-LND II	Gal(β1-4)GlcNAc(β1-6) \\ Gal(β1-4)GlcNAc(β1-6) \\ Gal(β1-3)GlcNAc(β1-3) / Gal(β1-4)Glc Gal(β1-3)GlcNAc(β1-3) / Fuc(α1-4) Fuc(α1-4)	68
52	DF-LND III	Fuc(α1-3) Gal(β1-4)GlcNAc(β1-6) \\ Gal(β1-4)GlcNAc(β1-6) \\ Gal(β1-3)GlcNAc(β1-3) / Gal(β1-4)Glc Gal(β1-3)GlcNAc(β1-3) / Fuc(α1-4)	68
53	DF-LND IV	Fuc(α1-3) Gal(β1-4)GlcNAc(β1-6) \\ Gal(β1-4)GlcNAc(β1-6) \\ Gal(β1-3)GlcNAc(β1-3) / Gal(β1-4)Glc Gal(β1-3)GlcNAc(β1-3) / Fuc(α1-4)	68
54	DF-LND V	Fuc(α1-3) Gal(β1-4)GlcNAc(β1-6) \\ Gal(β1-4)GlcNAc(β1-6) \\ Fuc(α1-2)Gal(β1-3)GlcNAc(β1-3) / Gal(β1-4)Glc Gal(β1-3)GlcNAc(β1-3) /	68

Table 4 (continued)

No.	*Oligosaccharide*	*Structure*	*References*
55	DF-LND VI	Gal(β1-4)GlcNAc(β1-6) ╲ ╱Gal(β1-4)GlcNAc(β1-6) ╲ Gal(β1-3)GlcNAc(β1-3) ╱ ╱Gal(β1-4)Glc | Fuc(α1-4) Fuc(α1-2)Gal(β1-3)GlcNAc(β1-3)	68
56		Gal(β1-4)GlcNAc(β1-6) ╲ ╱Gal(β1-4)GlcNAc(β1-6) ╲ Fuc(α1-2)Gal(β1-3/ 4)GlcNAc(β1-3) ╱ ╱Gal(β1-4)Glc | Fuc(α1-4/ 3) Gal(β1-3)GlcNAc(β1-3)	68
57		Gal(β1-4)GlcNAc(β1-6) ╲ ╱Gal(β1-4)GlcNAc(β1-6) ╲ Gal(β1-3)GlcNAc(β1-3) ╱ ╱Gal(β1-4)Glc Fuc(α1-2)Gal(β1-3/4)GlcNAc(β1-3) | Fuc(α1-4/ 3)	68
58	TriF-LND I	Fuc(α1-3) | Gal(β1-4)GlcNAc(β1-6) ╲ ╱Gal(β1-4)GlcNAc(β1-6) ╲ Fuc(α1-2)Gal(β1-3)GlcNAc(β1-3) ╱ ╱Gal(β1-4)Glc Gal(β1-3)GlcNAc(β1-3) | Fuc(α1-4)	68
59	TriF-LND II	Fuc(α1-3) | Gal(β1-4)GlcNAc(β1-6) ╲ ╱Gal(β1-4)GlcNAc(β1-6) ╲ Fuc(α1-2)Gal(β1-3)GlcNAc(β1-3) ╱ ╱Gal(β1-4)Glc Fuc(α1-2)Gal(β1-3)GlcNAc(β1-3)	68
60	TriF-LND III	Fuc(α1-3) | Gal(β1-4)GlcNAc(β1-6) ╲ ╱Gal(β1-4)GlcNAc(β1-6) ╲ Gal(β1-3)GlcNAc(β1-3) ╱ ╱Gal(β1-4)Glc | Gal(β1-3)GlcNAc(β1-3) Fuc(α1-4) | Fuc(α1-4)	68
61	TriF-LND IV	Fuc(α1-3) | Gal(β1-4)GlcNAc(β1-6) ╲ ╱Gal(β1-4)GlcNAc(β1-6) ╲ Gal(β1-3)GlcNAc(β1-3) ╱ ╱Gal(β1-4)Glc | Fuc(α1-4) Fuc(α1-2)Gal(β1-3)GlcNAc(β1-3)	68

Table 4 (continued)

No.	*Oligosaccharide*	*Structure*	*References*
62		Gal(β1-4)GlcNAc(β1-6) \\ Gal(β1-4)GlcNAc(β1-6) \\ Fuc(α1-2)Gal(β1-3/4)GlcNAc(β1-3) / Gal(β1-4)Glc | Fuc(α1-4/3) Fuc(α1-2)Gal(β1-3)GlcNAc(β1-3) /	68
63		Fuc(α1-3) | Gal(β1-4)GlcNAc(β1-6) \\ Gal(β1-4)GlcNAc(β1-6) \\ Fuc(α1-2)Gal(β1-3/4)GlcNAc(β1-3) / Gal(β1-4)Glc | Gal(β1-3)GlcNAc(β1-3) / Fuc(α1-4/3)	68
64	TetraF-LND I	Fuc(α1-3) | Gal(β1-4)GlcNAc(β1-6) \\ Gal(β1-4)GlcNAc(β1-6) \\ Fuc(α1-2)Gal(β1-3)GlcNAc(β1-3) / Gal(β1-4)Glc Fuc(α1-2)Gal(β1-3)GlcNAc(β1-3) / | Fuc(α1-4)	68
65	TetraF-LND II	Fuc(α1-3) | Gal(β1-4)GlcNAc(β1-6) \\ Gal(β1-4)GlcNAc(β1-6) \\ Gal(β1-3)GlcNAc(β1-3) / Gal(β1-4)Glc | Fuc(α1-4) Fuc(α1-2)Gal(β1-3)GlcNAc(β1-3) / | Fuc(α1-4)	68
66	TetraF-LND III	Fuc(α1-3) | Gal(β1-4)GlcNAc(β1-6) \\ Gal(β1-4)GlcNAc(β1-6) \\ Fuc(α1-2)Gal(β1-3)GlcNAc(β1-3) / Gal(β1-4)Glc | Fuc(α1-4) Fuc(α1-2)Gal(β1-3)GlcNAc(β1-3) /	68
67		Gal(β1-4)GlcNAc(β1-6) \\ Gal(β1-4)GlcNAc(β1-6) \\ Fuc(α1-2)Gal(β1-3/4)GlcNAc(β1-3) / Gal(β1-4)Glc | Fuc(α1-4/3) Fuc(α1-2)Gal(β1-3/4)GlcNAc(β1-3) / | Fuc(α1-4/3)	68

Table 4 (continued)

No.	*Oligosaccharide*	*Structure*	*References*
68	F-LN*n*D I	Fuc(α1-3) Gal(β1-4)GlcNAc(β1-6) \\ Gal(β1-4)GlcNAc(β1-3) / Gal(β1-4)GlcNAc(β1-6) \\ Gal(β1-3)GlcNAc(β1-3) / Gal(β1-4)Glc	68
69	F-LN*n*D II	Gal(β1-4)GlcNAc(β1-6) \\ Fuc(α1-2)Gal(β1-3)GlcNAc(β1-3) / Gal(β1-4)GlcNAc(β1-6) \\ Gal(β1-3)GlcNAc(β1-3) / Gal(β1-4)Glc	68
70	DF-LN*n*D	Fuc(α1-3) Gal(β1-4)GlcNAc(β1-6) \\ Gal(β1-4)GlcNAc(β1-3) / Gal(β1-4)GlcNAc(β1-6) \\ Fuc(α1-2)Gal(β1-3)GlcNAc(β1-3) / Gal(β1-4)Glc	68
Acidic oligosaccharides			
71	3'SL	Neu5Ac(α2-3)Gal(β1-4)Glc	28
72	6'SL	Neu5Ac(α2-6)Gal(β1-4)Glc	29
73	F-SL	Neu5Ac(α2-3)Gal(β1-4)Glc Fuc(α1-3)	54
74	LST a	Neu5Ac(α2-3)Gal(β1-3)GlcNAc(β1-3)Gal(β1-4)Glc	30
75	LST b	Neu5Ac(α2-6) Gal(β1-3)GlcNAc(β1-3)Gal(β1-4)Glc	30
76	LST c	Neu5Ac(α2-6)Gal(β1-4)GlcNAc(β1-3)Gal(β1-4)Glc	30
77	F-LST a	Neu5Ac(α2-3)Gal(β1-3)GlcNAc(β1-3)Gal(β1-4)Glc Fuc(α1-4)	250
78	F-LST b	Neu5Ac(α2-6) Fuc(α1-2)Gal(β1-3)GlcNAc(β1-3)Gal(β1-4)Glc	250
79	F-LST c	Neu5Ac(α2-6)Gal(β1-4)GlcNAc(β1-3)Gal(β1-4)Glc Fuc(α1-3)	251
80	S-LNH	Neu5Ac(α2-6)Gal(β1-4)GlcNAc(β1-6) \\ Gal(β1-3)GlcNAc(β1-3) / Gal(β1-4)Glc	44
81	S-LN*n*H I	Neu5Ac(α2-6)Gal(β1-4)GlcNAc(β1-6) \\ Gal(β1-4)GlcNAc(β1-3) / Gal(β1-4)Glc	45

Table 4 (continued)

No.	*Oligosaccharide*	*Structure*	*References*
82	S-LN*n*H II	Gal(β1-4)GlcNAc(β1-6) \\ Gal(β1-4)Glc Neu5Ac(α2-6)Gal(β1-4)GlcNAc(β1-3) /	252
83	FS-LNH	Neu5Ac(α2-6)Gal(β1-4)GlcNAc(β1-6) \\ Gal(β1-4)Glc Fuc(α1-2)Gal(β1-3)GlcNAc(β1-3) /	48
84	FS-LNH I	Fuc(α1-3) | Gal(β1-4)GlcNAc(β1-6) \\ Gal(β1-4)Glc Gal(β1-3)GlcNAc(β1-3) / | Neu5Ac(α2-6)	56
85	FS-LNH II	Fuc(α1-3) | Gal(β1-4)GlcNAc(β1-6) \\ Gal(β1-4)Glc Neu5Ac(α2-3)Gal(β1-3)GlcNAc(β1-3) /	56
86	FS-LNH III	Neu5Ac(α2-6)Gal(β1-4)GlcNAc(β1-6) \\ Gal(β1-4)Glc Gal(β1-3)GlcNAc(β1-3) / | Fuc(α1-4)	56
87	FS-LNH IV	Gal(β1-4)GlcNAc(β1-6) \\ Gal(β1-4)Glc Neu5Ac(α2-3)Gal(β1-3)GlcNAc(β1-3) / | Fuc(α1-4)	58
88	FS-LN*n*H I	Fuc(α1-3) | Gal(β1-4)GlcNAc(β1-6) \\ Gal(β1-4)Glc Neu5Ac(α2-6)Gal(β1-4)GlcNAc(β1-3) /	54
89	FS-LN*n*H II	Fuc(α1- [Neu5Ac(α2-6)Gal(β1-4)GlcNAc(β1-6) \\ Gal(β1-4)Glc Gal(β1-4)GlcNAc(β1-3) /]	45
90	DFS-LNH I	Neu5Ac(α2-6)Gal(β1-4)GlcNAc(β1-6) \\ Gal(β1-4)Glc Fuc(α1-2)Gal(β1-3)GlcNAc(β1-3) / | Fuc(α1-4)	56
91	DFS-LNH II	Fuc(α1-3) | Gal(β1-4)GlcNAc(β1-6) \\ Gal(β1-4)Glc Neu5Ac(α2-3)Gal(β1-3)GlcNAc(β1-3) / | Fuc(α1-4)	58

Table 4 (continued)

No.	*Oligosaccharide*	*Structure*	*References*
92	DFS-LN*n*H	Fuc(α1-3) Fuc(α1-2)Gal(β1-4)GlcNAc(β1-6) ⟍ Gal(β1-4)Glc Neu5Ac(α2-6)Gal(β1-4)GlcNAc(β1-3) ⟋	56
93	FS-LNO	Gal(β1-4)GlcNAc(β1-3)Gal(β1-4)GlcNAc(β1-6) ⟍ Gal(β1-4)Glc Neu5Ac(α2-3)Gal(β1-3)GlcNAc(β1-3) ⟋ Fuc(α1-4)	62
94	FS-*iso*-LNO	Gal(β1-3)GlcNAc(β1-3)Gal(β1-4)GlcNAc(β1-6) ⟍ Gal(β1-4)Glc Neu5Ac(α2-3)Gal(β1-3)GlcNAc(β1-3) ⟋ Fuc(α1-4)	60
95	DFS-*iso*-LNO I	Fuc(α1-3) Gal(β1-3)GlcNAc(β1-3)Gal(β1-4)GlcNAc(β1-6) ⟍ Gal(β1-4)Glc Neu5Ac(α2-3)Gal(β1-3)GlcNAc(β1-3) ⟋ Fuc(α1-4)	60
96	DFS-*iso*-LNO II	Fuc(α1-4) Gal(β1-3)GlcNAc(β1-3)Gal(β1-4)GlcNAc(β1-6) ⟍ Gal(β1-4)Glc Neu5Ac(α2-3)Gal(β1-3)GlcNAc(β1-3) ⟋ Fuc(α1-4)	60
97	DFS-LNO	Fuc(α1-3) Gal(β1-4)GlcNAc(β1-3)Gal(β1-4)GlcNAc(β1-6) ⟍ Gal(β1-4)Glc Neu5Ac(α2-3)Gal(β1-3)GlcNAc(β1-3) ⟋ Fuc(α1-4)	60
98	TFS-*iso*-LNO	Fuc(α1-3) Fuc(α1-2)Gal(β1-3)GlcNAc(β1-3)Gal(β1-4)GlcNAc(β1-6) ⟍ Gal(β1-4)Glc Neu5Ac(α2-3)Gal(β1-3)GlcNAc(β1-3) ⟋ Fuc(α1-4)	62
99	DS-LNT	Neu5Ac(α2-6) Neu5Ac(α2-3)Gal(β1-3)GlcNAc(β1-3)Gal(β1-4)Glc	253
	FDS-LNT I	Neu5Ac(α2-6) Neu5Ac(α2-3)Gal(β1-3)GlcNAc(β1-3)Gal(β1-4)Glc Fuc(α1-4)	61

Table 4 (continued)

No.	*Oligosaccharide*	*Structure*	*References*
101	DS-LNT II	Neu5Ac(α2-6) Neu5Ac(α2-3)Gal(β1-3)GlcNAc(β1-3)Gal(β1-4)Glc Fuc(α1-3)	55
102	DS-LNH I	Neu5Ac(α2-6)Gal(β1-4)GlcNAc(β1-6) \ Gal(β1-4)Glc Neu5Ac(α2-3)Gal(β1-3)GlcNAc(β1-3) /	61
103	DS-LNH II	Gal(β1-4)GlcNAc(β1-6) \ Gal(β1-4)Glc Neu5Ac(α2-3)Gal(β1-3)GlcNAc(β1-3) / Neu5Ac(α2-6)	61
	DS-LN*n*H	Neu5Ac(α2-6)Gal(β1-4)GlcNAc(β1-6) \ Gal(β1-4)Glc Neu5Ac(α2-6)Gal(β1-4)GlcNAc(β1-3) /	56
	FDS-LNH I	Fuc(α1-2)Gal(β1-4)GlcNAc(β1-6) \ Gal(β1-4)Glc Neu5Ac(α2-3)Gal(β1-3)GlcNAc(β1-3) / Neu5Ac(α2-6)	254
	FDS-LNH II	Fuc(α1-3) Gal(β1-4)GlcNAc(β1-6) \ Gal(β1-4)Glc Neu5Ac(α2-3)Gal(β1-3)GlcNAc(β1-3) / Neu5Ac(α2-6)	254
	FDS-LNH III	Neu5Ac(α2-6)Gal(β1-4)GlcNAc(β1-6) \ Gal(β1-4)Glc Neu5Ac(α2-3)Gal(β1-3)GlcNAc(β1-3) / Fuc(α1-4)	58
	FDS-LN*n*H	Neu5Ac(α2-3/α2-6)Gal(β1-4)GlcNAc(β1-6) \ Gal(β1-4)Glc Neu5Ac(α2-3/α2-6)Gal(β1-4)GlcNAc(β1-3) / Fuc(α1-3)	254
	TS-LNH	Neu5Ac(α2-6)Gal(β1-4)GlcNAc(β1-6) \ Gal(β1-4)Glc Neu5Ac(α2-3)Gal(β1-3)GlcNAc(β1-3) / Neu5Ac(α2-6)	255
	-	Fuc(α1-4) Neu5Ac(α2-3)Gal(β1-3)GlcNAc	59
	-	Fuc(α1-4) Neu5Ac(α2-3)Gal(β1-3)GlcNAc(β1-3)Gal	59

Table 4 (continued)

No.	*Oligosaccharide*	*Structure*	*References*
112	FS-*novo*-LNP I	Fuc(α1-3) I Gal(β1-4)GlcNAc(β1-6) \\ Gal(β1-4)Glc Neu5Ac(α2-3)Gal(β1-3) /	56
113	DF-*para*-LNH sulfate I	6S I Fuc (α1-2)Gal(β1-3)GlcNAc(β1-3)Gal(β1-4)GlcNAc(β1-3)Gal(β1-4)Glc I Fuc(α1-3)	256
114	DF-*para*-LNH sulfate II	6S I Gal(β1-3)GlcNAc(β1-3)Gal(β1-4)GlcNAc(β1-3)Gal(β1-4)Glc I I Fuc(α1-4) Fuc(α1-3)	256
115	TF-*para*-LNH sulfate	6S I Fuc (α1-2)Gal(β1-3)GlcNAc(β1-3)Gal(β1-4)GlcNAc(β1-3)Gal(β1-4)Glc I I Fuc(α1-4) Fuc(α1-3)	256

F, fucose; L, lactose; S, sialyl; DF, difucosyl; DS, disialyl; TS, trisialyl; FS, fucosyl sialyl; DFS, difucosyl sialyl; TFS, trifucosyl sialyl; FDS, fucosyl disialyl; DGal, digalactosyl; FL fucosyllactose; GL, galactosyllactose; LDFT, lacto difucotetraose; LNT, lacto-*N*-tetraose; LN*n*T, lacto-*N*-neotetraose; LNFP, lacto-*N*-fucopentaose; LNP, lacto-*N*-pentaose; LNDFH, lacto-*N*-difucohexaose; LNH, lacto-*N*-hexaose; LN*n*H, lacto-*N*-neohexaose; LNO, lacto-*N*-octaose;LN*n*O, lacto-*N*-neooctaose; LND, lacto-*N*-decaos; LN*n*D, lacto-*N*-neodecaose

It may be worth noting that lacto-*N*-octaose and lacto-*N*-*neo*octaose as such have failed to be detected [66], even though their fucosyl and/or sialyl derivatives have been characterized. Further studies may reveal whether this failure is due to a defect in the characterization technique or to a loss in a step in the fractionation of HMO.

3. Quantitative Aspects of Human Milk Oligosaccharides

Although the structures of HMO have been investigated for more than 50 years as described above, studies on the quantification of each of the oligosaccharides, in terms of their concentration in human colostrum and mature milk, have only recently been undertaken. Milk oligosaccharides can be quantified using reverse – phase or normal – phase HPLC subsequent to pre- or post – column labeling techniques. Derivatizations are often performed by condensation with 2-aminopyridine (PA) [74], 1-methyl-3-phenyl-5-pyrazolone (PMP) [74, 75], anthranilic acid [76, 77], or benzoic anhydride [78] to the reducing end of the sugar aldehyde. They can also be quantified by high pH anion exchange chromatography (HPEAC-

PAD) without derivatization [79, 80, 81]. Sialyl oligosaccharides can be quantified by capillary electrophoresis using phosphate buffer containing SDS [82, 83].

When the concentrations of representative neutral HMO including 2'-FL, Gal(β1-4)[Fuc(α1-3)]Glc (3-FL), LDFT, LNT, LN*n*T, LNFP-I, LNFP-II, LNFP-III, LNDFH-I or LNDFH-II and of representative acidic HMO including Neu5Ac(α2-6)Gal(β1-4)Glc (6'-SL), Neu5Ac(α2-3)Gal(β1-4)Glc (3'-SL), Neu5Ac(α2-3)Gal(β1-3)GlcNAc(β1-3)Gal(β1-4)Glc (LST-a), Gal(β1-3)[Neu5Ac(α2-6)]GlcNAc(β1-3)Gal(β1-4)Glc (LST-b), Neu5Ac(α2-6)Gal(β1-4)GlcNAc(β1-3)Gal(β1-4)Glc (LST-c), Neu5Ac(α2-3)Gal(β1-3)[Neu5Ac(α2-6)]GlcNAc(β1-3)Gal(β1-4)Glc (DSLNT), Fuc(α1-2)Gal(β1-3)[Neu5Ac(α2-6)]GlcNAc(β1-3)Gal(β1-4)Glc (sLNFP-I) and Neu5Ac(α2-3)Gal(β1-3)[Fuc(α1-4)]GlcNAc(β1-3)Gal(β1-4)Glc (sLNFP-II) were quantified in the colostra of Japanese women, the predominant oligosaccharides were shown to be 2'-FL, LNFP-I, LNDFH-I and LNT, in that order [74, 75]. Assuming that the total oligosaccharide concentration in colostrum was 22 ~ 24 g/L [3], then these four oligosaccharides can be calculated to constitute 25% to 33% of the total. It has been shown that the above four HMO are also predominant oligosaccharides in human mature milk [78, 79]. The concentrations of each of the acidic oligosaccharides in colostra or in mature milk were lower than those of the four neutral oligosaccharides; the most predominant acidic oligosaccharide was LST-c, followed by DSLNT, 6'-SL, 3'-SL and LST-a in that order [79].

The values determined for the concentration of each neutral [74, 78, 79, 80, 84] and acidic oligosaccharide [4, 75, 83, 85] vary between studies. These variations may be due to differences between the quantification methods used and the values can depend on the donors' ethnicity and on the lactation stage at which the milk samples were obtained. It is generally considered that an internal standard should be added to the milk samples prior to any separation and fractionation steps; comparisons can then be made between the peak area of the internal standard and that of each oligosaccharide in HPLC, HPEAC-PAD or capillary electrophoresis. It is of great interest to compare the different quantification methods in order to establish which yields the best reproducibility of estimation for each peak area, and which shows a reliable change in peak area in a dose dependent manner.

It is notable that three/fourth of the prominent oligosaccharides in human milk or colostra are of type I, which contain Gal(β1-3)GlcNAc (lacto-*N*-biose I) at their non-reducing termini [74]. The concentration of LNT, a type I oligosaccharide, was found to be 3 to 4 times higher than that of LN*n*T, which is a type II oligosaccharide containing the Gal(β1-4)GlcNAc (*N*-acetyllactosamine) group at its non-reducing terminus [74]. Urashima et al. have been characterizing the milk oligosaccharides of many mammalian species; they have found that the milk or colostra of most of these species contain only type II oligosaccharides [1], notable exceptions being chimpanzee, bonobo and orangutan whose milk or colostrum contains both types I and II. However, type II oligosaccharides predominate over type I in the milk or colostra of these apes [86], suggesting that the predominance of type I milk oligosaccharides is a characteristic feature of human milk. As will be described later, this predominance of type I is of interest in relation to the formation of the bifidus flora in the infant colon.

Differences between ethnic groups have been found with respect to the ratios between the concentrations of individual HMO's. Leo et al. have quantified the concentrations of representative oligosaccharides in the milk of Samoan women [76, 77]. They found no large differences between Samoan and Japanese women with respect to the concentrations of acidic oligosaccharides in milk/colostrum, but among the neutral oligosaccharides LNT was the most prominent in the milk of Samoan women, and most of their milk samples did not contain 2'-FL or LNFP-I. It was concluded that most Samoan donors are non-secretors, similar to only 15% of European and Asian donors, including Japanese. These observations could mean that a small group of non-secretors had traveled to Samoa from South East Asia a long time ago; later a small number of secretor people arrived and mixed with the Samoan ancestors. These considerations thus suggest that the quantification of milk oligosaccharide concentrations may be used as a possible approach for ethnology studies.

4. Gastrointestinal Digestion and Absorption of Milk Oligosaccharides

When infants consume milk, the free lactose therein is split into galactose and glucose by intestinal lactase (neutral β-galactosidase, lactose phlorizin hydrolase), an enzyme that is located in the membrane of the microvilli of the brush border of the small intestine. The two monosaccharides are transported into the enterocytes by a specific mechanism, whereupon the glucose enters the circulation and is used as an energy source while most of the galactose is converted to glucose in the liver, to be used as an energy source as well.

Much less is known about the exact metabolic fate of human milk oligosaccharides. They are resistant to enzymatic hydrolysis by the intestinal lactase of the brush border [8] and there is evidence that the major part of oligosaccharides through the small intestine without degradation and enters the colon where they are fermented by colonic bacteria [87, 88]. Evidently, the brush border of the small intestine does not contain enzymes, such as sialidase, fucosidase or N-acetylglucosaminidase that can remove sialic acid, fucose, N-acetylglucosamine residues, respectively, from the lactose or other core units of the milk oligosaccharides. A small fraction of human milk oligosaccharides is absorbed intact, perhaps by receptor-mediated endocytosis [9], some of which are excreted in the urine. It is unclear what proportion and exactly which of the consumed milk oligosaccharides are absorbed, but there is an evidence suggesting that circulating oligosaccharides may have immunological effects on endothelial cells [89].

It has been of interest to investigate whether the sialic acid of sialylated milk oligosaccharides can be absorbed and utilized as a precursor for the biosynthesis of brain gangliosides and sialoglycoproteins. Rat milk contains significant amounts of sialyllactose [90], which can be hydrolyzed to sialic acid and lactose by a very active small intestinal sialidase that is present in suckling rats. Since this enzyme has a low pH optimum and is absent from the brush border, it is probably of lysosomal, i.e., intracellular origin [91]. An intracellular location of this sialidase implies that the ingested sialyllactose has to be transferred into the enterocytes before it can be digested within lysosomes or supernuclear vacuoles; the most likely mechanism for this transfer is pinocytosis or endocytosis.

It would also be of interest to investigate whether, and to what extent, other monosaccharides, such as fucose, the major constituent of the neutral milk oligosaccharides, can similarly be absorbed and used as biosynthetic precursors. Rudloff et al. [92] recently fed 2 or 4 g of ^{13}C galactose to 11 breast-feeding women immediately after breakfast and determined the amount of ^{13}C incorporated into lactose and both neutral and acidic oligosaccharides, notably LNT and fucosyl LNT. Incorporation of ^{13}C was also observed into the fraction containing difucosyl LNT, fucosyl LNH and difucosyl LNH [92]. These results suggested that feeding ^{13}C-enriched milk oligosaccharides to infants would facilitate studies of their metabolic fate.

5. Bifidobacterium Growth Stimulation by Human Milk Oligosaccharides

As mentioned above, it is well known that HMO act as prebiotics that stimulate the growth of bifidobacteria in the infant colon. In breast-fed infants, bifidobacteria usually dominate the intestinal flora within 1 week after birth, constituting 95 to 99.9% of the bacterial population [93, 94]. By contrast, intestinal colonization by bifidobacteria is not rapid and predominant in bottle-fed infants who, prior to the early 20th Century, often experienced infection by pathogenic bacteria [95]. To improve the growth of intestinal bifidobacteria, saccharides such as lactulose [96] (Petuely, 1957) have been used as supplements to formula milk, resulting in improved health of bottle-fed infants.

It is not yet clear which oligosaccharides of more than 115 HMO are utilized and by which metabolic pathways they are digested by *Bifidobacteria*. Recently two series of reports based on different approaches have been published. One approach is to determine the enzymes for the metabolism of HMO by bifidobacteria. The other is to investigate the consumption of each component of HMO during *in vitro* fermentation by bifidobacteria.

The first approach is based on the bifidobacterial intracellular metabolic pathway specific for galacto-*N*-biose (GNB, Galβ1,3GalNAc) and lacto-*N*-biose I (LNB, Galβ1,3GlcNAc), the GNB/LNB pathway (Fig 4). The key enzyme of this pathway, β-1,3-galactosyl-*N*-acetylhexosamine phosphorylase (galacto-*N*-biose/lacto-*N*-biose I phosphorylase, GLNBP, EC 2.4.1.211) was found by Derensy-Dron et al. [97] in cell free extracts of *Bifidobacterium bifidum*. Kitaoka et al. [98] cloned the gene (lnpA) for this enzyme from *Bifidobacterium longum* subsp. *longum* JCM1217 and found that it was located in a putative operon (*lnpA-lnpD*) whose genes encode the GNB/LNB pathway. In this pathway LNB and GNB are finally converted to α-glucose 1-phosphate and to N-acetyl-α-glucosamine 1-phosphate, which are able to enter the glycolytic pathway [99]. The genes encoding the GNB/LNB specific transporter are locates just upstream of the operon for the GNB/LNB pathway [100].

The extracellular enzymatic system that liberates LNB from HMOs with the Type I structure has been isolated from *B. bifidum*. The key enzyme is lacto-*N*-biosidase, which catalyses the hydrolysis of LNT into LNB and lactose [101]. The genes encoding enzymes that generate the core structures of HMOs, α-1,2-fucosidase [102], α-1,3/4-fucosidase [103], and two α-sialidases [104, 105], were also cloned from the same strain. All these enzymes have N-terminal secretion signals and C-terminal membrane anchor domains. These extracellular enzymes may be specific to *B. bifidum* since genes encoding such extracellular enzymes are not found in the genomic sequences of *B. longum* subsp. *longum* NCC2705 [106] and *B. longum* subsp. *infantis* ATCC15697 [107]. It should be noted that *B. longum* subsp. *infantis* ATCC15697 has these, except for lacto-*N*-biosidase as intracellular enzymes.

Because of the presence of the intracellular GNB/LNB pathway, the abundance of Type I oligosaccharides in HMOs, and the extracellular enzymes that liberate LNB from HMOs, LNB is considered to be the key disaccharide in the utilization of HMOs. Its fermentability by various microorganism was examined using LNB that had been prepared on a large scale [108]. It was found that it was specific to several species of bifidobacteria (such as *B. longum* subsp. *longum*, *B. longum* susbsp. *infantis*, *B. bifidum*, and *Bifidobacterium breve*) that had been isolated from the feces of infants [109, 110].

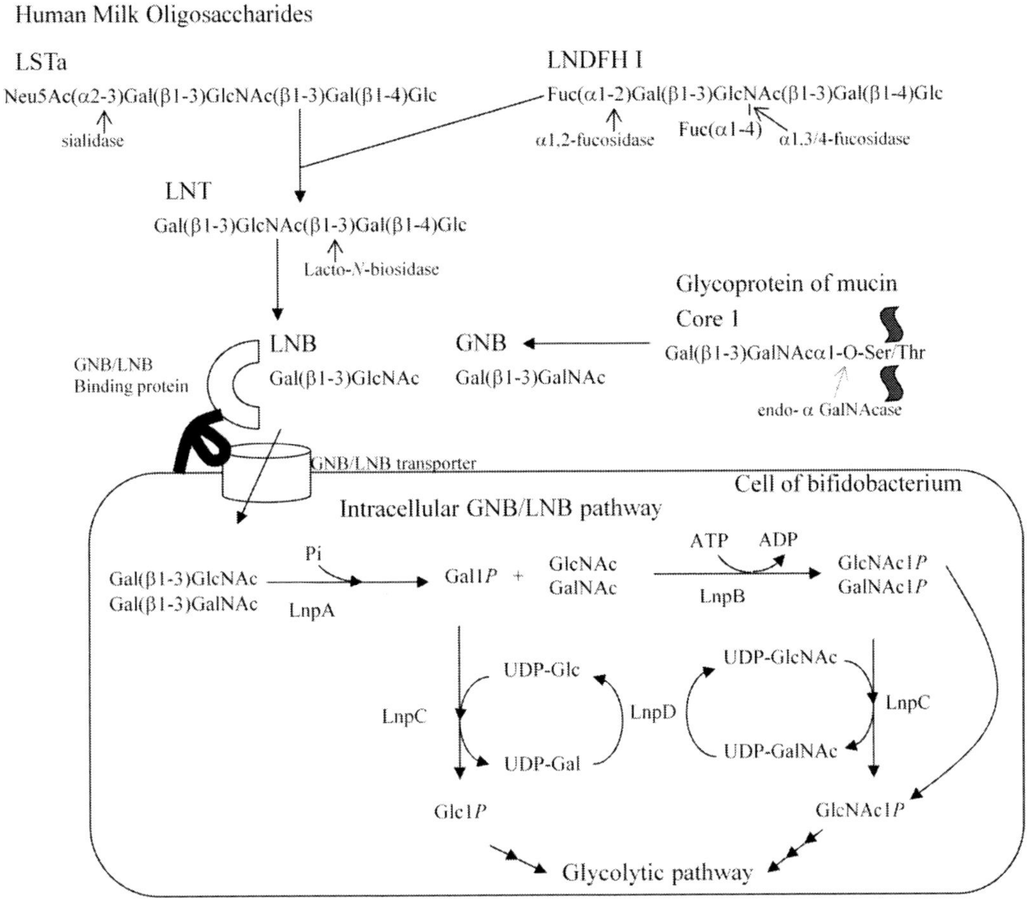

Figure 4. Metabolic pathway of the core portions of human milk oligosaccharides by *Bifidobacterium bifidum*.

The other approach is based on the quantification of each component of HMOs using mass spectrometry. Ward et al. [111] were the first to demonstrate that *B. longum* subsp. *infantis* ATCC15697 consumed components of HMOs. Ninonuevo et al. [112] developed a method designed to quantify the consumption of HMOs with each peak (m/z) using MALDI-FTICR MS. The residual HMO in the culture medium was reduced with $NaBH_4$, and then the same amount of HMO, which had been reduced with $NaBD_4$, was added into it. After MS spectrometry of this mixture, the D/H ratio of each peak was measured. From this calculation, they estimated how many percent of each oligosaccharide was consumed by a bifidobacterial strain. This method is very simple, but there is a disadvantage, because it is unclear which oligosaccharide isomer such as LNT or LN*n*T, LNH or LN*n*H was consumed predominantly by the strain. Using this method, LoCascio et al. [113] reported that *B. longum* subsp. *infantis* ATCC15697, grown in MRS medium, consumed HMOs that were smaller or equal to DP 8. Later they examined several strains of *B. longum* subsp. *infantis* and found that some strains did not consume fucosylated oligosaccharides [114]. Marcobal et al. [115] have reported a different glycoprofile for the consumption of HMOs by the same strain in ZMB1 medium, suggesting a difficulty in the interpretation of the glycoprofiling insofar as *in vitro* fermentation might be affected by the culture conditions.

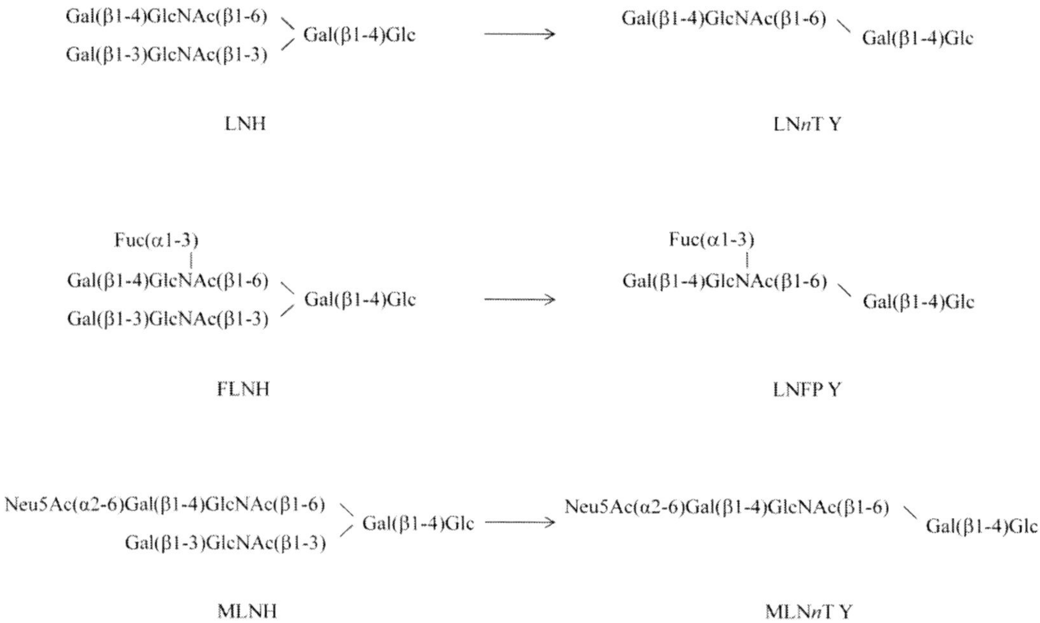

Figure 5. Possible partial degradations of lacto-*N*-hexaose, fucosyl lacto-*N*-hexaose or sialyl lacto-*N*-hexaose during passage through breast-fed infant colon.

To discover which of more than 115 HMO are actually digested by intestinal microflora dominated by *Bifidobacteria*, during passage through the infant colon, a comparison between the oligosaccharide profiles of human milk and those of the feces of breast-fed infants should provide important information. Recently Albecht et al. [116] extracted the oligosaccharides from human milk and from the feces of breast-fed infants and separated the oligosaccharides by capillary electrophoresis followed by characterization of each oligosaccharide by connected electro spray mass spectrometry. The relative concentration of each oligosaccharide was determined by calculation of the peak area in relation to that of the internal xylose standard in the electropherograms. From the comparison of peak areas, it was concluded that the LNT concentration in the feces was much smaller than that in the milk. In addition, two oligosaccharides, tentatively characterized as Gal(β1-4)GlcNAc(β1-6)Gal(β1-4)Glc (LN*n*T Y) and Gal(β1-4)[Fuc(α1-3)]GlcNAc(β1-6)Gal(β1-4)Glc (LNFP Y), were found in the feces, although these have not been found in human milk. In another study, Neu5Ac(α2-6)Gal(β1-4)GlcNAc(β1-6)Gal(β1-4)Glc as well as LNFP Y were separated from human infant feces and were characterized [117]. These results suggest that LNT, LNH, monofucosyl LNH or monosialyl LNH might have been partially digested by lacto-*N*-biosidase secreted by *Bifidobacterium bifidum* (see Figure 5), since this would reduce the concentration of LNT and produce the above two oligosaccharides. It has been hypothesized that type I HMO are digested in preference to the type II HMO by colonic *Bifidobacteria*, but the possibility that LNH or LNT were partially digested by colonic glycosidase cannot be excluded.

6. Milk Oligosaccharides as Anti-Pathogenic Agents

As indicated above, it is assumed that HMOs are synthesized by the actions of several glycosyltransferases, which usually catalyze the biosynthesis of carbohydrate moieties of glycoproteins and glycolipids. Therefore it is thought that the structures of the non-reducing ends of HMO are similar to those of the non-reducing ends of the sugar chains of glycoconjugates, which are found on the surface of epithelial cells. Actually, LNFP-III, and other HMO containing sialyl Le^x in their structures, were effectively used for the study of the binding specificities of siglecs and other recognition molecules on cell surfaces [46].

It was known that infection by many bacteria and viruses starts by binding to particular sugar chains of glycoconjugates on the surface of cells of the mucous epithelium of the digestive and respiratory tracts. Therefore, human milk oligosaccharides might be useful for elucidating the structures of the sugar chains on the surface of epithelial cells that are the targets of specific bacteria or viruses.

After entering into 1990th, several research groups began to report the phenomenon that some of the human milk oligosaccharides inhibit the attachment of intestinal bacteria to the surface of intestinal epithelial cells [118, 119, 120]. Based on these findings, Newburg started to investigate the possibility that human milk oligosaccharides may be useful for developing new drugs to protect suckling babies from bacterial and viral infections.

In his review, Newburg suggested that the addition of several oligosaccharides or glycoconjugates to milk formulae would protect infants against infectious diseases [121]. The toxic protein STa, produced by pathogenic *E. coli*, causes diarrhea by activation of guanylate cyclase, which is present on the surface of epithelial cells [122]. It was known that T84 cells, established by Dharmsathaphorn et al. [123, 124, 125], differentiated into colonic epithelial-like cells, equipped with microvilli containing guanylate cyclase. Crane et al. [126] found, using these cultured differentiated cells, that a human milk oligosaccharide fraction inhibits the activation of guanylate cyclase by STa protein. Subsequent to this finding, they observed that a fucosylated oligosaccharide fraction, which was obtained from human milk oligosaccharides by passing them through a column of insoluble *Ulex europaeus* lectin, inhibited this activation by the STa enterotoxin, whereas a non-fucosylated oligosaccharide fraction did not [126]. Recently Morrow [127] suggested that human milk oligosaccharides containing the Fuc(α1-2)Gal unit inhibit the attachment of STa to colonic epithelial cells. This field has recently expanded into a major research area.

Ruiz-Palacios et al. [128] found that HMO containing Fuc(α1-2)Gal(β1-4)GlcNAc (H-2 antigen epitope) at their non-reducing termini inhibit the attachment of *Campyrobacter jejuni* to colonic epithelial cells. This organism causes diarrhea and paralysis of motor nerves in infants. A universal correlation has been observed between the breast milk concentration of 2'-FL, which is usually the most prominent oligosaccharide in secretor donor's milk, and the frequency of diarrhea in breast fed infants, supporting the view that 2'-FL reduces the pathogenicity of *C. jejuni*. In addition, they reported that *Campyrobacter* did not attach to CHO cells, but did attach to α1,2-fucosyltransferase gene transfected CHO cells which expressed H-2 antigen. This attachment was inhibited by the addition of *Ulex europaeus* lectin, *Lotus tetragonolobus* lectin, anti H-2 monoclonal antibody, H-2 antigen related glycoconjugates, or a human milk oligosaccharides fraction, showing that the attachment to H-2 antigen expressed on colonic epithelial cells is a requirement for infection by *Campyrobacter*.

Le Pandu [129] reported that the frequency of infection by Norwalk virus (NV) is relatively low in non-secretor people, suggesting that carbohydrate moieties containing Fuc(α1-2)Gal(β1-3)GlcNAc (H-1 antigen epitope) at their non-reducing termini, expressed on colonic epithelial cells, are the receptors for infection by NV. It has also been reported that the frequency of infection by NV in infants was reduced by feeding secretor donor's milk, which contains oligosaccharides with the H-1 antigen, but not by feeding non-secretor donor's milk.

In 2005, Perret at al. [130] reported that *Pseudomonas aeruginasa* lectin specifically bind to 3-FL and to the human milk oligosaccharides containing Le^a and hypothesized that the receptors for *Pseudomonas aeruginosa* are Le^a and Le^x.

It is known that *E. coli* containing F18 pili causes diarrhea and oedema in piglets immediately after weaning and that this bacterial infection begins by binding of the adhesion FedF protein in the pili to the small intestine. Recently, Coddens et al. [131] found that the in vitro attachment of this bacterium to piglet small intestinal microvilli was inhibited by the addition of LNFP-I but not of LNT.

Further, an interaction of *Helicobacter pylori* with sialylated glycans has been reported. *H. pylori* is a Gram-negative bacterium, resides in the gastric mucosa and adheres to the epithelial cells lining the stomach. This organism infects around 50% of the world population, with a higher incidence in developing countries. *H. pylori* is associated with the development of peptic ulcers, mucosa-associated lymphoid-tissue (MALT) lymphoma and gastric adenocarcinoma. The preferred interaction is with α3-linked sialic acid; glycans having α6-linked Neu5Ac are non-binding. For example, 50% inhibition by *H. pylori* of hemagglutination of human erythrocytes was observed at a low concentration of some sialylated saccharides. The data showed that S-3 PG (Neu5Ac(α2-3)Gal(β1-4)GlcNAc(β1-3)Gal(β1-4)Glc-Cer), Neu5Ac(α2-3)Gal(β1-4)[Fuc(α1-3)]GlcNAc, 3'-*N*-acetylneuraminyl-*N*-acetyllactosamine (Neu5Ac(α2-3)Gal(β1-4)GlcNAc) as well as 3-*N*-acetylneuraminyl-lacto-*N*-neotetraose (Neu5Ac(α2-3)Gal(β1-4)GlcNAc(β1-3)Gal(β1-4)Glc) all bound to *H. pylori* CCUG17874 at similar strength. 3'-SL also bound to this organism but its binding ability was somewhat weaker than that of the above saccharides. It has also been reported that LST a, a human milk oligosaccharide, was able to bind to another strain, *H. pylori* J99 [132].

The binding of 3'-SL to *H. pylori* CCUG17874 is noteworthy because this saccharide is found in human milk and bovine colostrum. Asakuma et al. [75] found that at the start of lactation the concentration of 3'-SL in human colostrum was 360 mg/L, similar to that of 6'-SL. However, the concentration of 3'-SL decreased during the subsequent two days of lactation, whereas that of 6'-SL did not. This suggests that, very early in lactation, 3'-SL may be more significant in the prevention of transmission of *H. pylori* from mother to infant than later on.

Recent studies on the ability of various fractions of human milk oligosaccharides to inhibit the adhesion of three intestinal microorganisms (enteropathogenic *E. coli* serotype 0119, *Vibrio cholerae* and *Salmonella fyris*) to differentiated Caco-2 cells have shown that the acidic fraction had an anti-adhesive effect on all three pathogenic strains. The neutral high molecular weight fraction significantly inhibited the adhesion of *E. coli* 0119 and *V. cholerae*, but not that of *S. fyris*; the neutral low molecular weight fraction was effective toward *E. coli* 0119 and *S. fyris* but not *V. cholerae* [133]. This demonstrated that human milk oligosaccharides inhibit the adhesion to epithelial cells not only of common pathogens such as *E. coli* but also of other aggressive bacteria such as *V. cholerae and S. fyris*. Thus, oligosaccharides may be important factors in human milk that defend against acute diarrhea in breast-fed infants.

Furthermore, it has been reported that the human milk oligosaccharides fraction, at a concentration of 0.5 g/L, reduced the binding of the HIV-1 envelope glycoprotein gp120 to dendritic cell – specific ICAM3 – grabbing non – integrin (DC-SIGN) in human dendritic cells by more than 60%. In addition, the binding of gp120 to Raji cells, which expressed DC-SIGN, was reduced by more than 60%. It is worth noting that mother to child transmission accounts for more than 40% of all HIV-1 infections in children, with breast-feeding being the predominant postnatal transmission route, especially in developing countries. However, a majority of breast-fed infants born to HIV-positive mothers remain uninfected despite continuous exposure to the virus over many months. Viral entry across the infant's mucosal barrier is partially mediated by binding of gp120 to DC-SIGN on human dendritic cells (DC). It has been hypothesized that human milk oligosaccharides, when they reach DC in the colon, bind to DC-SIGN and inhibit the transfer of HIV to CD4 + T lymphocytes. It has also been suggested that this inhibition of HIV-gp120 binding to DC-SIGN is caused by oligosaccharides carrying multiple Lewis epitopes [134].

It is generally recognized that breastfed babies are less likely than bottle-fed babies to have asthma, lower or upper respiratory infections, and ear infections. A pilot study tested the relationship between the consumption of oligosaccharides, the oligosaccharide content of feces, and subsequent disease in breastfed infants [135]. In this study the concentration of LNFP-II was determined in breast milk and in the feces of breastfed babies, and the relationship between these concentrations and the frequencies of respiratory and gastrointestinal infection of the babies was evaluated. The results showed that the LNFP-II levels in the feces collected at 2 weeks, in those babies that had had symptoms of respiratory disease within 6 or 24 weeks after birth, were significantly lower than those of infants that had had no symptoms. Also, the levels of LNFP-II in the milk of mothers whose babies had had these symptoms at the same ages were significantly lower than those of other mothers whose babies had had no symptoms. Although the mechanism by which human milk oligosaccharides prevent respiratory disease in infants is unknown, an immuno modulating effect by a prebiotic mechanism was suggested; alternatively some of the milk oligosaccharides may have been absorbed into the systemic circulation and acted as inhibitory receptor analogs within the respiratory tract [135].

7. Immuno – Modulating Effect of Milk Oligosaccharides

It is thought that the immuno – modulating effect of human milk oligosaccharides in infants is acquired while the oligosaccharides are circulating after their absorption, or by immuno stimulation in the colon produced by growth-stimulated beneficial bacteria. Although unequivocal detection of human milk oligosaccharides in the blood of infants has not yet been reported [136], it nevertheless seems very likely on the basis of their observed urinary excretion [89, 137] that small amounts of intact milk oligosaccharides are normally absorbed from the gastrointestinal tract, and that they are transported into the systemic circulation.

It follows that they may alter protein-carbohydrate interactions also at a systemic level. For example, recent studies suggest that human milk oligosaccharides interfere with the adhesion of neutrophils to vascular endothelial cells [138] and platelets [139]. These effects appear to be based on the structural resemblance of some human milk oligosaccharides to the glycoprotein ligands of selectins. Selectins are trans-membrane proteins that are involved in cell-cell interactions in the immune system. P-selectin mediates leukocyte deceleration (rolling) on activated endothelial cells and intiates leukocyte extravasation at sites of inflammation. P-selectin is also involved in the formation of platelet-neutrophil complexes (PNC), a sub-population of highly activated neutrophils primed for adhesion, phagocytosis and enhanced production of reactive oxygen species. Recent studies suggest that oligosaccharides containing sialyl Le^x or its stereoisomer sialyl Le^a, which resemble the P-selectin ligand, inhibit the binding of selectin ligands to the surface of endothelial cells and platelets; this interferes with the formation of PNC, the effect of which is anti-inflammatory. The following oligosaccharide fractions were tested *in vitro* to establish whether they reduce leukocyte deceleration on U937 cells, which express the P-selectin ligand: total human milk oligosaccharides, neutral oligosaccharides, total acidic oligosaccharides, neutral oligosaccharides with a polymerization degree of 4, fucosylated oligosaccharides and disialyl lacto-*N*-tetraose. The acidic oligosaccharide fraction produced a slight but definite reduction of P-selectin ligand binding, similar to that of standard sialyl Le^x, whereas the total neutral oligosaccharides and neutral fucosylated oligosaccharides fractions did not [140]. These results support the notion of anti-inflammatory effects of acidic human milk oligosaccharides.

It has been reported that the incidence of necrotizing enterocolitis, a condition which is considered to be an exaggerated immune response, is about 85% lower in breast-fed than in formula-fed infants. This is consistent with an anti-inflammatory effect of absorbed human milk oligosaccharides [141].

The question of whether human milk oligosaccharides influence cytokine production and activation of cord blood T cells has recently been investigated [142]. Cord blood mononuclear cells from randomly chosen healthy newborns were co-cultured for 20 days with acidic or neutral oligosaccharides, and intracellular cytokine production and surface marker expression of T cells was studied using flow cytometry. The authors used concentrations of oligosaccharides (neutral human milk oligosaccharides, 10 μg/mL; acidic human milk oligosaccharides, 1 μg/mL) that were considered by them to mimic physiologic conditions, although these concentrations are considerably lower than the values of 100-200 mg/L for circulating human milk oligosaccharides estimated by Bode [136]. The acidic, but not the neutral oligosaccharide fraction, increased the percentage of interferon-γ-producing

CD3+CD4+ and CD3+CD8+ cells, of IL-13 production in CD3+CD8+ cells and significantly elevated CD25+ expression in CD3+CD4+ cells. These results showed that human milk oligosaccharides affect cytokine production and activation of cord blood-derived T cells *in vitro*. Oligosaccharides and, in particular, acidic milk oligosaccharides may therefore influence lymphocyte maturation in breast-fed newborns. The authors concluded that human milk oligosaccharides can modulate the immune system of the maturing infant.

8. MILK OLIGOSACCHARIDES OF THE COW AND OTHER DOMESTIC FARM ANIMALS AND THEIR FUTURE INDUSTRIAL UTILIZATION

It is known that bovine colostrum collected immediately after parturition contains more than 1 g/L of oligosaccharides, but the mature milk contains only small amounts [6,7]. Although 25 bovine oligosaccharides structures isolated from the colostrum have been completely characterized (Table 5), as many as 39 oligosaccharides have been detected using a combination of nanoelectrospray Fourier transform ion cyclotron resonance (nESI-FTICR) mass spectrometry and matrix-assisted laser desorption/ionization Fourier transform ion cyclotron resonance (MALDI-FTICR) mass spectrometry [143]. A similar analysis showed that cheese whey, which is a by-product in the production of cheese from mature bovine milk, contains at least 8 neutral and 7 acidic oligosaccharides [144]. Saito et al. [145] found Gal(β1-4)[Fuc(α1-3)]GlcNAc (3-fucosyl-*N*-acetyllactosamine) in bovine colostrum, whereas Tao et al. [143] did not detect any fucosyl oligosaccharide in the colostrum. Tao et al. [143] concluded from their results that bovine colostrum contains LN*n*T, LN*n*H and lacto-*N*-novopentaose I (Gal(β1-3)[Gal(β1-4)GlcNAc(β1-6)]Gal(β1-4)Glc), and their *N*-acetyl and *N*-glycolyl neuraminyl derivatives. Tao et al. [143] reported that sialyl oligosaccharides constitute 70% of the total oligosaccharides of bovine colostrum, of which those containing Neu5Gc constitute less than 5%. Bovine oligosaccharides include those that have *N*-acetyllactosamine (Gal(β1-4)GlcNAc) instead of lactose at their reducing ends, such as free *N*-acetyllactosamine or 6'-*N*-acetylneuraminyl-*N*-acetyllactosamine (Neu5Ac(α2-6)Gal(β1-4)GlcNAc), and others. There is a difference in this respect between bovine and human milk oligosaccharides, insofar as almost all HMO contain a lactose unit at their reducing ends.

Most of the acidic oligosaccharide fraction of bovine colostrum consists of 3'-SL, 6'-SL, 6'-SLN and DSL (Neu5Ac(α2-8)Neu5Ac(α2-3)Gal(β1-4)Glc), with 3'-SL constituting 70% of this total [143]. There is a difference between human and cow in the ratio of 3'-SL to 6'-SL, in that 6'-SL predominates over 3'-SL in human milk/colostrum. Changes in the levels of 3'-SL, 6'-SL and 6'-SLN in Holstein colostrum from immediately after parturition to 1 week post-partum are shown in Figure 6. The levels were maximal immediately after parturition, rapidly decreasing by 48 h post-partum [6]. In another study, the concentrations of 3'-SL, 6'-SL, 6'-SLN and DSL were found to be 681, 243, 239 and 201 mg/L, respectively, in Holstein colostrum and 867, 136, 220 and 283 mg/L, respectively in Jersey colostrum immediately after parturition [146].

Table 5. Bovine milk oligosaccharides

No.	*Oligosaccharide*	*Structure*	*Reference*
Neutral oligosaccharides			
1	*N*-Acetylgalactosaminylglucose	GalNAc(β1-4)Glc	257
2	LacNAc	Gal(β1-4)GlcNAc	257
3	Le^x trisaccharide	Gal(β1-4)GlcNAc Fuc(α1-3)	257
4	α3'-GalNAcL	GalNAc(α1-3)Gal(β1-4)Glc	258
5	Isoglobotriose	Gal(α1-3)Gal(β1-4)Glc	258
6	β3'-GL	Gal(β1-3)Gal(β1-4)Glc	145
7	β4'-GL	Gal(β1-4)Gal(β1-4)Glc	259
8	β6'-GL	Gal(β1-6)Gal(β1-4)Glc	145
9	*novo*-LNP I	Gal(β1-4)GlcNAc(β1-6) \\ Gal(β1-4)Glc Gal(β1-3) /	258
10	β3'-GalNAcL	GalNAc(β1-3)Gal(β1-4)Glc	260
11	β6'-GlcNAcL	GlcNAc(β1-6)Gal(β1-4)Glc	143
12	LN*n*T	Gal(β1-4)GlcNAc(β1-3)Gal(β1-4)Glc	143
13	LN*n*H	Gal(β1-4)GlcNAc(β1-6) \\ Gal(β1-4)Glc Gal(β1-4)GlcNAc(β1-3) /	143
Acidic oligosaccharides			
14	3'-*O*-Lac phosphate	Gal(β1-4)Glc-3'-PO_4	261
15	3'S Gal	Neu5Ac(α2-3)Gal	30
16	3'-SL	Neu5Ac(α2-3)Gal(β1-4)Glc	262
17	6'-SL	Neu5Ac(α2-6)Gal(β1-4)Glc	30
18	3'-Neu5GcL	Neu5Gc(α2-3)Gal(β1-4)Glc	30
19	6'-Neu5GcL	Neu5Gc(α2-6)Gal(β1-4)Glc	263
20	6'-SLacNAc	Neu5Ac(α2-6)Gal(β1-4)GlcNAc	30
21	6'-Neu5GcLacNAc	Neu5Gc(α2-6)Gal(β1-4)GlcNAc	263

No.	*Oligosaccharide*	*Structure*	*Reference*
22	-	Neu5Ac(α2-3)Gal(1-3)Gal(1-4)Glc	264
23	DSL	Neu5Ac(α2-8)Neu5Ac(α2-3)Gal(β1-4)Glc	30
24	6'-SLacNAc-1-*O*-phosphate	Neu5Ac(α2-6)Gal(β1-4)GlcNAc-1-PO_4	264
25	6'-SLacNAc-6-*O*-phosphate	Neu5Ac(α2-6)Gal(β1-4)GlcNAc-6-PO_4	264

Since bovine colostrum contains some oligosaccharides, which are mainly acidic, and mature bovine milk also contains them, although at lower concentrations, it can be expected that oligosaccharides derived from bovine colostrum or mature milk will be utilized by industry as functional foods, animal feeds or biomedical products. For example, several reports have suggested that bovine milk oligosaccharides can be used as anti infection materials.

It is thought that adhesion of *Neisseria meningitides*, a human-specific pathogen causing meningitis and speticemia, is mediated by type IV pili [147]. A mictotiter well pili-binding assay was used to investigate the binding of type IV pili isolated from *N. meningitides* to different glycoproteins. Inhibition of pili binding to bovine thyroglobulin and human salivary agglutinin by fractionated human and bovine milk oligosaccharides was demonstrated. The binding of *Neisseria pili* to bovine thyroglobulin was more effective and was clearly inhibited by neutral or acidic bovine milk oligosaccharides at concentrations of 1~2 g/L, suggesting that these fractions had the potential ability to inhibit the attachment of this bacterium to the colonic mucosa [147].

The bovine acidic oligosaccharide fraction, in which 3'-SL is dominant, can be expected to be utilized as an anti-adhesion agent against enteropathogenic *Escherichia coli* (EPEC). Angeloni et al. [148] showed that upon exposure to 3'-SL, EPEC adhesion to Caco-2 cells was reduced by 50% compared to control cells. They also showed that upon exposure to 3'-SL, Caco-2 cells changed their cell surface glycan profile in that the expression of α(2-3) and α(2-6) linked sialic acid residues was significantly reduced, while that of fucose and galactose residues was also diminished. These results suggest a novel mechanism by which milk oligosaccharides, such as 3'-SL, regulate bacteria-host interactions.

Fractions containing milk oligosaccharides, in the form of supernatants that had been separated from colostrum and from transitional, mature and late lactation milk of Spanish brown cows by ethanol precipitation and subsequent centrifugation, were used to investigate the inhibition of hemagglutination by seven *enterotoxigenic E. coli* strains (K99, FK, F41, F17, B16, B23 and B64). These strains had been isolated from diarrheal calves. The fractions from the transitional and mature milk inhibited hemagglutination by all of these strains, whereas those from colostrum and late lactation milk produced weaker inhibition [149]. It was suggested that this inhibition was due to 3'-SL, 6'-SL, 6'-SLN and DSL. The fractions from transitional and mature milk, in which the ratio of 6'-SL to 3'-SL was higher than in the fractions from colostrum and late lactation milk, had a stronger effect than the others.

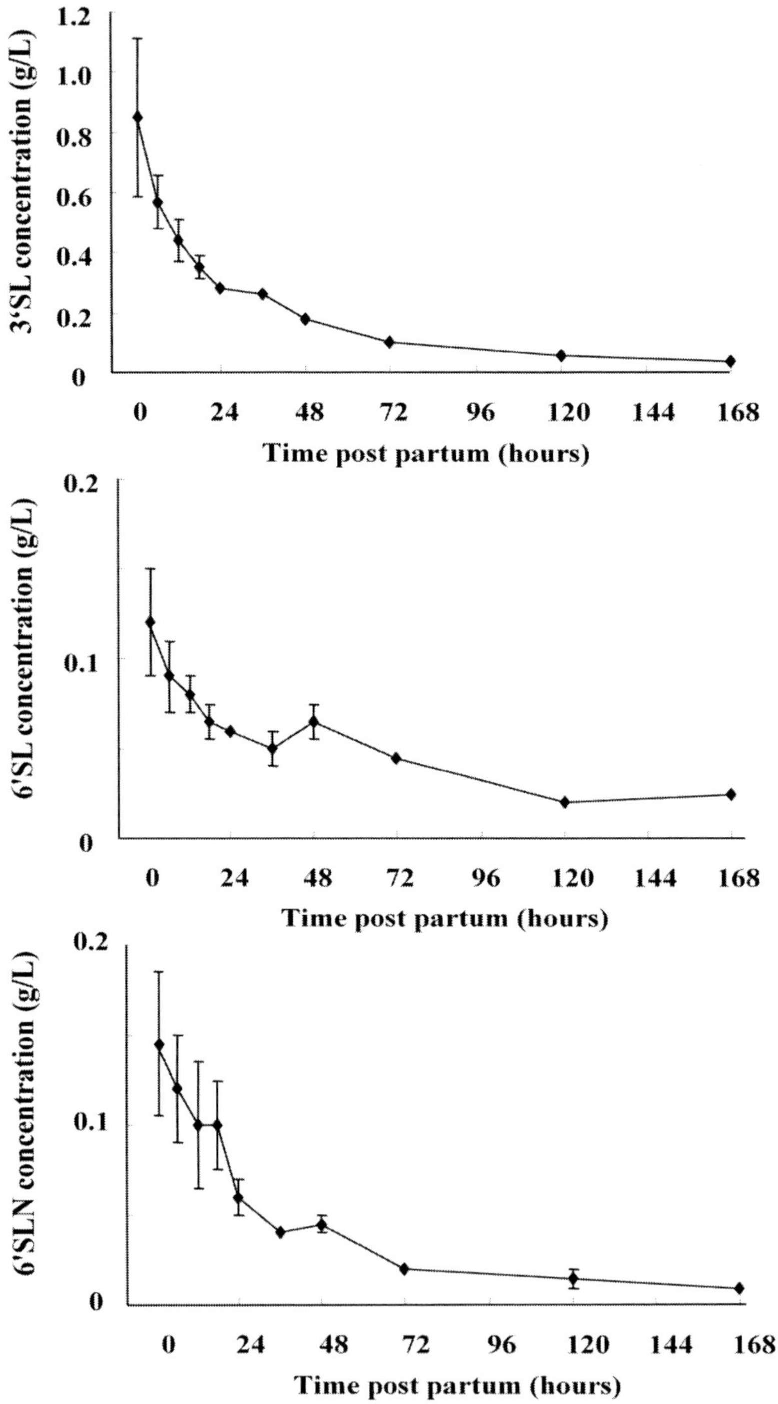

Reproduced from the figure in *J. Dairy Sci.* 86, 1315-1320 (2003) with permission.

Figure 6. Changes in the concentrations of 3'-SL, 6'-SL and 6'-SLN in Holstein bovine colostrum during early lactation. Values are indicated as means ±SD (n=4).

Terabayashi et al. [150] have been attempting to enhance the possible biological function of bovine colostrum sialyl oligosaccharides by chemical modification. The OH-1 at the reducing end of the oligosaccharide was substituted by an amino group, using an amination reaction with saturated ammonium hydrogen carbonate in water solution, with stirring for four days at room temperature. This amino derivative was then condensed with a long chain fatty acid using a reagent for amidation, DMT-MM (4-(4,6-dimethoxy-1,3,5-triazin)-2-yl)-4-methylmorpholinim chloride), in 2.5% methanol in water; the acyl amido derivative of the sialyl oligosaccharide was synthesized at high yield. It is well known that the epitope structures for infection to host cell with influenza virus are Neu5Ac(α2-3)Gal or Neu5Ac(α2-6)Gal. It was observed that this artificial ganglioside GM3 like component, which was synthesized by the condensation of 3'-SL with lauric acid (C12) using the above method, inhibited the infection to MDCK cells with human type influenza virus A/PR/8 (H1N1). The 50% inhibitory oncentration (IC_{50}) of the compound was comparable to that of the representative influenza medicine "Symmetrel (Amantadine Hydrochloride; Novartis Farma K.K.), while the cell toxicity was extremely low when compared with that by Symmetrel. On the other hand, free 3'-SL as such did not show such inhibition effect on infection with influenza virus. As the acyl amido derivative of 3'-SL is amphipathic, it associates as a micelle in water solution. The sugar chain of the 3'-SL of the derivative arranges itself on the micelle surface. It can be assumed that this localization of the 3'-SL epitope of the derivative should enhance its binding to the virus so that infection to the host cells with the virus would be competitively inhibited by this component. It can be expected that in future this approach may lead to the development of anti-infection drugs that utilize bovine colostrum sialyl oligosaccharides.

Further, it may be feasible to use sialyllactose, separated from cheese whey or bovine colostrum, as a functional food for brain activation. Sakai et al. [151] fed 1% of a fraction, which contained 87% 3'-SL and 13% 6'-SL, separated from cheese whey, or artificial galactosylated sialic acid, to rats at 8 weeks age and studied their swimming learning ability. Rats were grouped into the following; control group, 1% lactose feeding, 1% galactosyllactose feeding, 1% free sialic acid feeding, 1% sialyllactose feeding, and 1% galactosylated sialic acid feeding, and each group was supplied with food containing the above, for two weeks. The authors then subjected each group to the T maze test for 7~10 days followed by the Morris swimming test for 11~14 days, and measured the time that the rats took to reach the goal. They observed that there was a tendency for the two groups that had been fed with sialyllactose and galactosylated sialic acid to reach the goal faster than the other groups. In addition, they found a small increase in the total sialic acid, total ganglioside and GM_3 concentrations in the brains of these two groups compared with the other groups. These data suggest that sialic acid is absorbed during the consumption of sialyl glycoconjugates and is used for the biosynthesis of brain gangliosides, and that the feeding of sialylglycoconjugates could improve learning ability. In that case sialyllactose might be utilized as a functional food marketed as a brain stimulating factor.

It can be expected that milk oligosaccharides of other domestic farm animals, such as goats, sheep, and camels, will also be used as biofunctional materials. The milk oligosaccharide content of goat milk is 0.25~0.39 g/L; this is higher than that of bovine (0.03~0.06 g/L) or ovine (0.02~0.04 g/L) milk. In addition, the variety of oligosaccharides in goat milk is greater than that in bovine or ovine milk, as shown by the profiles from HPEAC analysis [152, 153]. Colostrum from the Japanese Saanen breed contains more 6'-SL than 3'-

SL; it also contains 6'-*N*-glycolylneuraminyllactose (Neu5Gc(α2-6)Gal(β1-4)Glc), Gal(α1-3)Gal(β1-4)Glc, Gal(β1-3)Gal(β1-4)Glc, Gal(β1-6)Gal(β1-4)Glc and 2'-FL [154, 155]. Another study has shown that mature milk from Spanish goats contains 6'-SL, 3'-SL, disialyllactose, *N*-glycolylneuraminyllactose, 3'-galactosyllactose, *N*-acetylglucosaminyllactose, LNH and additional high molecular oligosaccharides, as demonstrated by analysis with FAB-MS, but no fucosyl oligosaccharides [152]. Ovine colostrum contains more 3'-*N*-glycolylneuraminyllactose than 3'-SL and 6'-SL [156] and, notably, contains Neu5Gc in preference to Neu5Ac. Camel colostrum contains 3-FL, Gal(β1-3)Gal(β1-4)Glc, Gal(β1-6)Gal(β1-4)Glc, 3'-SL, 6'-SL, Neu5Ac(α2-3)Gal(β1-3)Gal(β1-4)Glc, LST-c, monosialyl LN*n*H (Neu5Ac(α2-6)Gal(β1-4)GlcNAc(β1-3)[Gal(β1-4)GlcNAc(β1-6)]Gal(β1-4)Glc and two monosialyl lacto-*N*-novopentaose I (Neu5Ac(α2-3)Gal(β1-3)[Gal(β1-4)GlcNAc(β1-6)]Gal(β1-4)Glc and Gal(β1-3)[Neu5Ac(α2-6)Gal(β1-4)GlcNAc (β1-6)]Gal(β1-4)Glc) [157]. As this colostrum contains greater amounts of sialyl oligosaccharides, other than 3'-SL, than bovine colostrum, it can be expected that it will be utilized as a source of biofunctional materials.

A recent study on rats showed that goat milk oligosaccharides have an anti-inflammatory effect in the colon [158]. In this study, colitis was induced by the hapten, trinitrobenzenesulfonic acid (TNBS). The experimental rats (OS) were fed a diet containing 500 mg/kg per day of goat milk oligosaccharides, from 2 days prior to the induction until day 6, after which all the rats were weighed and then killed, the entire colon was removed, opened and scored for visible damage and then divided into several pieces for biochemical determinations. When the OS rats were compared with control rats in which colitis had been induced by TNBS but that had not been treated with oligosaccharides, it was found that the OS rats showed decreased anorexia, reduced loss of body weight, reduced bowel wall thickening and less necrosis of the colon. Biochemically, the OS rats had lower colonic levels of inducible nitric oxide synthase (iNOS), cyclooxygenase 2 (COX2), interleukin-1β and mucin 3, as well as increased trefoil factor 3. These results showed that goat milk oligosaccharides are anti-inflammatory when administered as a pretreatment in the TNBS model of rat colitis, most likely due to their action as prebiotics resulting in favorable changes in the colonic bacterial flora. Since TNBS-induced colitis is widely used as a preclinical model of inflammatory bowel disease in humans, it was suggested that goat milk oligosaccharides may be useful in the management of this disease.

Although this trial using domestic farm animal milk oligosaccharides has only just begun, their industrial utilization can be expected in the near future.

9. Glycoproteins and Glycopeptides

Among caseins (α_{s0}-, α_{s1}-, α_{s2}-, α_{s3}-, α_{s4}-, α_{s6}, β-, and κ-) that account for about 50% (human) and 80% (bovine) of milk proteins, only κ-casein is glycosylated. It is well known that κ-casein shows microheterogeneity owing mainly to multiple post-translational modifications such as phosphorylation and glycosylation [159, 160, 161, 162]. There are at least six *O*-glycosylation sites, Thr142, Thr152, Thr154, Thr163, Thr166, and Thr186 (numbering from the N-terminal Met of the bovine κ-casein A precursor), on the surface of bovine κ-casein. Saito et al. [163] determined 5 chemical structures of sugar moieties of bovine mature milk κ-casein, viz., GalNAc, Gal(β1-3)GalNAc, NeuAc(α2-3)Gal(β1-3)GalNAc, NeuAc(α2-6)[Gal(β1-3)]GalNAc, and NeuAc(α2-6)[NeuAc(α2-3)Gal(β1-3)]GalNAc. Further structural analysis was performed by the same author on the sugar moieties of bovine colostrum κ-casein, viz., NeuAc(α2-6)[GlcNAc(β1-3)Gal(β1-3)]GalNAc, Gal(β1-4)GlcNAc(β1-6)[Gal(β1-3)]GalNAc, Gal(β1-4)GlcNAc(β1-6)[NeuAc(α2-3)Gal(β1-3)]GalNAc, NeuAc(α2-3)Gal(β1-4)GlcNAc(β1-6)[NeuAc(α2-3)Gal(β1-3)]GalNAc, and it was shown that these sugar moieties were totally different from those of mature milk, despite a common core structure (mucin type I: Gal(β1-3)GalNAc) [164, 165, 166, 167]. Otani et al. [168] reported that bovine κ-casein, in the range of 50-150 μg/ml, had a significantly inhibitory effect on the blastogenesis of mouse splenocytes induced by lipopolysaccharide, due to the glycomacropeptide region of the κ-casein molecule. Strömqvist et al. [169] claimed that not bovine but human milk κ-casein, especially its fucose containing sugar moiety, inhibits adhesion of *Helicobacter pyroli* to human gastric mucous cells. κ-Casein is hydrolyzed by chymosin (rennin), at Phe126-Met127, releasing an N-terminal-fragment, para-κ-casein, and a glycosylated C-terminal fragment, glycomacropeptide (or caseinomacropeptide). The latter fragment exhibits a wide variety of well documented bioactivities, presumably due mainly to its sugar moieties; these activities include anticariogenesis [170], anti-inflammatory [171] and immunomodulatory [172, 173, 174, 175] actions, inhibition of cholera toxin binding [176], inhibition of influenza-virus hemagglutination [177], and prebiotic effects [178].

With respect to the industrial utilization of milk derived bioactive glycoproteins, the most prominent one is lactoferrin, which is present in the whey fraction. The concentration of bovine lactoferrin in colostrum and mature milk is 6-8 and 2-4 g/L, respectively. The protein is produced commercially using cheese whey as source material. The molecular weight of bovine lactoferrin is 83 kDa, and its sugar content is about 11%. In bovine lactoferrin, there are five asparagine residues that *N*-glycans are potentially bound to; four of these, Asn233, Asn368, Asn476, and Asn545, had been believed to be glycosylated by high-mannose type *N*-glycan. On the other hand, Wei et al. [179] showed that bovine lactoferrin isoform-α is glycosylated by a complex type *N*-glycan at the fifth potential *N*-glycosylation site, Asn281. Lactoferrin exhibits a broad spectrum of biological functions, including anti-inflammatory, anti-tumor and bacteriostatic activities, actions on cell growth and differentiation and on iron homeostasis, and others [see reviews 180, 181, 182, 183, 184]. Its strong iron-binding property confers multifunctionality to lactoferrin; it is likely, however, that the binding ability of lactoferrin to a wide range of epithelial and immune cells via specific receptors is also involved in its anti-inflammatory activity and cancer protective functions by regulating cellular signaling pathways [180]. Although intensive as well as extensive investigations have been performed on lactoferrin, there is no evidence so far that sugar moieties of lactoferrin are relevant to its bioactivities. van Berkel et al. [185] suggested that the role of human lactoferrin *N*-glycan is protection of the protein portion against proteolytic degradation.

Lactophorin (identical to proteose-peptone component 3) is a 27kDa-phosphoglycoprotein, which has been found in the whey fraction of the milk of camel [186], bovine [187], goat [188], llama [189], and sheep [188]. There is so far no evidence for the presence of lactophorin in human milk. One *N*-glycosylation site (Asn77) and two *O*-glycosylation sites (Thr16 and Thr86) have been specified in this protein [190]. Girardet et al. [191] determined 8 chemical structures of ConA-bound *N*-glycans of bovine lactophorin using methylation analysis, mass spectrometry, and 400-MHz ^{1}H-NMR spectroscopy. All the structures determined had biantennary *N*-acetyllactosamine-type carbohydrate chains, and some of them were with a GalNAc(β1-4)GlcNAc or a NeuAc(α2-6)GalNAc(β1-4)GlcNAc group. Using the same analytical techniques as above, Coddeville et al. [192] determined 3 chemical structures of ConA-bound *O*-glycans of bovine lactophorin: GalNAcα1-*O*-Thr, Gal(β1-3)GalNAcα1-*O*-Thr, and Gal(β1-4)GlcNAc(β1-6)[Gal(β1-3)]GalNAcα1-*O*-Thr. Recently, Inagaki et al. [193] demonstrated further complexity of the lactophorin glycosylation pattern in that that heparin-bound *N*-glycans showed mono-sialylated bi-, tri-, and tetra-antennary complex-type carbohydrate chains carrying Gal-GlcNAc (LacNAc) or GalNAc-GlcNAc (LacdiNAc) with or without core-fucose. Being homologous to murine glycosylation-dependent cell-adhesion molecule 1 (GlyCAM-1), which is a mucin-like endothelial cell surface ligand for the leukocyte adhesion molecule, namely L-selectin, bovine lactophorin is presumed to be involved somehow in the immune response, but its *in vivo* physiological role still remains unknown. On the other hand, several bioactivities, such as emulsifying [194], immunostimulative [195], and inhibition against lipoprotein lipase [196, 197], have been reported for lactophorin. Furthermore, a synthetic 23-residue peptide mimicking the C-terminal region of bovine lactophorin, termed lactophoricin, showed anti-bacterial activity [198].

Lipid droplets secreted from mammary epithelial cells into milk are surrounded by an apical membrane, termed the milk fat globule membrane (MFGM). The MGFM contains several membrane proteins, which represent 1-4% of the total milk proteins [199]. Using proteomic approaches towards bovine [200, 201, 202], goat [203], human [204], and mouse MFGM [205], more than 100 MFGM proteins have been characterized. A large part of MFGM-associated proteins are glycosylated, and these show a great diversity of glycan structures that presumably ensure prevention of microbial and viral infection of the intestinal mucosa of infants by mimicking specific ligands of pathogenic microorganisms [206]. Other physiological activities of MFGM-associated glycoproteins, related to regulation of milk fat globule secretion and innate immunity or cellular development of neonates, are also considerable [199, 207]. MUC1 is a mucin-type glycoprotein, which is expressed in the apical membrane of mammary epithelial cells and is integrated into the MFGM. This highly glycosylated protein carries both *N*- and *O*-glycans that are composed of *N*-acetylgalactosamine, *N*-acetylglucosamine, galactose, mannose, sialic acid, and fucose, judging from its lectin binding affinities [208, 209, 210]. The structures of human MUC1 *O*-glycans significantly differ from those of bovine O-glycans [211]. Wilson et al. [212] have found a novel fucosylated *O*-glycan, Gal-GalNAc1-3Fuc-ol, in human MUC1. Sato et al. [212,] showed that the *N*-linked sugar chains of most bovine glycoproteins from milk fat globule membranes contain the GalNAc(β1-4GlcNAc) group. Since expression of the disaccharide structure is influenced by peptide sequences near the glycosylation sites [213], the site-specificity of the *N*-acetylgalactosaminylated sugar chains was investigated using bovine butyrophilin, a major MFGM glycoprotein [214]. Structural analyses of the oligosaccharides revealed that only complex-type sugar chain with the GalNAc(β1-4)GlcNAc structure are included in Asn-55 – linked oligosaccharides, while only hybrid-type sugar chains in bovine MFGM glycoproteins are included in Asn-215 – linked oligosaccharides. This result showed that the glycosylation of butyrophilin occurs in a site-specific manner [214]

There are many glycoproteins and glycopeptides, such as immunoglobulins and enzymes, in milk other than those described above. Although there can be no doubt that these are a potential source of beneficial bioactive substances, their low abundance and heterogeneity hamper their mass production. To date, either chemical or enzymatic synthesis of these bioactive glycoproteins and glycopeptides would still be a little too ambitious. Establishment of heterologous expression systems in appropriate hosts could be a promising methodology for the industrial use of milk glycoproteins and glycopeptides [216, 217].

10. Glycolipids

Bovine milk contains small amounts (around 40 μmol/L) of glycolipids, which are found mainly on the surface of fat globule membranes [218]. They consist only of sphingo-glycolipids, glycero-glycolipids being absent. Three neutral glycolipids have been characterized in bovine milk, while eleven gangliosides, which contain Neu5Ac, have been found [219, 220, 221]. It has been reported that bovine mature milk contains 8 and 17 μmol/L of glucosylceramide and lactosylceramide, respectively, while this milk contains 14 μmol/L of gangliosides, 80% of which are composed of GD3 (Neu5Ac(α2-8)Neu5Ac(α2-3)Gal(β1-4)Glc-1'Cer) and GM3 (Neu5Ac(α2-3)Gal(β1-4)Glc-1'Cer); the former predominates over the latter [222, 223]. As the concentration of gangliosides in bovine milk is lower than it is in human milk, its concentration in infant formula, which is produced from mature bovine milk, is lower than that in human milk [224]. In human colostrum, GD3 (disialyl) predominates over GM3 (monosialyl), while the latter predominates over the former in mature human milk [225, 226]; this supports the suggestion that the relatively high content of sialic acid in colostrum is biologically significant for the early development of infants.

For many years, the anti-pathogenic effects of milk glycolipids including inhibition of the attachment of pathogenic microorganisms and bacterial toxins to the colonic mucosa have received much attention [227]. In addition, other possibilities such as the significance of sialic acid for brain development [228] and of glycolipids as immunomodulating factors [218] have been recently explored.

As the carbohydrate moieties of sphingo-glycolipids are found on the outside of epithelial cell membranes, pathogenic microorganisms employ them as receptors for attachment to the cell surface during infections. Since milk gangliosides have similar carbohydrate structures to the receptor glycolipids on cell surfaces, it is assumed that they act as soluble receptor analogues that inhibit this attachment. Among milk gangliosides, GM1 (Gal(β1-3)GalNAc(β1-4)[Neu5Ac(α2-3)]Gal(β1-4)Glc-1'Cer), which is a minor ganglioside in milk, is thought to inhibit the attachment of cholera toxin [229] and of thermo-tolerant toxin produced by pathogenic *E. coli* [230], to the colon, and it has been reported that GM3 and GD3 bind to tetanus toxin and also to influenza A virus [150, 223]. It has been found that 9-O-acetyl-GD3, separated from bovine milk, binds to influenza C virus [231]. As 9-O-acetyl-GD3 has been detected in human tumor cells but not in normal cells [220], it is believed that a monoclonal antibody, prepared using bovine milk 9-O-acetyl-GD3, can be utilized for drugs of diagnosis and as a medical treatment for cancers [232, 233]. It has been found that GD3 does not bind to botulinum toxin, while GT1b (Neu5Ac(α2-3)Gal(β1-3)GalNAc(β1-4)[Neu5Ac(α2-8)Neu5Ac(α2-3)]Gal(β1-4)Glc-1'Cer), GQ1b (Neu5Ac(α2-8)Neu5Ac(α2-

3)Gal(β1-3)GalNAc(β1-4)[Neu5Ac(α2-8)Neu5Ac(α2-3)]Gal(β1-4)Glc-1'Cer) and GD1b (Gal(β1-3)GalNAc(β1-4)[Neu5Ac(α2-8)Neu5Ac(α2-3)Gal(β1-4)Glc-1'Cer], which had been prepared from caprine milk, does bind to this toxin [234]. These observations suggest that anti-pathogenic drugs will be developed using these gangliosides, separated from the milk or colostrum of domestic farm animals.

It has been observed that milk gangliosides fed to adult female rats were absorbed and then delivered to body tissues including the brain [235], and trace amounts were also incorporated into the fetus via the placenta. When one compared the ganglioside concentrations in the brains of breast-fed and bottle-fed babies, it was observed that they were higher in the former than in the latter [236]. From these observations, it can be expected that orally fed gangliosides may in the future be utilized as materials that promote the in vivo synthesis of brain gangliosides and that they may improve learning and cognitive abilities of infants [228].

It has been shown that when rats were fed with fodder containing 1% milk gangliosides, the total ganglioside concentration increased in the mucus of the small intestine accompanied by an increase or decrease in GD3 and GM3 levels, respectively, compared with the control [237]. In addition, the cholesterol level as well as the ratio of cholesterol to gangliosides remarkably decreased in the mucus of the small intestine, compared with the control [237]. It is well known that sphingo-glycolipids including gangliosides, as well as cholesterol, are constituents of a microdomain in cell membranes. Several signaling proteins are found in the microdomain area, which are involved in biological functions such as metabolism, internal secretion, immunity, inflammation and cell growth. It is known that the decrease in the cholesterol level in the microdomain of epithelial cells of the small intestine controls phospholipase A2 activity [238]. As this enzyme releases arachidonic acid from membrane phopholipids to form eicosanoid, which induces inflammation, it is thought that the feeding of milk gangliosides controls inflammation in the intestine [237]. In addition, it has been reported that the levels of platelet activating factor (PAF) and of diacyl glycerol, which are the biofunctional lipid and second messenger, respectively, in this microdomain, decreased when the experimental animals were fed with milk gangliosides [237]. From these observations, it seems possible that a biofunctional food, that has anti-inflammatory actions in the gastrointestinal tract, could be developed through the incorporation of milk gangliosides.

As the whey concentrate, which is a by-product in cheese manufacturing, contains a fairly high concentration of glycolipids amounting to 162 mg and 40 mg per 100 g of total solid neutral glycolipids and GD3, respectively, it can be expected that their separation and utilization will in future be developed on the industrial scale. The following technique for the separation of glycolipids and gangliosides from the whey concentrate has been developed; proteolysis of the whey proteins using proteases and membrane ultrafiltration followed by concentration and extraction of the glycolipids, and separation of neutral glycolipids and gangliosides using ion exchange chromatography [239]. It seems likely that milk glycolipids will in future be utilized as materials in biofunctional foods and medicines.

Concluding Remarks

As described above, studies on milk oligosaccharides, which have a history of over 50 years, were begun in order to find the bifidobacterium growth stimulating factor in human milk. To date, 115 varieties of HMO have been characterized together with development of techniques for their purification, and it seems likely that more oligosaccharides will be characterized, including those with higher molecular mass and others that are present at low concentrations. During this fifty year history, there have been unique findings such as the absence of oligosaccharides containing non-reducing H antigen in non-secretor milk, by Kobata et al., and the inhibition, by HMO that contain Fuc(α1-2)Gal(β1-4)GlcNAc unit of the attachment of *Campylobacter jejuni* to colonic epithelial cells. There is evidence suggesting that part of the HMO are absorbed unchanged by the small intestine of breast-fed infants and act as immuno modulators within the systemic circulation. The biosynthetic pathways of carbohydrate moieties of glycoproteins or glycolipids can be assumed to be based on those of milk oligosaccharides; these studies therefore contribute to the development of glycobiology. Along with the development of techniques for the quantification of oligosaccharides, the biological significance of individual HMO's will more readily be assessed. In recent years the metabolism of HMO by several bifidobacterial strains has been clarified in studies based on their genomes. The data suggests that type I HMOs are utilized more predominantly than type II HMOs by the colonic bacteria present in breast-fed infants. In addition, it has been found that the predominance of type I oligosaccharides in milk or colostrum is a feature that is specific to humans, not being found in other mammals, including other primates. These observations suggest the possibility of symbiotic evolution between humans and infant colonic bifidobacteria.

During the last few years, the idea that it may be possible to utilize bovine milk or colostrum oligosaccharides, or chemically modified BMO's by industry has spread, even though the concentration of oligosaccharides in bovine milk is low and their variety is small compared with that in human milk. In addition, glycoproteins or glycolipids derived from bovine milk, buttermilk or cheese whey may in future be utilized on the industrial scale as biofunctional materials in foods or medicines.

Although techniques for the large industrial scale preparation of oligosaccharides, with chemical structures similar to those of HMO's, have not so far been developed, that for lacto-*N*-biose I, which is a basic unit of type I HMO's, has been established, suggesting that this disaccharide may soon be available for utilization by industry. It can also be expected that in the near future basic research will develop to a stage in which the industrial utilization of other oligosaccharides derived from bovine milk, or the milk of other farm animals, will be possible.

ACKNOWLEDGMENTS

The work was supported by fundings from the Global COE Program "Frontier Program for Animal Global Health and Hygiene", Ministry of Education, Culture, Sports, Science, and Technology, Japan, and Yotsuba Milk Products Co.

REFERENCES

[1] Urashima, T., Saito, T., Nakamura, T., Messer, M. 2001. Oligosaccharides of milk and colostrum in non-human mammals. *Glycoconj. J.* 18, 357-371.

[2] Messer, M., Urashima, T. 2002. Evolution of milk oligosacchride and lactose. *Trends Glycosci. Glycotech.* 14, 153-176.

[3] Newburg, D.S., Naubauer, S.H. 1995. Carbohydrates in milks: Anlysis, quantities and significance. In: *Handbook of Milk Composition* (R, G. Jensen ed.), pp. 273-349, Academic Press, San Diego.

[4] Kunz, C., Rudloff, S., Baier, W., Klein, N., Strobel, S. 2000. Oligosaccharides in human milk: Structural, functional, and metabolic aspects. *Annu. Rev. Nutr.* 20, 699-722.

[5] Boehm, G., Stahl, B. 2003. Oligosaccharides. In: *Functional Dairy Products* (T. Mattila-Sandholm, M. Saarela, eds.), pp. 203-243, Woodhaed Publishing Ltd, Cambridge.

[6] Nakamura, T., Kawase, H., Kimura, K., Watanabe, Y., Ohtani, M., Arai, I., Urashima T. 2003. Concentrations of sialyl oligosaccharides in bovine colostrum and milk during the prepartum and early lactation. *J. Dairy Sci.* 86, 1315-1320.

[7] Gopal, P.K., Gill, H.S. 2000. Oligosaccharides and glycoconjugates in bovine milk and colostrums. *Br. J. Nutr.* 84, S69-S74.

[8] Engfer, M.B., Stahl, B., Finke, B., Sawatzki, G., Daniel, H. 2000. Human milk oligosaccharides are resistant to enzymatic hydrolysis in the upper gastrointestinal tract. *Am. J. Clin. Nutr.* 71, 1589-1596.

[9] Gnoth, M.J., Rudloff, S., Kunz, C., Kinne, R.K.H. 2001. Investigations of the *in vitro* transport of human milk oligosaccharides by Caco-2 monolayer using a novel high performance liquid chromatography-mass spectrometry technique. *J. Biol. Chem.* 276, 34363-34370.

[10] Polonovski, M., Lespagnol, A. 1933. Nouvelles acquisitions sur les composes glucidiues du lai de femme. *Bull Soc. Chem. Biol.* 15, 320-349.

[11] Polonovski, M., Montreuil, J. 1954. Etude chromatographique des polyosides du lai de femme. *C.R.Acad.Sci.* 238, 2263-2264.

[12] Tissier, H. Recherches sur la flora intestinale de nourissons, PhD thesis. Univ. Paris, Paris France.

[13] Moro, E. 1900. Morphologi und bakteriologisch Untersuchungen uber die Darmbakterien des Sauglings: die Bakterium-flora des normalen Frauenmilchstuls. *Jahrbuch Kinderh.* 61, 686-734.

[14] Grulee, C., Sanford, H., Schwartz, H. 1935. Breast and artificially fed infants; study of the age incidence in morbidity and mortality in 20,000 cases. *J.A.M.A.* 104, 1986-1988.

[15] Schönfeld, H. 1929. Uber die Beziehungen der einzelnen Bestandteile der Fauenmilch zur Bifidusflora. *Jahrbuch Kinderh.* 113, 19-69.

[16] György, P., Kuhn, R., Rose, C.S., Zilliken, F. 1954. Bifidus factor I. A variant of *Lacobacillus bifidus* requiring a special growth factor. *Arch. Biochem. Biophys.* 48, 193-201.

[17] György, P., Kuhn, R., Rose, C.S., Zilliken, F. 1954. Bifidus factor III. The rate of dialysis. *Arch. Biochem. Biophys.* 48, 209-213.

[18] György, P., Kuhn, R., Rose, C.S., Zilliken, F. 1954. Bifidus factor IV. Preparations obtained from human milk. *Arch. Biochem. Biophys.* 48, 214-224.

[19] Kuhn, R., Baer, H.H., Gauhe, A. 1956. Kristallisierte fucoside-lactose. *Chem. Ber.* 89, 2513.

[20] Montreuil, J. 1956. Structure de deux triholosides isoles du lai de femme. *C.R.Acad.Sci.* 242, 192-193.
[21] Kuhn, R., Gauhe, A. 1958. Über die lacto-difuco-tetraose der Frauenmilch. *Justus Liebig's Ann. Chem.* 611, 249-252.
[22] Kuhn, R., Baer, H.H. 1956. Die Konstitution der Lacto-*N*-tetraose. *Chem. Ber.* 89, 504-511.
[23] Kuhn, R., Gauhe, A. 1962. Die Konstitution der Lacto-*N*-neotetraose. *Chem. Ber.* 95, 518-522.
[24] Kuhn, R., Baer, H.H., Gauhe, A. 1956. Kristallisation und Konstitutionsermittlung der Lacto-*N*-fucopentaose I. *Chem. Ber.* 89, 2514-2523.
[25] Kuhn, R., Baer, H.H., Gauhe, A. 1958. Die Konstitution der Lacto-*N*-fucopentaose II. *Chem. Ber.* 91, 364.
[26] Kuhn, R., Baer, H.H., Gauhe, A. 1958. 2-α-L-Fucopyranosyl-D-galaktose und 2-α-L-fucopyranosyl-D-talose. *Justus Liuebig's Ann. Chem.* 611, 242-249.
[27] Kuhn, R., Gauhe, A. 1960. Uber ein Kristallisiertes, Le^a-aktives Hexasaccharid aus Frauenmilch. *Chem. Ber.* 93, 647-651.
[28] Kuhn, R., Brossmer, R. 1959. Uber das durch Viren der Influenza-Gruppe spaltbare Trisaccharid der Milch. *Chem. Ber.* 92, 1667-1671.
[29] Kuhn, R. 1959. Biochemie der Rezeptoren und Resistenzfaktoren. Von der Widerstandsfähigkeit der Lebewesen gegen Einwirkungen der Umwelt. *Naturwissenschaften* 46, 43-50.
[30] Kuhn, R., Gauhe, A. 1965. Bestimmung der Bindungsstelle von sialinsäureresten in Oligosaccharides mit Hilfe von Periodat. *Chem. Ber.* 98, 395-413.
[31] Kuhn, R., Gauhe, A. 1962. Uber drei saure pentasaccharide aus frauenmilch. *Chem. Ber.* 95, 513-517.
[32] Watkins, W.M. 1966. Blood group specific substances. In: *Glycoproteins* 2nd ed., Part B (A. Gottschalk, ed.), pp. 462-515, Elsevier, Amserdam.
[33] Watkins, W.M. 1972. Blood group specific substances. In: *Glycoproteins* (A. Gottschalk, ed.), Elsevier, pp 830-891, Amsterdam.
[34] Grollman, E.F., Ginsburg, V. 1967. Correlation between secretor status and occurrence of 2'-fucosyllactose in human milk. *Biochem. Biophys. Res. Commun.* 28, 50-53.
[35] Kobata, A., Tsuda, M., Ginsburg, V. 1969. Oligosaccharides of human milk. I. Isolation and characterization. *Arch. Biochem. Biophys.* 130, 509-513.
[36] Shen, L., Grollman, E.F., Ginsburg, V. 1968. An enzymatic basis for secretor status and blood group substanve specificity in humans. *Proc. Natl. Acad. Sci. U. S.* 59, 2124-230.
[37] Grollman, E.F., Kobata, A., Ginsburg, V. 1969. An enzymatic basis for Lewis blood types in man. *J. Clin. Invest.* 48, 1489-1494.
[38] Kobata, A., Grollman, E.F., Ginsburg, V. 1968. An enzymatic basis of blood type A in humans. *Arch. Biochem. Biophys.* 124, 609-612.
[39] Kobata, A., Grollman, E.F., Ginsburg, V. 1968. An enzymatic basis of blood type B in humans. *Biochem. Biophys. Res. Commun.* 32, 272-277.
[40] Kobata, A., Gindburg, V. 1970. UDP-*N*-acetyl-D-galactosamine: D-Galactose α-3-*N*-acetyl-D-galactosaminyltransferase, a product of gene that determines blood type A in man. *J. Biol. Chem.* 245, 1484-1490.

[41] Yamashita, K., Kobata, A. 1974. Oligosaccharides of human milk V. Isolation and characterization of a new trisaccharides, 6'-galactosyllactose. *Arch. Biochem. Biophys.* 161, 164-170.

[42] Ginsburg, V., Zopf, D.A., Yamashita, K., Kobata, A. 1976. Oligosaccharides of human milk: Isolation of a new pentasaccharide, lacto-*N*-fucopantaose V. *Arch. Biochem. Biophys.* 175, 565-568.

[43] Kobata, A., Ginsburg, V. 1969. Oligosaccharides of human milk. II. Isolation and characterization of a new pentasaccharide, lacto-*N*-fucopentaose III. *J. Biol. Chem.* 244, 5496-5502.

[44] Kobata, A., Ginsburg, V. 1972. Oligosaccharides of human milk. III. Structure of a new hexasaccharide, lacto-*N*-hexaose. *J. Biol. Chem.* 247, 1525-1529.

[45] Kobata, A., Ginsburg, V. 1972. Oligosaccharides of human milk. IV. Isolation and characterization of a new hexasaccharide, lacto-*N*-neohexaose. *Arch. Biochem. Biophys.* 150, 273-281.

[46] Fenderson, B.A., Zehavi, U., Hakomori, S. 1984. A multivalent lacto-*N*-fucopentaose III-lysyllysine conjugate decompacts preimplantation mouse embryos, while the free oligosaccharide is ineffective. *J. Exp. Med.* 160, 1591-1596.

[47] Yamashita, K., Tachibana, Y., Takasaki, S., Kobata, A. 1976. ABO blood group determinants with branched cores. *Nature* 262, 702-703.

[48] Yamashita, K., Tachibana, Y., Kobata, A. 1977. Oligosaccharides of human milk: Structures of three lacto-*N*-hexaose derivatives with H-haptenic structure. *Arch. Biochem. Biophys.* 182, 546-555.

[49] Takasaki, S., Kobata, A. 1974. Microdetermination of individual neutral and amino sugars and *N*-acetylneuraminic acid in complex saccharides, *J. Biochem. Tokyo* 76, 783-789.

[50] Kobata, A. 1979. Use of endo- and exoglycosidases for structural studies of glycoconjugates. *Anal. Biochem.* 100, 1-14.

[51] Kobata, A., Takasaki, S. 1992. Structural characterization of oligosaccharides from glycoproteins. 4A Glycosidase treatment and other methods, including methylation analysis. In: *Glycobiology: A Practical Approach* (M. Fukuda, and A. Kobata, eds.), pp165-185, IRL Press, Oxford.

[52] Tai, T., Yamashita, K., Kobata, A. 1975. Synthesis and mass fragmentographic analysis of partially *O*-methylated 2-*N*-methylglucosamines. *J. Biochem. Tokyo* 78, 679-686.

[53] Donald, A.S.R., Feeney, J. 1988. Separation of human milk oligosaccharides by recycling chromatography. First isolation of lacto-*N*-neo-difucohexaose II and 3'-galactosyllactose from this source. *Carbohydr. Res.* 178, 79-91.

[54] Grönberg, G., Lipniunas, P., Lundgren, T., Erlansson, K., Lindh, F., Nilsson, B. 1989. Isolation of monosialylated oligosaccharides from human milk and structural analysis of three new compounds. *Carbohydr. Res.*, 191, 261-278.

[55] Grönberg, G., Lipniunas, P., Lundgren, T., Lindh, F., Nilson, B. 1990. Isolation and structural analysis of three new disialylated oligosaccharides from human milk. *Arch. Biochem. Biophys.* 278, 297-311.

[56] Grönberg, G., Lipniunas, P., Lundgren, T., Lindh, F., Nilsson, B. 1992. Structural analysis of five new monosialylated oligosaccharides from human milk. *Arch. Biochem. Biophys.* 296, 597-610.

[57] Kitagawa, H., Nakada, H., Numata, Y., Kurosaka, A., Fukui, S., Funakoshi, I., Kawasaki, T., Shimada, K., Inagaki, F., Yamashina, I. 1988. Immunoaffinity isolation os a sialyl-Le^a oligosaccharides from human milk. *J. Biochem.* 104, 591-594.
[58] Kitagawa, H., Nakada, H., Kurosaka, A., Hiraiwa, N., Numata, Y., Fukui, S., Funakoshi, I., Kawasaki, T., Yamashina, I., Shimada, I., Inagaki, F. 1989. Three novel oligosaccharides with the sialyl-Le^a structure in human milk: Isolation by immonoaffinity chromatography. *Biochemistry* 28, 8891-8897.
[59] Kitagawa, H., Nakada, H., Numata, T., Kurosaka, A., Fukui, S., Funakoshi, I., Kawasaki, T., Shimada, I., Inagaki, F., Yamashina, I. 1990. Occurrence of tetra- and penta-saccharides with the sialyl-Le^a structure in human milk. *J. Biol. Chem.* 265, 4859-4862.
[60] Kitagawa, H., Nakada, H., Fukui, S., Funakoshi, I., Kasawaki, T., Yamashina, I. 1991. Novel oligosaccharides with the sialyl-Le^a structure in human milk. *Biochemistry* 30, 2869-2876.
[61] Kitagawa, H., Takaoka, M., Nakada, H., Fukui, S., Funakoshi, I., Kawasaki, T., Tate, S., Inagaki, F., Yamashina, I. 1991. Isolation and structural studies of human milk oligosaccharides that are reactive with a monoclonal antibody MSW 113. *J. Biochem.* 110, 598-604.
[62] Kitagawa H., Nakada, H., Fukui, S., Funakoshi, I., Kawasaki, T., Yamashina, I., Tate, S., Inagaki, F. 1993. Novel oligosaccharides with the sialyl-Le^a structure in human milk. *J. Biochem.* 114, 504-508.
[63] Yamashita, K., Tachibana, Y., Kobata, A. 1977. Oligosaccharides of human milk: Structural studies of two new octasaccharides, difucosyl derivatives of *para*-lacto-*N*-hexaose and *para*-lacto-*N*-*neo*hexaose. *J. Biol. Chem.* 252, 5408-5411.
[64] Tachibana, Y., Yamashita, K., Kobata, A. 1978. Oligosaccharides of human milk: Structural studies of di- and trifucosyl derivatives of lacto-*N*-octaose and lacto-*N*-*neo*octaose. *Arch. Biochem. Biophys.* 188, 83-89.
[65] Strecker, G., Fievre, S., Wieruszeski, J.M., Michalski, C., Montreuil, J. 1992. Primary structure of four human milk octa-, nona- and undeca-saccharides established by ^{1}H- and ^{13}C-nuclear magnetic resonance spectrometry. *Carbohydr. Res.* 226, 1-14.
[66] Haeuw-Fiévre, S., Wieruszeski, J.-M., Plancke, Y., Strecker, G. 1993. Primary structure of human milk octa-, dodeca-, and tridecasaccharides determined by a combination of ^{1}H-NMR spectroscopy and fast-atom-bombardment mass spectrometry. Evidence for a new core structure, the *para*-lacto-*N*-octaose. *Eur. J. Biochem.* 215, 361-371.
[67] Chai, W., Piskarev, V.E., Zhang, Y., Lawson, A.M., Kogelberg, H. 2005. Structural determination of a novel lacto-*N*-decaose and its monofucosylated analogue from human milk by electrospray tandem mass spectrometry and ^{1}H NMR spectroscopy. *Arch. Biochem. Biophys.* 434, 116-127.
[68] Amano, J., Osanai, M., Orita, T., Sugahara, D., Osumi, K. 2009. Structural determination by negative-ion MALDI-QIT-TOFMS^n after pyrene derivatization of variously fucosylated oligosaccharides with branched decaose cores from human milk. *Glycobiology* 19, 601-614.
[69] Sasaki, K., Kurata-Miura, K., Ujita, M., Angata, K., Kitagawa, S., Sekine, S., Nishi, T., Fukuda, M. 1997. Expression cloning of cDNA encoding a human β-1,3-*N*-acetylglucosaminyltransferase that is essential for poly-*N*-acetyllactosamine synthesis. *Proc. Natl. Acad. Sci. USA.* 94, 14294-14299.

[70] Almeida, R., Amado, M., David, L., Levery, S.B., Holmes, E.H., Merkx, G., vam Kessel, G., Rygaard, E., Hassa, H., Bennertt, E., Clausen, H. 1997. A family of human β4-galactosyltransferase: cloning and expression of two novel UDP-galactose:β-*N*-acetylglucosamine β1,4-galactosyltransferase, β4Gal-T2 and β4Gal-T3. *J. Biol. Chem.* 272, 31979-31991.

[71] Issiki, S., Togayachi, A., Kudo, T., Nishihara, S., Watanabe, M., Kubota, T., Kitajima, M., Shiraishi, N., Sasaki, K., Andoh, T., Narimatsu, H. Cloning, expression, and charactetization of a novel UDP-galactose: β-*N*-acetylglucosamine, β1,3-galactosyltransferase (β3GalT-V) responsible for synthesis of type-I chain in colorectal and pancreatic epithelia and tumor cells derived therefrom. *J. Biol. Chem.* 274, 12499-12507.

[72] Ropp, P.A., Little, M.R., Cheng, P.W. 1991. Mucin biosynthesis: purification and characterization of a mucin β6 *N*-acetylglucosaminyltransferase. *J. Biol. Chem.* 266, 23863-23871.

[73] Ninonuevo, M.R., Park, Y., Yin, H., Zhang, J., Ward, R.E., Clowers, B.H., German, J.B., Freeman, S.L., Killeen, K., Grimm, R., Lebrilla, C.B. 2006. A strategy for annotating the human milk glycome. *J. Agric. Food Chem.* 54, 7471-7480.

[74] Asakuma, S., Urashima, T., Akahori, M., Ohbayashi, M., Nauamura, T., Kimura, K., Watanabe, Y., Arai, I., Sanai, Y. 2008. Variation in neutral oligosaccharide levels in human colostrums. *Eur. J. Clin. Nutr.* 62, 488-494.

[75] Asakuma, S., Akahori, M., Kimura, K., Watanabe, Y., Nakamura, T., Tsunemi, M., Arai, I., Sanai, Y., Urashima, T. 2007. Sialyl oligosaccharides of human colostrum: Changes in concentration during the first three days of lactation. *Biosci. Biotech. Biochem.* 71, 1447-1451.

[76] Leo, F., Asakuma, S., Nakamura, T., Fukuda, K., Senda, S., Urashima, T. 2009. Improved determination of milk oligosaccharides using a single derivatization with anthranilic acid and separation by revsersed-phase high-performance liquid chromatography. *J. Chromatogr. A.* 1216, 1520-1523.

[77] Leo, F., Asakuma, S., Fukuda, K., Senda, A., Urashima, T. 2010. Determination of sialyl and neutral oligosaccharide levels in transition and mature milks of Samoan women, using anthranilic derivatization followed by reverse phase high performance liquid chromatography. *Biosci. Biotech. Biochem.* 74, 298-303.

[78] Chaturvedi, P., Warren, C.D., Altaye, M., Morrow, A.L., Ruiz-Palacios, G., Pickering, L.K., Newburg, D.S. 2001. Fucosylated human milk oligosaccharides vary between individuals and over the course of lactation. *Glycobiology* 11, 365-370.

[79] Thurl, S., Muller-Wermer, B., Sawatzki, G. 1996. Quantification of individual oligosaccharide compounds from human milk using high-pH anion exchang chromatography. *Anal. Biochem.* 235, 202-206.

[80] Coppa, G.V., Pierani, P., Zampini, L., Carloni, I., Carlucci, A., Catacci, C., Gabrielli, O. 1999. Oligosaccharides in human milk during different phases of lactation. *Acta Paediat. Suppl.* 430, 89-94.

[81] Kunz, C., Rudloff, S., Hintelmann, A., Pohlentz, G., Egge, H. 1996. High-pH anion-exchange chromatography with pulsed amperometric detection and molar response factors of human milk oligosaccharides. *J. Chromatogr. B* 685, 311-221.

[82] Shen, Z., Warren, C.D., Newburg, D.S. 2000. High-performance capillary electrophoresis of sialylated oligosaccharides of human milk. *Anal. Biochem.* 279, 37-45.

[83] Bao, Y., Zhu, L., Newburg, D.S. 2007. Simultaneous quantification of sialyloligosaccharides from human milk by capillary electrophoresis. *Anal. Biochem.* 370, 206-214.

[84] Kunz, C., Rudloff, S., Schad, W., Braun, D. 1999. Lactose-derived oligosaccharides in the milk of elephants: Comparison with human milk. *Br. J. Nutr.* 82, 391-399.

[85] Martin-Sosa, S., Martin, M.J., Garcia-Pardo, L.A., Hueso, P. 2003. Sialyloligosaccharides in human and bovine milk and infant formulas: variations with the progression of lactation. *J. Dairy Sci.* 86, 52-59.

[86] Urashima, T., Odaka, G., Asakuma, S., Uemura, Y., Goto, K., Senda, A., Saito, T., Fukuda, K., Messer, M., Oftedal, O.T. 2009. Chemical characterization of oligosaccharides in chimpanzee, bonobo, gorilla, orangutan, and siamang milk or colostrum. *Glycobiology* 19, 499-508.

[87] Brand-Miller, J.C., McVeagh, P., McNeil, Y., Messer, M. 1998. Digestion of human milk oligosaccharides by healthy infants evaluated by the lactulose hydrogen breath test. *J. Pediatr.* 133, 95-98.

[88] Newburg, D.S. 2000. Oligosaccharides in human milk and bacterial colonization. *J. Pediatr. Gastroenterol. Nutr.* 30, S8-S17.

[89] Rudloff, S., Pohlentz, G., Diekmann, L., Egge, H., Kunz, C. 1996. Urinary excretion of lactose and oligosaccharides in preterm infants fed human milk or infant formula. *Acta Paediatr.* 85, 598-603.

[90] Kuhn, N.J. 1972. The lactose and neuraminyllactose content of rat milk and mammary tissue. *Biochem. J.* 130, 177-180.

[91] Dickson, J.J., Messer, M. 1978. Intestinal neuraminidase activity of suckling rats and other mammals. Relationship to the sialic acid content of milk. *Biochem. J.* 170, 407-413.

[92] Rudloff, S., Obermeier, S., Borsch, C., Pohlenz, G., Hartmenn, R., Brösicke, H., Lentze, M.J., Kunz, C. 2006. Incorporation of orally applied ^{13}C-galactose into milk lactose and oligosaccharides. *Glycobiology* 16, 477-487.

[93] Rotimi, V.O., Duerden, B.I. 1891. The development of the bacterial flora in normal neonates. *J. Med. Microbiol.* 14, 51-58.

[94] Benno, Y., Mitsuoka, T. 1986. The development of gastrointestinal micro-flora in humans and animals. *Bifidofact. Microflora* 5, 13-25.

[95] Bezkorovainy, A. 1989. Ecology of bifidobacteria. In: *Biochemistry and Physiology of Bifodobacteria* (A. Bezkorovainy, R. Miller-Catchpole, eds.), pp. 29-72, CRC Press, Cleveland.

[96] Petuely, F. 1957. Bifidusflora bei Flaschenkindern durch bifidogene Substanzen (Bifidusfaktor). *Z. Kinderheikd.* 79, 174-179.

[97] Derensy-Dron, D., Krzewinski, F., Brassart, C., Bouquelet, S. 1999. β-1,3-Galactosyl-*N*-acetylhexosamine phosphorylase from *Bifidobacterium bifidum* DSM 20082: characterization, partial purification and relation to mucin degradation. *Biotechnol. Appl. Biochem.* 29, 3-10.

[98] Kitaoka, M., Tian, J., Nishimoto, M. 2005. Novel putative galactose operon involving lacto-*N*-biose phosphorylase in *Bifidobacterium longum*. *Appl. Environ. Microbiol.* 71, 3158-3162.

[99] Nishimoto, M., Kitaoka, M. 2007. Identification of *N*-acetylhexosamine 1-kinase in the complete lacto-*N*-biose I/galacto-*N*-biose metabolic pathway in *Bifidobacterium longum*. *Appl. Environ. Microbiol.* 73, 6444-6449.

[100] Suzuki, R., Wada, J., Katayama, T., Fushinobu, S., Wakagi, T., Shoun, H., Sugimoto, H., Tanaka, A., Kumagai, H., Ashida, H., Kitaoka, M., Yamamoto, K. 2008. Structural and thermodynamic analyses of solute-binding Protein from *Bifidobacterium longum* specific for core 1 disaccharide and lacto-*N*-biose I. *J. Biol. Chem.* 283, 13165-13173.

[101] Wada, J., Ando, T., Kiyohara, M., Ashida, H., Kitaoka, M., Yamaguchi, M., Kumagai, H., Katayama, T., Yamamoto, K. 2008. *Bifidobacterium bifidum* lacto-*N*-biosidase, a critical enzyme for the degradation of human milk oligosaccharides with a type 1 structure. *Appl. Environ. Microbiol.* 74, 3996-4004.

[102] Katayama, T., Sakuma, A., Kimura, T., Makimura, Y., Hiratake, J., Sakata, K., Yamanoi, T., Kumagai, H., Yamamoto, K. 2004. Molecular cloning and characterization of *Bifidobacterium bifidum* 1,2-α-L-fucosidase (AfcA), a novel inverting glycosidase (glycoside hydrolase family 95). *J. Bacteriol.* 186, 4885-4893.

[103] Ashida, H., Miyake, A., Kiyohara, M., Wada, J., Yoshida, E., Kumagai, H., Katayama, T., Yamamoto, K. 2009. Two distinct α-L-fucosidases from Bifidobacterium bifidum are essential for the utilization of fucosylated milk oligosaccharides and glycoconjugates. *Glycobiology* 19, 1010-1017.

[104] Yamamoto, K., Katayama, T., Kitaoka, M., Fushinobu, S. 2010. Analyses of bifidobacterial glycosidases involved in the metabolism of oligosaccharides. *Biosci. Microflora*, 29, 23-30.

[105] Kiyohara, M., Tanigawa, K., Chaiwangsri, T., Katayama, T., Ashida, H., Yamamoto, K. 2010. An exo-α-sialidase from bifidobacteria involved in the degradation of sialyloligosaccharides in human milk and intestinal glycoconjugates. *Glycobiology, in press*.

[106] Schell, M.A., Karmirantzou, M., Snel, B., Vilanova, D., Berger, B., Pessi, G., Zwahlen, M.C., Desiere, F., Bork, P., Delley, M., Pridmore, R.D., Arigoni, F. 2002. The genome sequence of *Bifidobacterium longum* reflects its adaptation to the human gastrointestinal tract. *Proc. Natl. Acad. Sci. USA* 99, 14422-14427.

[107] Sela, D.A., Chapman, J., Adeuya, A., Kim, J.H., Chen, F., Whitehead, T.R., Lapidus, A., Rokhsar, D.S., Lebrilla, C.B., German, J.B., *et al.* 2008. The genome sequence of *Bifidobacterium longum* subsp. *infantis* reveals adaptations for milk utilization within the infant microbiome. *Proc. Natl. Acad. Sci. U S A* 105, 18964-18969.

[108] Nishimoto, M., Kitaoka, M. 2007b. Practical preparation of lacto-*N*-biose I, a candidate for the bifidus factor in human milk. *Biosci. Biotechnol. Biochem.* 71, 2101-2104.

[109] Kiyohara, M., Tachizawa, A., Nishimoto, M., Kitaoka, M., Ashida, H., Yamamoto, K. 2009. Prebiotic effect of lacto-*N*-biose I on bifidobacterial growth. *Biosci. Biotechnol. Biochem.* 73, 1175-1179.

[110] Xiao, J., Takahashi, S., Nishimoto, M., Odamaki, T., Yaeshima, T., Iwatsuki, K., Kitaoka, M. 2010. Distribution of in vitro fermentation ability of lacto-*N*-biose I, the major building block of human milk oligosaccharides, in bifidobacterial strains. *Appl. Environ. Microbiol.* 76, 54-59.

[111] Ward, R.E., Ninonuevo, M., Mills, D.A., Lebrilla, C.B., German, J.B. 2006. In vitro fermentation of breast milk oligosaccharides by *Bifidobacterium infantis* and *Lactobacillus gasseri*. *Appl. Environ. Microbiol.* 72, 4497-4499.

[112] Ninonuevo, M.R., Ward, R.E., LoCascio, R.G., German, J.B., Freeman, S.L., Barboza, M., Mills, D.A., Lebrilla, C.B. 2007. Methods for the quantitation of human milk oligosaccharides in bacterial fermentation by mass spevtroscopy. *Anal. Biochem*. 261, 15-23.

[113] LoCascio, R.G., Ninonuevo, M.R., Freeman, S.L., Sela, D.A., Grimm, R., Lebrilla, C.B., Mills, D.A., German, J.B. 2007. Glycoprofiling of bifidobacterial consumption of human milk oligosaccharides demonstrates strain specific, preferential consumption of small chain glycans secreted in early human lactation. *J. Agric. Food. Chem*. 55, 8914-8919.

[114] LoCascio, R.G., Ninonuevo, M.R., Kronrwitter, S.R., Freeman, S.L., German, J.B., Lebrilla, C.B., Mills, D.A. 2009. A versatile and scalable strategy for glycoprofiling bifidobacterial consumption of human milk oligosaccharides. *Microbial Biotechnol*. 2, 333-342.

[115] Marcobal, A., Barboza, M., Froehlich, J.W., Block, D.E., German, J.B., Lebrilla, C.B., Mills D.A. 2010. Consumption of human milk oligosaccharides by gut-related microbes. *J. Agric. Food. Chem*., *58*, 5334–5340.

[116] Albrecht, S., Schols, van den Heuvel, E.G.H.M., Voragen, A.G.J., Gruppen, H. 2010. CE-LIF^n profiling of oligosaccharides in human milk and feces of breast-fed babies. *Electrophoresis* 31, 1264-1273.

[117] Bailey, D., Davies, M.J., Routier, F.H., Bauer, C., Feeney, J., Hounsell, E.F. 1997. 1H NMR analysis of novel sialylated and fucosylated lactose-based oligosaccharides having linear GlcNAc(β1-6)Gal and Neu5Ac(α2-6)GlcNAc sequence. *Carbohydr. Res*. 300, 289-300.

[118] Ramphal, R., Carnoy, C., Fievre, S., Michalski, J.C., Houndret, N., Lamblin, G., Strecker, G., Roussel, P. 1991. *Pserumonas aeruginosa* recognizes carbohydrate chains containing type 1 (Galβ1-3GlcNAc) or type 2 (Galβ1-4GlcNAc) disaccharide units. *Infect. Immun*. 59, 700-704.

[119] Cravioto, A., Tello, A., Villafan, H., Ruiz, J., del Vedovo, S., Neeser, J.R. 1991. Inhibition of localized adhesion of enteropathogenic *Escherichia coli* to HEp-2 cells by immunoglobulin and oligosaccharides of human colostrum and breast milk. *J. Infect. Dis*. 163, 1247-1255.

[120] Brassart, D., Woltz, A., Golliard, M., Neeser, J.R. 1991. In vitro inhibition of adhesion of *Candida albicans* clinical isolates to human buccal epithelial cells by Fucα1-2Galβ-bearing complex carbohydrates. *Infect. Immun*. 59, 1605-1613.

[121] Newburg, D.S. 1997. Do the binding properties of oligosaccharides in milk protect human infants from gastrointestinal bacteria? *J. Nutr*. 127, 980S-984S.

[122] Schulz, S., Green, C.K., Yuen, P.S.T., Garbers, D.L. 1990. Guanylyl cyclase is a heat-stable enterotoxin receptor. *Cell* 179, 941-948.

[123] Dharmsathaphorn, K., McRoberts, J.A., Mandel, K.G., Tisdae, L.D., Masui, H. 1984. A human colonic tumor cell line that maintains vectorial electrolyte transport. *Am. J. Physiol*. 246, G204-G208.

[124] Guarino, A., Cohen, M., Thompson, M., Dharmsathaphorn, K., Giannella, R. 1987. T84 cell receptor binding and guanyl cyclase activation by *Escherichia coli* heat-stable toxin. *Am. J. Physiol.* 253, 6775-6780.

[125] Weikel, C.S., Spann, C.L., Chambers, C.P., Crane, J.K., Linden, J., Hewlett, E.L. 1990. Phorbol esters enhance the cyclic GMP response of T84 cells to the heat-stable enterotoxin of Escherichia coli (STa). *Infect. Immunol.* 58, 1402-1407.

[126] Crane, J.K., Azar, S.S., Stam, A., Newburg, D.S., 1994. Oligosaccharides from human milk block binding and activity of the Escherichia coli heat-stable enterotoxin (Sta) in T84 intestinal cells. *J. Nutr.* 124, 2358-2364.

[127] Morrow, A.L., Ruiz-Palacios, G.M., Jiang, X., Newburg, D.S. 2005. Human-milk glycans that inhibit pathogen binding protect breast-feeding infants against infectious diarrhea. *J. Nutr.* 135, 1304-1307.

[128] Ruiz-Palacios, G.M., Cervantes, L.E., Romas, P., Chavez-Munguia, B., Newburg, D.S. 2003. *Campylobacter jejuni* binds intestinal H(O) antigen (Fucα1,2Galβ1,4GlcNAc), and fucosyloligosaccharides of human milk inhibit its binding and infection. *J. Biol. Chem.* 278, 14112-14120.

[129] Le Pandu, J. 2004. Histo-blood group antigen and human milk oligosaccharides: genetic polymorphism and risk of infectious disease. *Adv. Exp. Med. Biol.* 554, 135-143.

[130] Perret, S., Sabin, C., Dumon, C., Pokorna, M., Gautier, C., Galanina, O., Ilia, S., Bovin, N., Nicaise, M., Desmadril, M., Gilboa-Garber, N., Wimmerova, M., Mitchell, E.P., Imberty, A. 2005. Structural basis for the interaction between human milk oligosaccharides and the bacterial lectin PA-IIL of *Pseudomonas aeruginosa*. *Biochem. J.* 389, 325-332.

[131] Coddens, A., Diswall, M., Amgstrom, J., Breimer, M.E., Goddeers, B., Cox, E., Teneberg, S. 2009. Recognition of blood group ABH type I determinants by the FedF adhesin of F18-fimbriated *Escherichia coli*. *J. Biol. Chem.* 284, 9713-9726.

[132] Johansson, P., Nilsson, J., Angstrom, J., Miller-Pandraza, H. 2005. Interaction of *Helicobacter pylori* with sialylated carbohydrates: the dependence on different parts of the binding trisaccharide Neu5Acα3Galβ4GlcNAc. *Glycobiology* 15, 625-636.

[133] Coppa, G.V., Zampini, L., Galeazzi, T., Facinelli, B., Ferrante, L., Carpetti, R., Orazio, G. 2006. Human milk oligosaccharides inhibit the adhesion to caco-2 cells of diarrhea pathogens: *Escherichia coli, Vibrio cholerae*, and *Salmonella fyris*. *Pediatr. Res.* 59, 377-382.

[134] Hong, P., Ninonuevo, M.R., Lee, B., Lebrilla, C., Bode, L. 2009. Human milk oligosaccharides reduce HIV-1-gp120 binding to dendritic cell-specific ICAM3-grabbing non-integrin (DC-SIGN). *Br. J. Nutr.* 101, 482-486.

[135] Stepans, M.B.F.S., Wilhelm, S.L., Hertzog, M., Rodehorst, T.K.C., Blany, S., Clemens, B., Polak Ⅲ, J.J., Newbirg, D.S. 2006. Early consumption of human milk oligosaccharides is inversely related to subsequent risk of respiratory and enteric disease in infant. *Brestfeeding Medicine*, 1, 207-215.

[136] Bode, L. 2006. Recent advances on structure, metabolism and function of human milk oligosaccharides. *J. Nutr.* 136, 2127-2130.

[137] Obermeier, S., Rudloff, S., Pohlentz, S., Lentze, M.J., Kunz, C. 1999. Secretion of ^{13}C-labelled oligosaccharides into human milk and infants urine after an oral [^{13}C]galactose load. *Isotopes Environ. Haelth Stud.* 35, 119-125.

[138] Klein, A., Schwertman, A., Peters, M., Kunz, C., Strobel, S. 2000. Immunomodulatory effects of breast milk oligosaccharides. In: *Short and Long Term Effects of Breast Feeding on Child Health* (B. Koletzki, ed.), pp. 251-259, Kluwer Academic/Plenum Publishers, New York.
[139] Bode, L., Rudloff, S., Kunz, C., Strobel, S., Klein, N. 2004. Human milk oligosaccharides reduce platelet-neutrophil complex formation leading to a decrease in neutropil β2 integrin expression. *J. Leukoc Biol.* 76, 820-826.
[140] Schumacher, G., Bendas, G., Stahl, B., Beermann, C. 2006. Human milk oligosaccharides affect P-selectin binding capacities: *In vitro* investigation. *Nutr.* 22, 620-627.
[141] Lucas, A., Cole, T.J. 1990. Breast milk and neonatal necrotizing enterocolitis. *Lancet* 336, 1519-1523.
[142] Eiwegger, T., Stahl, B., Schmitt, J., Boehm, G., Gerstmayr, M., Pichler. J., Dehink, E., Loibichler, C., Urbanek, R., Szepfalnsi, I. 2004. Human milk-derived oligosaccharides and platelet-derived oligosaccharides stimulate cytokine production of cord blood T-cells in vitro. *Pediatr. Res.* 56, 536-540.
[143] Tao, N., DePeters, E.J., Freeman, S., German, J.B., Grimm, R., Lebrilla, C.B. 2008. Bovine milk glycome. *J. Dairy Sci.* 91, 3768-3778.
[144] Barile, D., Tao, N., Lebrilla, C.B., Coisson, J.D., Arlorio, M., German, J.B. 2009. Permete from cheese whey ultrafiltration in a source of milk oligosaccharides. *Int. Dairy J.* 19, 524-530.
[145] Saito, T., Itoh, T., Adachi, S. 1987. Chemical structures of three neutral trisaccharides isolated from in free forms from bovine colostrum. *Carbohydr. Res.* 165, 43-51.
[146] McJarrow, P., van Amelsfort-Schoonbeek, J. 2004. Bovine sialyl oligosaccharides: seasonal variations in their concentrations in milk, and a comparison of the colostrum of Jersey and Friesian cows. *Int. Dairy J.*, 14, 571-579.
[147] Hakkarainen, J., Toivanen, M., Leinonen, A., Frangsmyr, L., Stronberg, N., Lapinjoki, S., Nassif, X., Tikkanen-Kaukanen, C. 2005. Human and bovine milk oligosaccharides inhibit *Neisseria meningitidis* pili attachment *in vitro*, *J. Nutr.* 135, 2445-2448.
[148] Angeloni, S., Ridet, J.L., Kusy, N., Gao, H., Crevoisier, F., Guinchard, D., Kochhar, S., Sigrist, H., Sprenger, N. 2005. Glycoprofiling with micro-arrays of glycoconjugates an lectins. *Glycobiology* 15, 31-41.
[149] Martin, M.J., Martin-Sosa, S., Hueso, P. 2002. Binding of milk oligosaccharides by several enterotoxigenic *Escherichia coli* strains isolated from calves. *Glycoconj. J.* 19, 5-11.
[150] Terabayashi, T., Morita, M., Ueno, M., Nakamura, T., Urashima, T. 2006. Inhibition of influenza-virus-induced cytopathy by sialylconjugates, *Carbohydr. Res.*, 341, 2246-2253.
[151] Sakai, F., Ikeuchi, F., Urashima, T., Fujiwara, M., Ohtsuki, K., Yanahira, S. 2006. Effects of feeding sialyllactose and galactosylated *N*-acetylneuraminic acid on swimming learning ability and brain lipid composition in adult rats. *J. Appl. Glycosci.* 53, 249-254.
[152] Martinez-Ferez, A., Rudloff, S., Guadix, A., Henkel, C.A., Pohlentz, G., Boza, J.J., Guadix, E.M., Kunz, C. 2006. Goat's milk as a natural source of lactose-derived oligosaccharides isolation by membrane technology. *Int. Dairy J.* 16, 173-181.

[153] Mehra, R., Kelly, P. 2006. Milk oligosaccharides: Structural and technological aspects. *Int. Dairy J.* 16, 1334-1340.

[154] Urashima, T., Bubb, W.A., Messer, M., Tsuji, Y., Taneda, Y. 1994. Studies of the neutral trisaccharides of goat (*Capra hircus*) colostrum and of the one- and two-dimensional ^{1}H and ^{13}C NMR spectra of 6'-*N*-acetylglucosaminyllactose. *Carbohydr. Res.* 262, 173-184.

[155] Urashima, T., Murata, S., Nakamura, T. 1997. Structural determination of monosialyl trisaccharide obtained from caprine colostrum. *Comp. Biochem. Physiol.* 116B, 431-435.

[156] Nakamura, T., Urashima, T., Nakagawa, M., Saito, T. 1998. Sialyllactose occurs as free lactones in ovine colostrum. *Biochim. Biophys. Acta* 1381, 286-292.

[157] Fukuda, K., Yamamoto, A., Ganzorig, K., Khuukhenbaatar, J., Senda, A., Saito, T., Urashima, T. 2010. Chemical characterization of the oligosaccharides in Bactrian camel (*Camelus bactrianus*) milk and colostrum. *J. Dairy Sci.* in press.

[158] Daddaowa, A., Puerta, V., Requena, P., Martinez-Ferez, A., Guadix, E., Sanchez de Mediza, Zarzuelo, A., Dolerez Svarez, M., Josa Boya, J., Martinez-Augustin, O. 2006. Goat milk oligosaccharides are anti-inflammatory in rats with hapten-induced colitis. *J. Nutr.* 136, 672-675.

[159] Clowers, B.H., Dodds, E.D., Seipert, R.R., Lebrilla C.B. 2007. Site determination of protein glycosylation based on digestion with immobilized nonspecific proteases and Fourier transform ion cyclotron resonance mass spectrometry. *J. Proteome Res.* 6, 4032-4040.

[160] Holland, J.W., Deeth, H.C., Alewood, P.F. 2004. Proteomic analysis of κ-casein micro-heterogeneity. *Proteomics* 4, 743-752.

[161] Holland, J.W., Deeth, H.C., Alewood, P.F. 2005. Analysis of *O*-glycosylation site occupancy in bovine κ-casein glycosylation separated by two-dimensional gel electrophoresis. *Proteomics* 5, 990-1002.

[162] Holland, J.W., Deeth, H.C., Alewood, P.F. 2006. Resolution and characterisation of multiple isoforms of bovine κ-casein by 2-DE following a reversible cysteine-tagging enrichment strategy. *Proteomics* 6, 3087-3095.

[163] Saito, T., Itoh, T., Adachi, S. 1980. The chemical structure of the main sugar moiety isolated from bovine whole casein. *Agric. Biol. Chem.* 44, 1023-1030.

[164] Saito, T., Itoh, T., Adachi, S. 1981. The chemical structure of a tetrasaccharide containing *N*-acetyl-glucosamine obtained from bovine colostrum κ-casein. *Biochem. Biophys. Acta* 673, 487-494.

[165] Saito, T., Itoh, T., Adachi, S., Suzuki, T., Usui, T. 1981. The chemical structure of neutral and acidic sugar chains obtained from bovine colostrum κ-casein. *Biochem. Biophys. Acta* 678, 257-267.

[166] Saito, T., Itoh, T., Adachi. S., Suzuki, T., Usui, T. 1982. A new hexasaccharide chain isolated from bovine colostrum κ-casein taken at the time of parturition. *Biochem. Biophys. Acta* 719, 309-317.

[167] Saito, T., Itoh, T. 1992. Variations and distributions of *O*-glycosidically linked sugar chains in bovine κ-casein. *J. Dairy Sci.* 75, 1768-1774.

[168] Otani, H., Monnai, M., Hosono, A. 1992. Bovine k-casein as inhibitor of the proliferation of mouse splenocytes induced by lipopolysaccharide stimulation. *Milchwissenschaft* 47, 512-515.
[169] Strömqvist, M., Falk, P., Bergström, S., Hansson, L., Lönnerdal, B., Normark, S., Hernell, O. 1995. Human milk κ-casein and inhibition of *Helicobacter pylori* adhesion to human gastric mucosa. *J. Pediatr. Gastroenterol. Nutr.* 21, 288-296.
[170] Neeser, J.R., Chambaz, A., Delvedovo, S., Prigent, M.J., Guggenheim, B. 1988. Specific and nonspecific inhibition of adhesion of oral actinomyces and streptococci to erythrocytes and polystyrene by caseinoglycopeptide derivatives. *Infect. Immun.* 56, 3201-3208.
[171] Daddaoua, A., Puerta, V., Zarzuelo, A., Suarez, M.D., de Medina, F.S., Martinez-Augustin, O. 2005. Bovine glycomacropeptide is anti-inflammatory in rats with hapten-induced colitis. *J. Nutr.* 135, 1164-1170.
[172] Otani, H., Monnai, M., Kawasaki, Y., Kawakami, H., Tanimoto, M. 1995. Inhibition of mitogen-induced proliferative responses of lymphocytes by bovine κ-caseinoglycopeptides having different carbohydrate chains. *J. Dairy Res.* 62, 349-357.
[173] Monnai, M., Horimoto, Y., Otani, H. 1998. Immunomodificatory effect of dietary bovine κ-caseinoglycopeptide on serum antibody levels and proliferative responses of lymphocytes in mice. *Milchwissenschaft* 53, 129-132.
[174] Li, E.W.Y., Mine, Y. 2004. Immunoenhancing effects of bovine glycomacropeptide and its derivatives on the proliferative response and phagocytic activities of human macrophagelike cells, U937. *J. Agric. Food Chem.* 52, 2704-2708.
[175] Requena, P., Gonzalez, R., Lopez-Posadas, R., Abadia-Molina, A., Suarez, M.D., Zarzuelo, A., de Medina, F.S., Martinez-Augustin, O. 2010. The intestinal antiinflammatory agent glycomacropeptide has immunomodulatory actions on rat splenocytes. *Biochem. Pharm.* 79, 1797-1804.
[176] Kawasaki Y, Isoda H, Tanimoto M, Dosako S, Idota T, and Ahiko K. 1992. Inhibition by lactoferrin and κ-casein glycomacropeptide of binding of Cholera-toxin to its receptor. *Biosci. Biotechnol. Biochem.* 56, 195-198.
[177] Kawasaki Y, Isoda H, Shinmoto H, Tanimoto M, Dosako S, Idota T, and Nakajima I. 1993. Inhibition by κ-casein glycomacropeptide and lactoferrin of influenza-virus hemagglutination. *Biosci. Biotechnol. Biochem.* 57, 1214-1215.
[178] Azuma, N., Yamauchi, K., Mitsuoka, T. 1984. Bifidus growth-promoting activity of a glycomacropeptide derived from human κ-casein. *Agric. Biol. Chem.* 48, 2159-2162.
[179] Wei, Z., Nishimura, T., Yoshida, S. 2000. Presence of a glycan at a potential *N*-glycosylation site, Asn-281, of bovine lactoferrin. *J. Dairy Sci.* 83, 683-689.
[180] Ward, P.P., Paz, E., Conneely, O.M. 2005. Multifunctional roles of lactoferrin: a critical overview. *Cell Mol. Life Sci.* 62, 2540-2548.
[181] Weinberg, E.D. 2007. Antibiotic properties and applications of lactoferrin. *Curr. Pharm. Des.* 13, 801-811.
[182] Actor, J.K., Hwang, S.A., Kruzel, M.L. 2009. Lactoferrin as a natural immune modulator. *Curr. Pharm. Des.* 15, 1956-1973.
[183] Kanwar, J.R., Kanwar, R.K., Sun, X., Punj, V., Matta, H., Morley, S.M., Parratt, A., Puri, M., Sehgal, R. 2009. Molecular and biotechnological advances in milk proteins in relation to human health. *Curr. Protein Pept. Sci.* 10, 308-338.

[184] Rodrigues, L., Teixeira, J., Schmitt, F., Paulsson, M., Månsson, H.L. 2009. Lactoferrin and cancer disease prevention. *Crit. Rev. Food Sci. Nutr.* 49, 203-217.

[185] van Berkel, P.H.C., Geerts, M.E.J., Vanveen. H.A., Kooiman, P.M., Pieper, F.R., de Boer, H.A., Nuijens, J.H. 1995. Glycosylated and unglycosylated human lactoferrins both bind iron and show identical affinities towards human lysozyme and bacterial lipopolysaccharide, but differ in their susceptibilities towards tryptic proteolysis. *Biochem J.* 15, 107-114.

[186] Girardet, J.M., Saulnier, F., Gaillard, J.L., Ramet, J.P., Humbert, G. 2000. Camel (*Camelus dromedarius*) milk PP3: evidence for an insertion in the amino-terminal sequence of the camel milk whey protein. *Biochem. Cell Biol.* 78, 19-26.

[187] Kanno, C. 1989. Purification and separation of multiple forms of lactophorin from bovine milk whey and their immunological and electrophoretic properties. *J. Dairy Sci.* 72, 883-891.

[188] Sørensen, E.S., Rasmussen, L.K., Møller, L., Petersen, T.E. 1997. The localization and multimeric nature of component PP3 in bovine milk: purification and characterization of PP3 from caprine and ovine milks. *J. Dairy Sci.* 80,3176-3181.

[189] Cantisani, A., Napolitano, L., Giuffrida, M.G., Conti, A. 1990. Direct identification and characterization of llama (*Lama glama* L.) whey proteins by microsequencing after western blotting. *J. Biochem. Biophys. Methods* 21, 227-236.

[190] Girardet, J.M., Linden, G. 1996. PP3 component of bovine milk: a phosphorylated whey glycoprotein. *J. Dairy Res.* 63, 333-350.

[191] Girardet, J.M., Coddeville, B., Plancke, Y., Strecker, G., Campagna, S., Spik, G., Linden, G. 1995. Structure of glycopeptides isolated from bovine milk component PP3. *Eur. J. Biochem.* 234, 939-946.

[192] Coddeville, B., Girardet, J.M., Plancke, Y., Campagna, S., Linden, G., Spik, G. 1998. Structure of the *O*-glycopeptides isolated from bovine milk component PP3. *Glycoconj. J.* 15, 371-378.

[193] Inagaki, M., Nakaya, S., Nohara, D., Yabe, T., Kanamaru, Y., Suzuki, T. 2010. The multiplicity of *N*-glycan structures of bovine milk 18 kDa lactophorin (milk GlyCAM-1). *Biosci. Biotechnol. Biochem.* 74, 447-450.

[194] Shimizu, M., Yamauchi, K., Saito M. 1989. Emulsifying properties of the proteose-peptone fraction obtained from bovine milk. *Milchwissenschaft* 44, 497-500.

[195] Sugahara, T., Onda, H., Shinohara, Y., Horii, M., Akiyama, K., Nakamoto, K., Hara K. 2005. Immunostimulation effects of proteose-peptone component 3 fragment on human hybridomas and peripheral blood lymphocytes. *Biochim. Biophys. Acta* 1725, 233-240.

[196] Cartier, P., Chilliard, Y., Paquet, D. 1990. Inhibiting and activating effects of skim milks and proteose-peptonefractions on spontaneous lipolysis and purified lipoprotein-lipase activity in bovine-milk. *J. Dairy Sci.* 73, 1173-1177.

[197] Girardet, J.M., Linden, G. 1993. Study of mechanism of lipolysis inhibition by bovine milk proteose-peptone component 3. *J. Dairy Sci.* 76, 2156-2163.

[198] Campagna, S., Mathot, A.G., Fleury, Y., Girardet, J.M., Gaillard, J.L. 2004. Antibacterial activity of lactophoricin, a synthetic 23-residues peptide derived from the sequence of bovine milk component-3 of proteose peptone. *J. Dairy Sci.* 87, 1621-1626.

[199] Cavaletto, M., Giuffrida, M.G., Conti, A. 2008. Milk fat globule membrane components-a proteomic approach. *Adv. Exp. Med. Biol.* 606, 129-141.

[200] Reinhardt, T.A., Lippolis, J.D. 2006. Bovine milk fat globule membrane proteome. *J. Dairy Res.* 73, 406-416.
[201] Fong, B.Y., Norris, C.S., MacGibbon, A.K.H. 2007. Protein and lipid composition of bovine milk-fat-globule membrane. *Int. Dairy J.* 17, 275-288.
[202] Smolenski, G., Haines, S., Kwan, F.Y., Bond, J., Farr, V., Davis, S.R., Stelwagen, K., Wheeler, T.T. 2007. Characterisation of host defence proteins in milk using a proteomic approach. *J. Proteome Res.* 6, 207-215.
[203] Cebo, C., Caillat, H., Bouvier, F., Martin, P. 2010. Major proteins of the goat milk fat globule membrane. *J. Dairy Sci.* 93, 868-876.
[204] Quaranta, S., Giuffrida, M.G., Cavaletto, M., Giunta, C., Godovac-Zimmermann, J., Cañas, B., Fabris, C., Bertino, E., Mombrò, M., Conti, A. 2001. Human proteome enhancement: high-recovery method and improved two-dimensional map of colostral fat globule membrane proteins. *Electrophoresis* 22, 1810-1818.
[205] Wu, C.C., Howell, K.E., Neville, M.C., Yates, J.R. 3rd, McManaman, J.L. 2000. Proteomics reveal a link between the endoplasmic reticulum and lipid secretory mechanisms in mammary epithelial cells. *Electrophoresis* 21, 3470-3482.
[206] Peterson, J.A., Hamosh, M., Scallan, C.D., Ceriani, R.L., Henderson, T.R., Mehta, N.R., Armand, M., Hamosh, P. 1998. Milk fat globule glycoproteins in human milk and in gastric aspirates of mother's milk-fed preterm infants. *Pediatr. Res.* 44, 499-506.
[207] Robenek, H., Hofnagel, O., Buers, I., Lorkowski, S., Schnoor, M., Robenek, M.J., Heid, H., Troyer, D., Severs, N.J. 2006. Butyrophilin controls milk fat globule secretion. *Proc. Natl. Acad. Sci. U. S. A.* 103, 10385-10390.
[208] Shimizu, M., Yamauchi, K., Miyauchi, Y., Sakurai, T., Tokugawa, K., McIlhinney, R.A. 1986. High-Mr glycoprotein profiles in human milk serum and fat-globule membrane. *Biochem. J.* 233, 725-730.
[209] Patton, S., Huston, G.E., Jenness, R., Vaucher, Y. 1989. Differences between individuals in high-molecular weight glycoproteins from mammary epithelia of several species. *Biochim. Biophys. Acta* 980, 333-338.
[210] Liu, C., Erickson, A.K., Henning, D.R. 2005. Distribution and carbohydrate structures of high molecular weight glycoproteins, MUC1 and MUCX, in bovine milk. *J. Dairy Sci.* 88, 4288-4294.
[211] Pallesen, L.T., Andersen, M.H., Nielsen, R.L., Berglund, L., Petersen, T.E., Rasmussen, L.K., Rasmussen, J.T. 2001. Purification of MUC1 from bovine milk-fat globules and characterization of a corresponding full-length cDNA clone. *J. Dairy Sci.* 84, 2591-2598.
[212] Sato, T., Furukawa, K., Greenwalt, D.E., Kobata, A. 1993. Most bovine milk fat globule membrane glycoproteins contain Asparagine-linked sugar chains with GalNAcβ1→5GlcNAc groups. *J. Biochem.* 114, 890-900.
[213] Smith, P.L., Baeziger, J.U. 1990. Recognition by the glycoprotein hormone-specific *N*-acetylgalactosaminetransferase is independent of hormone native conformation. *Proc. Natl. Acad. Sci. USA* 87, 7275-7279.
[214] Sato, T., Takio, K., Kobata, A., Greenwalt, D.E., Furukawa K. 1995. Site-specific glycosylation of bovine butyrophilin. *J. Biochem.* 117, 147-157.
[215] Wilson, N.L., Robinson, L.J., Donnet, A., Bovetto, L., Packer, N.H., Karlsson, N.G. 2007. Glycoproteomics of milk: differences in sugar epitopes on human and bovine milk fat globule membranes. *J. Proteome Res.* 7, 3687-3696.

[216] Ward, P.P., Lo, J.Y., Duke, M., May, G.S., Headon, D.R., Conneely, O.M. 1992. Production of biologically active recombinant human lactoferrin in *Aspergillus oryzae*. *Biotechnology (N. Y.)* 10, 784-789.

[217] González-Chávez, S.A., Arévalo-Gallegos, S., Rascón-Cruz, Q. 2009. Lactoferrin: structure, function and applications. *Int. J. Antimicrob. Agents* 33, 301.e1-8.

[218] Sprong, R.C., Hulstein, M.F.E., van der Meer, R. 2001. Bactericidal activities of milk lipids. *Antimicrob Agents Chemother.* 45, 1298-1301.

[219] Takamizawa, K., Iwamori, M., Mutai, M., Y. Nagai, Y. 1986. Gangliosides of Bovine Buttermilk. *J. Biol. Chem.* 261, 5625-5630.

[220] Hanagata, G., Kobayashi, T., Yanahira, S., Deya, E., Gasa, S. 1991. Characterization of ganglioside 9-O-acetyl-GD_3 from Bovine cheese whey and human milk. *Glycoconj. J.* 8, 254-255.

[221] Rombaut, R., Dewettinck, K. 2006. Properties, analysis and purification of milk polar lipids. *Int. Dairy J.* 16, 1362-1373.

[222] Martin, M.J., Martin-Sora, S., Hueso, P. 2001. Bovine Milk Gangliosides: Changes in Ceramide Moiety with Stage of Lactation. *Lipids* 36, 291-298.

[223] Rueda, R. 2007. The role of dietary gangliosides on immunity and the prevention of infection. *Br. J. Nutr.* 98, Suppl. 1, S68-S73.

[224] Pan, X.L., Izumi, T. 2000. Variation of the ganglioside compositions of human milk, cow's milk and infant formulas. *Early Hum. Dev.* 57, 25-31.

[225] Takamizawa, K., Iwamori, M., Mutai, M., Nagai, Y. 1986. Selective changes in gangliosides of human milk during lactation: a molecular indicator for the period of lactation. *Biochim. Biophys. Acta* 879, 73-77.

[226] Pan, X.L., Izumi, T. 1999. Chronological changes in the ganglioside composition of human milk during lactation. *Early Hum. Dev.* 55, 1-8.

[227] Otnaess, A.-B.K., Legreid, A., Ertresvag, K. 1983. Inhibition of Enterotoxin from *Escherichia coli* and *Vibrio cholera* by Gangliosides from Human Milk. *Infect. Immun.* 40, 563-569.

[228] McJarrow, P., Schnell, N., Jumpsen, J., Clandinin, T. 2009. Influence of dietary gangliosides on neonatal brain development. *Nutr. Rev.* 67, 451-463.

[229] Otnaess, A.-B.K., Laegreid, A. 1986. Detection of Femtomolar Quantities of the Ganglioside G_{M1} on Thin-layer Chromatography Plates by Native Choleratoxin and Labeled Antisera. *Curr. Microbiol.* 13, 323-326.

[230] Laegreid, A., Otnaess, A.-B.K. 1987. Trace amounts of ganglioside GM1 in human milk inhibit enterotoxins from *Vibrio cholera* and *Escherichia coli*. *Life Sci.* 40, 55-62.

[231] Suzuki, Y., Nakao, T., Ito, T., Watanabe, N., Toda, Y., Guiyu, X., Suzuki, Kobayashi, T., Kimura, Y., Yamada, A., Sugawara, K., Nishimura, H., Kitame, F., Nakamura, K., Deya, E., Kiso, M., Hasegawa, A. 1992. Structural determination of gangliosides that bind to influenza A, B, and C viruses by an improved binding assay: Strain-specific receptor epitopes in sialo-sugar chains. *Virology* 189, 121-131.

[232] Houghton, A.N., Mintzer, D., Cordon-Cardo, C., Welt, S., Fligel, B., Vadhan, S., Carswell, E., Melamed, M.R., Oettgen, H.F., Old, L.J. 1985. Mouse monoclonal IgG3 antibody detecting G_{D3} ganglioside: A phase I trial in patients with malignant melanoma. *Proc. Natl. Acad. Sci. USA* 82, 1242-1246.

[233] Ritter, G., Boosfeld, E., Markstein, E., Yu, R.K., Ren, S., Stallcup, W.B., Oettgen, H.F., Old, L.J., Livingston, P.O. 1990. Biochemical and Serological Characteristics of

Natural 9-*O*-Acetyl GD3 from Human Melanoma and Bovine Buttermilk and Chemically *O*-Acetylated GD3. *Cancer. Res.* 50, 1403-1410.
[234] Takamizawa, K., Iwamori, M., Kozaki, S., Sakaguchi, G., Tanaka, R., Takayama, H., Nagai, Y. 1986. TLC immunostaining characterization of *Clostridium botulinum* type A neurotoxin binding to gangliosides and free fatty acids. *FEBS Lett.* 201, 229-232.
[235] Park, E.J., Suh, M., Ramanujam, K., Steiner, K., Begg, D., Clandinin, M.T. 2005. Diet-Induced Changes in Membrane Gangliosides in Rat Intestinal Mucosa, Plasma and Brain. *J. Pediatr. Gastroenterol. Nutr.* 40, 487-495.
[236] Wang, B., McVeagh, P., Petocz, P., Brand-Miller, J. 2003. Brain ganglioside and glycoprotein sialic acid in breastfed compared with formula-fed infants. *Am. J. Clin. Nutr.* 78, 1024-1029.
[237] Park, E.J., Suh, M., Thmson, B., Thomson, A.B.R., Ramanujam, K.S., Clandinin, M.T. 2005. Dietary ganglioside decreases cholesterol content, caveolin expression and inflammatory mediators in rat intestinal microdomains. *Glycobiology*, 15, 935-942.
[238] Koumanov, K., Wolf, C., Bereziat, G. 1997. Modulation of human type II secretory phosholipase A_2 by sphingomyelin and annexin VI. *Biochem. J.* 326, 227 -233.
[239] Deya, E. 1994. Feature and utilization of bovine milk glycolipids. *Jpn J. Dairy Food Sci.*, 43, A61-A64.
[240] Sugawara, M., Idota, T. 1995. *A new oligosaccharide; 4'-galactosyl-lactose in human milk.* Proceedings of the Annual Meeting of Japan Society for Bioscience, Biotechnology, and Agrochemistry. Sappro, Japan, p.132.
[241] Dabrowski, U., Egge, H., Dabrowski, J. 1983. Proton-nuclear magnetic resonance study of peracetylated derivatives of ten oligosaccharides isolated from human milk. *Arch. Biochem. Biophys.* 224, 254-260.
[242] Dua, V.K., Goso, K., Dube, V.E., Bush, C.A. 1985. Characterization of lacto-*N*-hexaose and two fucosylated derivatives from human milk by high-performance liquid chromatography and proton NMR spectroscopy. *J. Chromatogr.* 328, 259-269.
[243] Sabharwal, H., Nilsson, B., Grönberg, G., Chester, M.A., Dakour, J. Sjöblad, S., Lundblad, A. 1988. Oligosaccharides from faeces of preterm infants fed on breast milk. *Arch. Biochem. Biophys.* 265, 390-406.
[244] Bruntz, R., Dabrowski, U., Dabrowski, J., Ebershold, A., Peter-Katalinic, J., Egge, K. 1988. Fucose containing oligosaccharides from human milk from a donor of blood group O Le^a non secretor. *Biol Chem. Hoppe-Seyler*, 369, 257-273.
[245] Sabharwal, H., Nilsson, B., Chester, M.A., Lindh, F., Grönberg, G., Sjöblad, S., Lundblad, A. 1988. Oligosaccharides from feces of a blood-group B, breast-fed infants. *Carbohydr. Res.* 178, 145-154.
[246] Strecker, G., Wieruszeski, J.M., Michlski, J.C., Montreuil, J. 1988. Structure of a new nonasaccharide isolated from human milk; IV^2Fuc, V^4Fuc, III^3Fuc-*p*-Lacto-*N*-hexaose. *Glycoconj. J.* 5, 385-396.
[247] Yamashita, K., Tachibana, Y., Kobata, A. 1976. Oligosaccharides of human milk. Isolation and characterization of two new nonasaccharides, monofucosyllacto-*N*-octaose and monofucosyllacto-*N*-neooctaose. *Biochemistry* 15, 3950-3955.
[248] Kogelberg, H., Piskarev, V.E., Zhang, V., Lawson, A.M., Chai, W. 2004. Determination by electrospray mass spectrometry and 1H-NMR spectroscopy of primary structures of variously fucosylated neutral oligosaccharides based on the iso-lacto-*N*-octaose core. *Eur. J. Biochem.* 271, 1172-1186.

[249] Strecker, G., Wieruszeski, J.M., Michalski, J.C., Montreuil, J. 1989. Primary structure of human milk nona- and deca-saccharides determined by a combination of fast atom bombardment masd spectrometry and ^{1}H-/^{13}C-niclear magnetic resonance spectroscopy. Evidence for a new core structure, iso-lacto-*N*-octaose. *Glycoconj. J.* 6, 169-182.

[250] Wieruszeski, J.M., Chekkor, A., Bouquelot, S., Montreuil, J., Stercker, G., Peter-Katalinic, J. Egge, H. 1985. Structure of two new oligosaccharides isolated from human milk: Sialylated lacto-*N*-fucopentaose I and II. *Carbohydr. Res.* 137, 127-133.

[251] Smith, D.F., Prieto, P.A., Meerumb, D.K., Wang, W.C., 1987. A novel sialylfucopentaose in human milk. *J. Bol. Chem.* 262, 12040-12047.

[252] Tarrago, M.T., Tucker, K.H., van Halbeek, H., Smith, D.F. 1988. A novel sialylhaxasaccharide from human milk: Purification by affninity chromatography on immobilized wheat germ agglutinin. *Arch. Biochem. Biophys.* 267, 353-362.

[253] Grimmonorez, L., Montreuil, J. 1968. Etude physico-chemique de sex nouveaux oligosaccharides isolees du lait de femme. *Bull. Soc. Chim. Biol.* 50, 843-855.

[254] Yamashita, K., Tachibana, Y., Kobata, A. 1976. Oligosaccharides of human milk. Isolation and characterization of three new disialylfucosylhexasaccharides. *Arch. Biochem. Biophys.* 174, 582-591.

[255] Fiévre, S., Wieruszeski, M., Michalski, J.C., Lemoine, J., Montreuil, J., Strecker, G., 1991. Primary structure of a trisialylated oligosaccharide from human milk. *Biochem. Biophys. Res. Commun.* 177, 720-725.

[256] Guerardel, Y., Morelle, W., Plancke, Y., Lemoine, J., Strecker, J. 1999. Structural analysis of three sulfated oligosaccharides isolated from human milk. *Carbohydr. Res.* 320, 230-238.

[257] Saito, T., Itoh, T., Adachi, S. 1984. Presence of two neutral disaccharides containing *N*-acetylhexosamine in bovine colostrum as free forms. *Bochim. Biophys. Acta* 801, 147-150.

[258] Urashima, T., Saito, T., Ohmisya, K., Shimazaki, K. 1991. Structural determination of three neutral oligosaccharides in bovine (Holstein-Friesian) colostrum, including the novel trisaccharide: GalNAcα1-3Galβ1-4Glc. *Biochim. Biophys. Acta* 1073, 225-229.

[259] Kimura, K., Watanabe, Y., Matsumoto, K., Miyagi, A. 1997. Studies on the neutral galactooligosaccharides in commercial cow's milk. *Yakult Reports*, 17, 1-7.

[260] Watanabe, Y., Kimura, K., Nakamura, T., Kume, S., Urashima, T. 2006. *Concentrations and compositions of the bovine milk oligosaccharides*. Proceedings of the 23rd International Carbohydrate Symposium. Whistler, Canada, p. 78.

[261] Cumar, F.A., Ferchmin, P.A., Caputto, R. 1965. Isolation and identification of a lactose phosphate ester from cow colostrum. *Biochim. Biophys. Res. Comm.* 20, 60-62.

[262] Schneir, M.L., Rafelson, M.E. 1966. Isolation of two structural isomers of *N*-acetylneuraminyllactose from bovine colostrum. *Biochim. Biophys. Acta* 130, 1-11.

[263] Veh, R.W., Michalski, J.C., Corfield, A.P., Sander-Wewer, M., Gies, D., Schauer, R. 1981. New chromatographic system and the rapid analysis and preparation of colostrum sialyloligosaccharides. *J. Chromatogr.* 212, 313-322.

[264] Parkkinen, J., Finne, J. 1987. Isolation of sialyl oligosaccharides and sialyl oligosaccharide phosphates from bovine colostrum and human urine. *Methods Enzymol.* 138, 289-300.

In: Oligosaccharides: Sources, Properties and Applications ISBN 978-1-61122-193-0
Editor: Nicole S. Gordon

Chapter 2

PREBIOTIC OLIGOSACCHARIDES: ORIGINS AND PRODUCTION, HEALTH BENEFITS AND COMMERCIAL APPLICATIONS

Santad Wichienchot*[*] *and Pavinee Chinachoti

Nutraceutical and Functional Food Research and Development Center, Faculty of Agro-Industry, Prince of Songkla University, Hat Yai, Songkhla, 90112, Thailand

ABSTRACT

In recent years, oligosaccharides and their derivatives have become useful for health applications in various fields because of their specific biological activities. In the food industries, several oligosaccharides have received increasing attention as key components for functional foods and nutraceutical products. Prebiotics are non-digestible oligosaccharides which have been shown to have properties that can modulate gastrointestinal problems and improve general health and well being. The benefits of prebiotic include relief of constipation, reduced risk of colon cancer, inhibition of pathogens in gastrointestinal tract, increased minerals absorption, immune modulation, short chain fatty acid and vitamin production, reduced blood cholesterol and lipids and improved microbial balance in the gut. Prebiotics are distinguished from other dietary fibers due to its ability to selectively promote fermentation by bifidobacteria and/or lactobacilli within the gut. The "prebiotics" was named by Gibson and Roberfroid in 1995 as "a non digestible food ingredient that beneficially affects the host by selectively stimulating the growth and/or activity of one or a limited number of bacteria in the colon, that can improve the host health". It was redefined several times until 2007 when FAO defined as a "non-viable food component that confers a health benefit on the host associated with modulation of the microflora.

Currently, commercial prebiotics are non-digestible but fermentable oligosaccharides which are consisted of sugar moieties between 2 and 10 saccharide units, with an exception of inulin and beta-glucans which are polysaccharides. Prebiotic oligosaccharides can be obtained from natural sources such as chicory and artichoke and

---

* E-mail: santad.w@psu.ac.th

produced through chemical synthesis (lactulose) or by enzymatic synthesis (hydrolysis or transfer reaction). Currently commercialized prebiotic oligosaccharides for food use include fructo-oligosaccharides, galacto-oligosaccharides, lactosucrose, isomalto-oligosaccharides, gentio-oligosaccharides, xylo-oligosaccharides and soybean oligosaccharides (whereas lactulose has been used as a laxative). Recommendation on prebiotics consumption is 3-10 g per day. Effectiveness on each type of prebiotic varies; for example, lactulose, lactosucrose and xylo-oligosaccharides have been reported more effective at low doses. Prebiotics are popular food ingredients used in functional foods and nutraceutical products such as low calories and mind sweetness for weight control or diabetes and for well being of gut health. Combination of prebiotics and probiotics (synbiotic products) offer synergistic benefits to the GI tract, such as in infant powder formula, beverage, yoghurt and dairy products. Functional claims of prebiotic depend on local law of each country, however, most country accepted as a source of dietary fiber.

1. INTRODUCTION

Today consumer interest in self-care and integrative medicine and also it has strong evidence scientific support on understanding of the relationship between health and diet. Thus the market for functional foods and nutraceuticals is continuing to expand rapidly. Among these products, prebiotics containing products have been much interest today because of their scientifically supported health promoting properties. The prebiotics currently in use are non-digestible, but fermentable carbohydrates, especially oligosaccharides, by selective strains of human microflora in the gut. A substantial amount of research on human clinical trials indicated that prebiotics may indeed prove to be a clinically beneficial dietary supplement (Charalampopoulos & Rastall, 2009). The combination of selective effect on the microflora and fermentation should lead to several benefits for health such as modulate the gut flora, to affect different gastrointestinal activities and lipid metabolism, to enhance immunity, to reduce risk to metabolic diseases (e.g., diabetes, obesity and cardiovascular diseases) and to protect against colorectal cancer (Qiang *et al.*, 2009). Quantities of carbohydrate entering the gut vary with diet, but it has been estimated that about 20-60 g of carbohydrate become available for microbial fermentation in the colon daily (Macfarlane & Gibson, 1997). However, minimum effective levels are 8-15 g/d for FOS, 20-40 g/d for inulin and 3 g/d for lactulose (Rycroft *et al.*, 2001). Thus, to bring about a change in the composition or relative population levels of selected groups of bacteria within this metabolically diverse ecosystem, the prebiotic must be selectively and differentially fermented by the bifidobacteria and/or lactobacilli.

The term prebiotic was first defined by Gibson and Roberfroid (1995) as "a non-digestible food ingredient that beneficially affects the host by selectively stimulating the growth and/or activity of one or a limited number of bacteria in the colon that can improve the host health". However, the definition was re-defined as "a selectively fermented ingredient that allows specific changes, both in the composition and/or activity in the gastrointestinal microbiota that confers benefits upon host wellbeing and health." (Gibson, 2004) until FAO (2007) defined as "a non-viable food component that confers a health benefit on the host associated with modulation of the microbiota" (Yanbo, 2009). The prebiotic concept considers that many potentially health-promoting microorganisms, such as bifidobacteria and lactobacilli, are already resident in the human colon. Prebiotics must be stable in the acid of the stomach, reach the colon where they are then selectively fermented by positive bacteria and they must not be absorbed in the small intestine (Roberfroid, 2001).

The traditional means of manipulating the gut flora was by the use of probiotics. The term probiotic was first defined by Parker (1974) as 'organisms and substances which contribute to intestinal microbial balance'. Later, Fuller (1989) re-defined as 'a live microbial feed supplement which beneficially affects the host animal by improving its intestinal microbial balance'. A synbiotic is defined as 'the mixtures of pro- and prebiotic, which beneficially affect the host, by improving the survival and implantation of live microbial dietary supplements in the gastrointestinal tract' (Gibson & Roberfroid, 1995). Examples of health promoting effects and mechanism of probiotics, prebiotics and synbiotics are summarized in Table 1.

Table 1. Health promoting effects and mechanisms of probiotic bacteria and prebiotic substances

Beneficial Effect	Mechanism	References
Reduce risk of pathogen infection	Decreased pH and produced bacteriocins-inhibited growth of undesirable bacteria	Lawhon *et al.* (2002) Tzortzis *et al.* (2004)
Improve mineral absorption	Decreased pH- increased mineral (esp. calcium) solubility	Scholz-Ahrens *et al.* (2001) Gudiel-Urbano and Goni (2002)
Decrease lipids and cholesterol	Increased short-chain fatty acid (SCFA)-modulated lipogenesis Lowering of pH & biles precipitation Suppression of hepatic triglyceride and very low density lipoprotein (VLDL) synthesis	Taylor and Williams (1998) Delzenne and Kok (2001)
Decrease risk of cancer	Increased butyrate- fuel for colonocytes and cell differentiation Decreased bile acid formation Decreased genotoxic metabolites & enzymes and carcinogens	Reddy (1998), Brady *et al.* (2000) Sanderson *et al.* (2004)
Immune system	Direct contact of lactic acid bacteria or bacterial product with immune cells in the intestine Production of SCFA	Salminen *et al.* (1998) Schley and Field (2002)
Reduce constipation	Faecal bulking & fibre-like effects	Hara *et al.* (1994), Mizota (1996)
Reduce diarrhoea	Reduction in pathogens causing infection	Gibson and Wang (1994a)
Aid in irritable bowel syndrome (IBS)	Lactobacilli inhibited *Candida albicans*, thought to be the IBS causative organism	Madden and Hunter (2002)
Weight management	Prebiotics could be relevant in the management of obesity by controlling gut microbiota	Delzenne and Cani (2010)

Source: Adapted from Gibson & Roberfroid (2008).

2. PREBIOTIC OLIGOSACCHARIDES

There are three main types of carbohydrates that are non-digestible in the human small intestine: non-starch polysaccharides (NSP), resistant starch (RS) and non-digestible oligosaccharides (NDO) (Voragen, 1998). Any dietary material that resists digestion in the upper gut and enters the large intestine is a candidate prebiotic (Crittenden *et al.*, 2001), however at the present time the only molecules known to act as prebiotics are carbohydrates, particularly oligosaccharides (Charalampopoulos & Rastall, 2009). Many oligosaccharides are not metabolised in the human small intestine (i.e. they are NDO), which is the first requirements of a prebiotic, and pass to the colon quantitatively. When the prebiotic arrives in the colon, certain members of the indigenous microflora must ferment it selectively which is the most important property of a prebiotic (Gibson & Roberfroid, 1995).

One of the biggest blocks to the development of prebiotics is our lack of knowledge of the structure-function relationships in these molecules. A useful prebiotic would; easily incorporate into appropriate food vehicles, possess multiple functions, have activity at low dosages, be non-carcinogenic, have a low calorific value and target the distal colon. It is possible to identify certain properties that the oligosaccharides should possess in order to achieve these attributes (Table 2).

Table 2. Design parameters for enhanced activity prebiotics

Desirable attribute in prebiotic	Properties of oligosaccharides
Active at low dosage	Highly selectively and efficiently metabolised by *Bifidobacterium* and/or *Lactobacillus* sp.
Lack of side effects	Highly selectively metabolised by beneficial bacteria but not by gas producers, putrefactive organisms, etc.
Persistence through the colon	High molecular weight, correct choice of glycosyl residue, chemical modification
Varying viscosity	Available in different molecular weights and linkages
Good storage and processing stability	Possess 1-6 linkages and pyranosyl sugar rings
Fine control of microflora modulation	Selectively metabolised by restricted species
Varying sweetness	Different monosaccharide composition
Inhibits adhesion of pathogens	Possess receptor sequence

Source: Charalampopoulos & Rastall (2009).

Typically, oligosaccharides are sugars consisting of between 2 and 10 saccharide units. Oligosaccharides can be commercially produced through the hydrolysis of polysaccharides or through enzymatic transfer reactions from lower molecular weight sugars. Various aspects of the production and properties of food grade oligosaccharides and their prebiotic effects have been tested using *in vitro* methods (i.e. batch and three-stage continuous culture) and *in vivo* methods (i.e. animal models and human clinical trials). Prebiotic oligosaccharides currently commercialized as food ingredients include fructo-oligosaccharides, galacto-oligosaccharides, lactosucrose, isomalto-oligosaccharides, gentio-oligosaccharides and xylo-oligosaccharides

(Gibson & Rastall, 2006). The current commercial available prebiotic oligosaccharides are as follow (Charalampopoulos & Rastall, 2009; Gibson & Rastall, 2006) which is summarized in Figure 1.

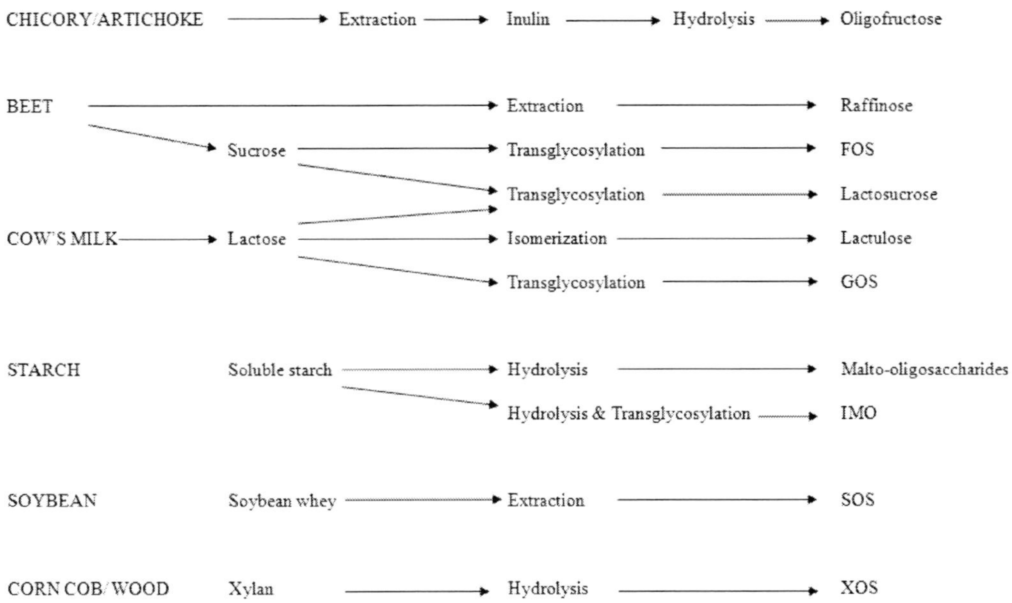

Adapted from Rastall, 2000.

Figure 1. Schematic process representation of prebiotic oligosaccharides.

2.1. Lactulose

Lactulose (4-O-β-D-galactopyranosyl-D-fructose) is a synthetic disaccharide in the form of Gal β1-4 Fru, produced by catalytic isomerization from lactose. Since the 1950s, lactulose has been used for the treatment of specific medical conditions in humans (Petuely, 1957 cited by Schumann, 2000). Lactulose is generally classified as a drug and it is a prebiotic experimentally but is not used commercially as a food ingredient.

Lactulose effectively escapes degradation in the stomach and small intestine and reaches the colon intact. When fed to patients with chronic constipation at doses greater than 20 g/day, lactulose increases faecal output. At lower doses, lactulose acts as a prebiotic within the colonic microflora, increasing numbers of bifidobacteria (Tuohy *et al.,* 2005). A number of human feeding studies have confirmed the prebiotic nature of lactulose in healthy volunteers. Terada and co-workers (1992) fed 8 volunteers lactulose at 3 g/day for 14 days and it was found a significant increase in numbers of bifidobacteria recovered from stool samples. Conversely, numbers of *Clostridium perfringens*, *Bacteroides* spp., *Streptococcus* spp. and the *Enterobacteriaceae* decreased. In addition, a number of potentially toxic microbial metabolites also decreased significantly upon lactulose ingestion and the activity of microflora associated enzymes involved in the production of carcinogens also decreased. Similar results were observed by Ballongue and co-workers (1997) when 36 volunteers were fed lactulose at 2 x 10 g/day for 4 weeks. Numbers of faecal bifidobacteria increased with lactulose ingestion and a concomitant increase in numbers of lactobacilli was reported. Numbers of *Bacteroides* spp., *Clostridium* spp. and coliforms were reduced where reduction on carcinogenic microbial enzyme activities and toxic faecal metabolites. The faecal concentration of acetate, also increased significantly upon lactulose ingestion. These studies used traditional culture methodology for the microbial analyses.

The prebiotic nature of lactulose at sub-clinical doses (10 g/day) has been recently confirmed using both traditional microbiological techniques and direct molecular enumeration of faecal microbial populations using FISH (Tuohy *et al.,* 2005). Ten volunteers were fed lactulose at 10 g/day and 10 volunteers were fed a placebo (glucose/lactose). Faecal water genotoxicity was assessed using the "Comet" assay to investigate the ability of lactulose to protect against DNA damage to colonocytes. Numbers of faecal bifidobacteria increased significantly during lactulose intake, whilst genetic probing showed a concomitant decrease in *Clostridium perfringens/histolyticum.* For inulin and FOS, the degree of bifidogenesis observed in volunteers fed lactulose was related to initial *Bifidobacterium* population size. Lactulose ingestion did not show elevated levels of DNA protection when compared to control samples. Lactulose is producing by Chephassar Chem-pharm (Germany), Danipharm (Denmark), Melei GmbH (Japan), Moringa Milk Industry (Japan), Resolution (UK) (Angus, 2000).

2.2. Inulin and Fructo-Oligosaccharides (FOS)

Inulin and fructo-oligosaccharides are β(2-1) fructans, which contain a mixture of oligo- and polysaccharides. They are almost all linear chains of fructose having the structure GF_n (with G=glucosyl unit, F=fructosyl unit and n=number of fructosyl units linked to one another) (De Leenheer & Hoebregs, 1994). In native chicory inulin, the number of fructose units linked together ranges from 2 to more than 60, with average degree of polymerization (DP) in the order of 10-12. Inulin having an average DP of about 25 is also available commercially from Orafti (Tienen, Belgium) as Raftiline HP, for "High Performance". Oligofructose obtained from inulin contains GF_n and F_n chains with number of fructosyl units ranging from 2 to 9, whereas oligofructose as produced from sucrose only has GF_n forms with number of fructosyl unit between 2 and 4 (Bornet, 1994; De Leenheer & Hoebregs, 1994).

Fructans are, after starch, the most abundant non-structural natural polysaccharides. They are present in a wide variety of plants and some bacteria and fungi. Plants containing inulin and oligofructose primarily belong to either Liliales, e.g. leek, onion, garlic and asparagus, or the Compositae, such as Jerusalem artichoke, dahlia, yacon and chicory (Praznik *et al*., 2004). Some candidate prebiotics derived from tropical fruits and vegetables was recently reported (Thammarutwasik *et al*., 2009; Wichienchot *et al*., 2010).

Fructo-oligosaccharide are probably the most extensively studied prebiotics and their ability to selectively stimulate *Bifidobacterium* spp. within the gut microflora is now well established and has been confirmed by numerous *in vitro* and *in vivo* studies (Roberfroid *et al.,* 1998). In pH controlled co-cultures, using FOS as sole carbon source, showed that *Bifidobacterium infantis* not only grew well on FOS but also inhibited *Escherichia coli* and *Clostridium perfringens* (Gibson & Wang, 1994).

Inulin and FOS have also been shown to selectively stimulate numbers of bifidobacteria in complex *in vitro* systems closely mimicking the microbial diversity of the human gut (Kolida *et al.,* 2002). Although *in vitro* models of the human gut microflora are useful for the early development of potentially prebiotic oligosaccharides and in comparing prebiotic capabilities of different oligosaccharides under controlled laboratory conditions, the final step in the validation of prebiotic activity can only be derived from human feeding studies. Tuohy and co-workers (1995) fed 15 g of FOS or inulin for 15 days to 8 healthy male volunteers in a 45 day placebo controlled, double-blind, cross-over trial with 8 healthy male volunteers. They reported that consumption of either FOS or inulin resulted in a significant increase in numbers of bifidobacteria recovered from stool samples. During consumption of FOS, numbers of *Bacteroides* spp., *Clostridium* spp. and fusobacteria decreased, while numbers of Gram positive cocci decreased during inulin consumption. Similarly, Roberfroid and co-workers (1997) confirmed the bifidogenic nature of chicory FOS and also observed a concomitant reduction in numbers of *Bacteroides* spp. upon prebiotic dietary supplementation in 8 healthy volunteers. Reviewing both *in vitro* and *in vivo* studies carried-out on the prebiotic nature of FOS and inulin concluded that at doses between 4 and 15g/day both substrates give rise to significant increases in numbers of bifidobacteria within the gut microflora, and that the degree of bifidogenesis is related to the initial numbers of faecal bifidobacteria (Roberfroid *et al.,* 1998).

The prebiotic nature of FOS, combined with the dietary fiber PHGG (partially hydrolyzed guar gum) was confirmed in a different food product (short bread biscuits) using culture-independent molecular techniques for bacterial enumeration (Tuohy *et al.,* 2001). A double-blind, placebo controlled, cross-over study was conducted with 31 healthy volunteers. The volunteers were given active biscuits containing FOS (6.6 g/day) and PHGG (3.4 g/day) for 21 day periods and faecal bacteriology carried out using fluorescent *in situ* hybridization (FISH). Upon ingestion of the active biscuits, a significant increase in numbers of bifidobacteria was observed with a small decrease in numbers of *Clostridium perfringens/histolyticum*. It was also found that the degree of bifidogenesis was related to the initial population level of *Bifidobacterium* spp. in volunteers before ingestion of the prebiotic. Large increases in numbers of bifidobacteria were observed in volunteers with relatively low starting levels of bifidobacteria (about 10^7 cells/g faeces), while little bifidogenesis was observed in volunteers with starting levels of bifidobacteria greater than 10^9 cells/g faeces. Companies producing inulin and FOS are as followed. Inulin is produced by Orafti Belgium and CoSucra Belgium, FOS is produced by Beghin Meiji Industies France, CoSucra Belgium and Orafti Belgium (Angus, 2000).

2.3. Galacto-Oligosaccharides (GOS)

Galacto-oligosaccharides, also called transgalacto-oligosaccharides (TOS), are non-digestible, galactose-containing oligosaccharides of the form Glu β 1-4 [β Gal 1-6]$_n$ where n= 2-5. They are produced from lactose by means of an enzymatic conversion (Crittenden & Playne, 1996). Two types of enzyme can theoretically be used to produce GOS from lactose: glycosyltransferases, which need a sugar nucleotide as a donor and glycosidases, which utilise galactose as a donor. The first type of enzyme is not available on an industrial scale and commercially only the glycosidases, namely β-galactosidases derived from bacteria (*Bacillus circulans*), moulds (*Aspergillus niger* or *oryzae*) and yeasts (*Kluyveromyces lactis* or *fragilis*) are used. The enzyme catalyses both hydrolysis and polymerization with the equilibrium depends on the biological sources of enzyme and the reaction conditions.

Galacto-oligosaccharides are not broken down in the stomach or small intestine and, thus reach the colon largely intact (Bouhnik *et al.,* 1997). *In vitro*, pure culture studies have shown that GOS were readily utilized by bifidobacteria and lactobacilli (Tanaka *et al.*, 1983). Similarly, GOS have been shown to selectively stimulate numbers of bifidobacteria within the human gut microflora in complex *in vitro* models of the human colon (Rycroft *et al.,* 2001). Rowland and Tanaka (1993) were able to demonstrate that ingestion of transgalacto-oligosaccharides (TOS) not only resulted in increased numbers of bifidobacteria and lactobacilli, but also in decreased numbers of enterobacteria. Transgalacto-oligosaccharides additionally have been shown to offer a level of protection against colon cancer by reducing microbial enzyme activities involved in the production of carcinogens within the gut (i.e. nitrate reductase and β-glucuronidase).

In human feeding studies, Tanaka and co-workers (1983) showed that although 3 g/day for 7 days resulted in little change in numbers of faecal bacteria, ingestion of 10 g/day TOS for 7 days resulted in a significant increase in numbers of bifidobacteria and lactobacilli recovered from stool samples. Conversely, numbers of faecal *Bacteroides* spp. decreased significantly. Ingestion of GOS by healthy volunteers resulted in an increase in faecal bifidobacteria and lactobacilli in a dose-dependent manner (Ito *et al*., 1990). The same group later showed that ingestion of 15 g/day of GOS also had a positive effect on potentially detrimental faecal microflora-associated characteristics, p-cresol, NH_3, propionate, valerate, isovalerate and isobutyrate (Ito *et al.,* 1993). Bouhnik and co-workers (1997) showed that the bifidogenic nature of GOS ingestion persisted over longer periods of time (21 days) while dietary supplementation continued and that ingestion of GOS reduced breath H_2. Conversely, Alles and co-workers (1999) found that ingestion of GOS at either 7.5 g/day or 15 g/day for a 3 week period had little effect upon numbers of faecal bacteria or faecal microbial metabolites. However, in that study, volunteers showed high initial stool bifidobacterial levels ($>10^9$ CFU/g faeces). As observed with FOS and lactulose, the degree of bifidogenesis was related to initial faecal population levels of bifidobacteria and this may have accounted for the lack of prebiotic effect observed. The GOS used in the previous study was only 62% oligosaccharide, the remainder being made up of lactose (20%) and glucose (18%), both of which are absorbed in the upper gut and, therefore, do not reach the colon. This may have had an effect on the prebiotic parameters monitored by the investigators. Galacto-oligosaccharides are produced by Borculo Domo Ingredients (Netherlands), Nihon Shokuhin Kako (Japan), Nissan Sugar (Japan) and Yakult Honsha (Japan) (Angus, 2000).

2.4. Soybean Oligosaccharides (SOS)

Soybean oligosaccharides are non-reducing α-galactosyl sucrose derivatives (Crittenden & Playne, 1996). They were developed by Calpis Food Industry Co., Japan. The predominant oligosaccharides in soybeans are raffinose (a trisaccharide, α-D-Gal-(1-6)-α-D-Glu-(1-2)-β-D-Fru) and the tetrasaccharide stachyose (α–D-Gal-(1-6)-α -D-Gal-(1-6)-α-D-Glu-(1-2)-β-D-Fru). Soybean oligosaccharides are able to reach the colon and are thought to stimulate bifidobacteria *in vivo* (Oku, 1994). Benno and co-workers (1987) fed 7 healthy volunteers raffinose at 15 g/day for 4 weeks and observed a significant increase in numbers of faecal bifidobacteria. A concomitant decrease in numbers of *Bacteroides* spp. and clostridia was observed. On feeding 6 volunteers 10 g/day SOS for 3 weeks, Hayakawa and co-workers (1990) reported a significant increase in numbers of faecal bifidobacteria and lactobacilli and a large decrease in clostridia and peptostreptococci. Soybean oligosaccharides are producing by Calpis Co. (Japan) (Angus, 2000).

2.5. Lactosucrose (LS)

Lactosucrose is a non-reducing oligosaccharide, in a mixture of lactose and sucrose which produced by the transfer of a fructosyl residue from sucrose to the glucose moiety of lactose by action of β-fructofuranosidase (Tamura, 1983). Lactosucrose was developed by Hayashibara Biochemical Laboratory Co., Japan. A pure culture study compared lactosucrose with lactulose, FOS, SOS, raffinose and glucose for its utilization by various intestinal bacteria. Six bifidobacteria and three lactobacilli strains grew to the same extent on lactosucrose and glucose, whereas all the other organisms tested preferred glucose (Hara *et al.*, 1994).

2.6. Isomalto-Oligosaccharides (IMO)

Isomalto-oligosaccharides (IMO) are composed of glucose monomers linked by α(1-6) glucosidic linkages. IMO are derived from starch by a two-step enzymatic process and are mixtures of α(1,6) glucosides such as isomaltose, isomaltotriose, panose and isomaltotetraose. Firstly, starch is hydrolyzed to α(1-4) malto-oligosaccharides by α-glucosidase. Secondly, glucosyl residues are transferred to produce α(1-6) linked IMO using the same enzyme. IMO were developed by Hayashibara Biochemical Laboratory Co, Japan in 1982 and marketed by Hayashibara as well as Showa Sangyo and Nikken Chemicals. The effective daily dose of IMO is 13 g. A commercial mixture known as Isomalto-900 has been produced by incubating α-amylase, pullulanase and α-glucosidase with corn starch. The major oligosaccharides in this mixture are isomaltose (Glu α1-6 Glu), isomaltotriose (Glu α1-6 Glu α1-6 Glu) and panose (Gluα1-6 Glu α1-4 Glu).

Commercially available products contain a mixture of isomaltose, isomaltotriose and panose. All have been shown, in pure culture, to support the growth of most species of bifidobacteria at least as well as raffinose. Isomalto-oligosaccharides fermentation also resulted in a significant increase in concentrations of lactic acid and acetate after 24 h compared to initial levels. A significant increase in bifidobacteria was observed at doses of 9.8 g/day over 10 days. In a further group of 9 male volunteers, ingestion of IMO at 10 g/day over a 2 week period also resulted in a significant increase in bifidobacteria. The minimum effective dose of pure IMO was around 9-10 g/day. The degree of polymerization of IMO played a role in the bifidogenic capabilities of this prebiotic *in vivo*. Unlike FOS, lactulose and GOS, certain moieties of commercially available IMO, especially those of low DP, may be digested in the small intestine by human isomaltase.

2.7. Gluco-Oligosaccharides

Gluco-oligosaccharides with differing structures can be made using glycosyltransferases or β-glucosidases (Crittenden & Playne, 1996). Gentio-oligosaccharides (GEOS) consisted of β(1-6)-D-glucosidic linkages and are made by Nihon Shokuhin Kako Co. (Playne & Crittenden, 1996) and sold under the trade name Gentose. Mixed linkages gluco-oligosaccharides produced by *Gluconobacter oxydans* NCIMB4943 shown prebiotic effect by colonic fermentation in batch culture (Wichienchot *et al.*, 2006a) and artificial colon system (Wichienchot *et al.*, 2006b).

2.8. Xylo-Oligosaccharides (XOS)

Xylo-oligosaccharides are low molecular weight reducing oligosaccharides, chains of xylose residues linked by β(1-4) bonds and mainly consist of xylobiose, xylotriose and xylotetraose (Hopkins *et al.*, 1998). Xylo-oligosaccharides can be manufactured by enzymatic hydrolysis of xylan from corncobs (Crittenden & Playne, 1996), oats (Jaskari *et al.*, 1998) and wheat arabinoxylan (Yamada *et al.*, 1993). Typical raw materials for XOS production are hardwoods (birchwood), corn cobs, straws, bagasses, hulls, malt cakes and bran (Vazquez *et al.*, 2000). Commercially available XOS are manufactured and marketed by Suntory Ltd., Japan.

Table 3. Minimum effective doses for a prebiotic effect in humans for various oligosaccharides

Oligosaccharides	Minimum effective dose (g d^{-1})
Raftilose L60	8
Raftilose P95	15
Neosugar	10
Inulin	20 for significant increase in bifidobacteria 40 for significant increase in bifidobacteria and decrease in bacteroides, coliforms and enterococci
Lactulose	3
95% Lactosucrose	3
Oligomate	10
Transgalactosylated oligosaccharides	10
Transgalactosylated disaccharides	15
Soybean oligosaccharide extract	10
Raffinose	15
Isomalto-900P	13
Xylo-oligosaccharides	1

Source: Rycroft (2001).

To compare oligosaccharides of all the different groups, the minimum effective doses needed to elicit a prebiotic effect in human may be compared (Table 3). The data suggested that there are large differences in the minimum dose needed for different oligosaccharides. More studies are needed for lactulose, lactosucrose and XOS as these were effective at lowest doses of all the substrates. As the studies all used different subjects, diets and methods, substrate effect is difficult to compare.

3. ORIGINS AND PRODUCTION

Oligosaccharides are increasingly being recognized as useful tools for the modulation of the colonic microflora, especially for the level of bifidobacteria and lactobacilli. The structures of commercial available prebiotic oligosaccharides are summarized in Table 4.

Table 4. Structure of commercial available prebiotic oligosaccharides

Oligosaccharides	Structure
Lactulose	Galβ1-4Fru
Fructo-oligosaccharides (FOS)	Fruβ2-1Fru$_n$, n=1-3 Gluα1-2[βFru1-2]$_n$, n = 2-9, average 4-5
Inulin, Raftiline LS	Fruβ2-1Fru$_n$ n=1->60 Gluα1-2[βFru1-2]$_n$, n > 10, average 10-12
Galacto-oligosaccharides (GOS)	Gluα1-4[βGal1-6]$_n$, n= 1-4, average 2
Soybean oligosaccharides (SOS)	Raffinose (Galα1-6Glu1-2βFru) Stachyose (Galα1-6Galα1-6Gluα1-2βFru)
Lactosucrose	Galβ1-4Gluα1-4Gluα-2βFru
Isomalto-oligosaccharides (IMO)	Glcα1-6[Gluα1-6]$_n$, n ≥ 1, average 1-2
Gentio-oligosaccharides (GEOS)	Gluβ1-6Glu
Xylo-oligosaccharides (XOS)	Xylβ1-4[Xyl]$_n$, n= 2-7

Source: Gibson & Rastall (2006).

3.1. Extraction

The simplest approach to the manufacture of prebiotics is to extract them from a biological material. This is currently commercially performed with inulin extracted from chicory (De Leenheer, 1994), and with raffinose and starchyose extracted from soybeans (Koga *et al.*, 1993), but the extraction yields are quite low.

Inulin is a β2↔1 linked fructan found in a variety of plants. The commercial source in Europe is chicory, with the largest manufacturer being Orafti in Tienen, Belgium. The chicory extraction process is similar to the sugar beet process (De Leenheer, 1994). Chicory chips are extracted with hot water and the extracted chip sold as animal feed. Protein, peptides, colloids and phosphates are then removed by liming and carbonatation at alkaline pH. The inulin is then demineralised by anion and cation exchange chromatography and decolourised by chromatography on activated carbon. The final products are then sterilized, concentrated and spray-dried.

Soybean oligosaccharides are also extracted from their biological source with no further modification. They are isolated from soybean whey and then concentrated to a 75% (w/v) syrup containing 35% oligosaccharides (Crittenden & Playne, 1996). SOS are extracted commercially from soybean whey by using reverse osmosis (RO) and nanofiltration (NF) membrane. Concentrations of the total oligosaccharides at 10% (w/v) and 22% (w/v) were obtained in a batch operation conditions (Matsubara *et al.*, 1996).

3.2. Chemical Approach

Oligosaccharides can be synthesized chemically (Garegg, 1990) but the processes are very complicated due to the many protection and deprotection steps that are necessary for regioselective synthesis, processes are also labour intensive as well as producing unwanted colour and flavour compounds that have to be removed in additional refining steps. The number of steps increases with the size of the oligosaccharides, so that, while synthesis of a disaccharide may require 5-7 steps, a trisaccharide may require more than ten steps. Total yields are often low and large-scale production is not applicable. In addition, steriospecific reactions giving the correct anomer (α or β) are often difficult (Bucke, 1996; Wong *et al.*, 1995). Lactulose is the only prebiotic manufactured using chemical rather than enzymatic approaches (Timmermans, 1994).

3.3. Enzymatic Approach

The manufacturer of oligosaccharides depends rather heavily on the use of enzymes to bring about either hydrolysis of polysaccharides or the synthesis of oligosaccharides.

3.3.1. Polysaccharide Hydrolysis

Polysaccharide degradation has much potential to manufacture prebiotic oligosaccharides from various biological sources. The aim of this approach is a controlled partial hydrolysis to give oligosaccharides of specific molecular weights, resulting in useful rheological properties and technological applications. Currently, only two prebiotics are commercially manufactured by this method, inulin-derived fructo-oligosaccharides (De Leenheer, 1994) and xylan derived xylo-oligosaccharides (Playne & Crittenden, 1996).

3.3.2. Enzymatic Synthesis

Future developments in the large-scale synthesis of oligosaccharides will most likely be in the area of enzymatic synthesis. The use of enzymes in synthesis of complex carbohydrates offers several advantages over chemical methods. A wide variety of regiospecific reactions can be catalyzed very efficiently without protection of the hydroxyl groups. These take place under mild conditions, often at room temperature and close to neutral pH, and organic solvents and hazardous chemicals or catalysts can be avoided (Rastall & Bucke, 1992). Two main types of enzymes are used to catalyze oligosaccharide synthesis: hydrolases (glycosidases E.C. 3.2) and transferases (glycosyltransferases E.C. 2.4). Most of these enzymes are of plant or microbial origin.

Some prebiotics such as IMO, TOS, FOS and lactosucrose are manufactured by enzymatic transfer reaction (Nakakuki, 1993). The reactions utilized cheap sugars as a donors and acceptors. The simplest reactions are the manufacture of FOS and lactosucrose. Fructo-oligosaccharides are made commercially by Meiji Seika in Japan and in Europe by collaboration between Meiji Seika and Eridania Beghin Say (Beghin-Meiji Industries). Fructosyltransferase from *Aureobasidium pullulans* or *Aspergillus niger* is used to build up higher fructo-oligosaccharides from a 60% (w/v) sucrose solution at 50-60°C in an immobilized cell-based reactor (Kono, 1993; Yun, 1996). All of the sucrose-derived FOS terminates in a non-reducing glucose residue. Higher product purities can be achieved by removal of the glucose and sucrose by ion-exchange chromatography (Kono, 1993).

Production of hetero-oligosaccharides can be promoted over the production of homo-oligosaccharides by increasing the percentage of acceptor sugar in the mixture, but total yield of oligosaccharide product is decreased (Rastall & Bucke, 1992). In addition, different linkages may be obtained by using different sources of the enzyme (Suwasono & Rastall, 1998). Generally glycosidases are used for the synthesis of shorter oligosaccharides while glycosyltransferases are suitable for synthesis of higher oligosaccharides (Rastall & Bucke, 1992).

4. Health Benefits

4.1. Reduction of Acute Infections and Inflammatory Reactions

Disturbances in the normal intestinal microbial community can result in a proliferation of pathogens. Acute inflammatory reactions cause diarrhea and sometimes vomiting, and can be associated with a number of bacteria, viruses, and protozoa including *E.coli*, *Campylobacter* spp., *Vibrio cholerae*, *S. aureus*, *B. cereus*, *Clostridium perfringens*, *Salmonella* spp., *Shigella* spp., *Yersinia* spp., *Giardia lamblia*, *Entamoeba histolytica,* and *Cryptosporidium parvum* (Casci *et al.,* 2006). Bacteria can also be linked to more chronic diseases in the colon. For example, *C. difficile* has been targeted as the primary causative agent of pseudomembranous colitis.

The GI tract functions as a barrier against antigens from microorganisms and food. Among the possible mechanisms of probiotic therapy is promotion of a non-immunological gut defense barrier, normalization of increased intestinal permeability and altered gut microecology. Another possible mechanism is improvement of the intestinal immunological barrier and alleviation of gut inflammatory responses. Many probiotic effects are mediated through immune regulation, particularly from the balance control of pro-inflammatory and anti-inflammatory cytokines (Isolauri, 2001).

4.2. Prevention of Antibiotic Associated Diarrhea (AAD)

Diarrhea occurs in about 20% of patients who receive antibiotics. Antibiotic associated diarrhea (AAD) results from microbial imbalance that leads to a decrease in the fermentation capacity of the colon. Invasion with *C. difficile*, *Klebsiella oxytoca,* and *C. septicum* are causes of AAD (Salminen *et al.,* 1998; Marteau and Boutron-Ruault, 2002). Studies have shown that oral administration of *S. boulardii* can decrease the risk of AAD. Other studies have shown the efficacy of *S. boulardii*, *Enterococcus faecium* SF68, and *L. rhamnosus* GG in shortening the duration of AAD (Casci *et al.,* 2006). Several trials have suggested a preventive effect of some fermented products on the risk of diarrhea in children (Gibson & Rastall, 2006). Saavedra and co-workers (1994) showed that feeding *B. bifidum* and *S. thermophilus* to infants significantly reduced the risk of diarrhea and the shedding of rotavirus.

4.3. Immune Stimulation

In a human trial, administration of probiotic yogurt gave an increase in the production of β-interferon (Halpern *et al.*, 1991). In animal models, probiotics have been shown to stimulate the production of antibodies and increase the concentration of natural killer cells (Fooks *et al.,* 1999). An enhancement in the circulating IgA antibody secreting cell response was observed in infants supplemented with a strain of *L. casei*, and this correlated with shortened duration of diarrhea in the study group when compared to a placebo group (Kaila *et al.,* 1992). Other studies have reported an enhancement in the non-specific immune phagocytic activity of granulocyte populations in the blood of human volunteers after consumption of *L. acidophilus* and *B. bifidum* (Marteau *et al.,* 1997). It is possible that stimulation of intestinal IgA antibody responses induced by LAB may be explained partly by an effect on phagocyte cell functions. Ingestion of yogurt has been reported to stimulate cytokine production and β-interferon in human blood mononuclear cells (Solis-Pereira & Lemonnier, 1996). It has been shown in experimental animals that the postnatal maturation of small intestinal brush border membranes was associated with increased food protein binding capacity (Bolte *et al.,* 1998). The capacity of antigens to bind to epithelial cells is related to the rate and route of antigen transfer and shown to influence the intensity of mucosal immune responses (Van der Heijden *et al.*, 1991). Oral administration of *Lactobacillus* species can enhance non-specific host resistance to microbial pathogens, thereby facilitating the exclusion of pathogens in the gut (Perdigon *et al.*, 1998). Several strains of live LAB have been shown *in vitro* to induce the release of the pro-inflammatory cytokines tumor necrosis factor α and interleukin, reflecting stimulation of non-specific immunity (Miettinen *et al.*, 1996).

4.4. Lowering Lactose Intolerance

Some lactic acid bacteria, including *L. acidophilus* and *B. bifidum,* produce β-D-galactosidase to increase tolerance to dairy products (Kim & Gilliland, 1983). There is good evidence for the alleviation of symptoms of lactose intolerance and milk hypersensitivity by specific probiotics (Pelto *et al.*, 1998).

It is well established that persons with lactose intolerance experience improved digestion and tolerance of the lactose contained in yogurt than of that contained in milk (De Vrese *et al.,* 2001). At least two mechanisms have been reported to explain this: digestion of lactose in the gut lumen by lactase contained in the yogurt bacteria, and slower intestinal delivery or transit time of yogurt compared to milk (Lin *et al.,* 1998). Kim and Gilliland (1983) suggested that improved lactose digestion was not caused by lactose hydrolysis prior to consumption, but rather by the action of the enzyme in the digestive tract.

4.5. Lowering Risk of Colon Cancer

Development of colon cancer consists of 3 consequence steps. First is an initiating step, in which a carcinogen produces an alteration in DNA. It is believed that several mutations must occur for a tumor to develop. Second is post-initiation step, involve changes in signal transduction pathways.

Third step is an overgrowth in colonic crypts, which can be seen morphologically in an aberrant crypt. Aberrant crypts, which are considered preneoplastic structures, have a serpentine growth pattern. Aberrant crypts may occur singly or as groups of aberrant crypts within a single focus and then the crypts will progress to polyps and eventually to tumors (Brady *et al*., 2000). Prebiotics increasing the portion of bifidobacteria and lactobacilli at the expense of bacteriods and clostridia may also decrease genotoxic enzyme production, as the former produce lower levels of such enzymes than latter, caused to reduce of colon cancer (Rowland, 1995). It can induce apoptosis, a process which is deactivated in cancer cell which would normally lead to their elimination and an increase in immunogenicity of cancer cells (Maciorkowska *et al.,* 2010). Inulin-type fructans and corresponding fermentation products reduced the risks for colon cancer (Munjal *et al*., 2009).

4.6. Cholesterol Lowering Effect and Lipid Metabolism

Prebiotics in the form of fermented milk are thought to have cholesterol lowering properties in humans. However, studies have given equivocal findings. Even recent studies fail to provide convincing evidence that fermented milk products containing live microorganisms have any cholesterol lowering efficacy in humans. In animal studies, dietary FOS caused a suppression of hepatic triglyceride and VLDL (very low density lipoprotein) synthesis, resulting in marked reduction in triglycerides and cholesterol levels. Propionate metabolism has also been studied in ruminants, where it is a major gluconeogenic precursor (Bergman *et al.,* 1966). In rats, it has been shown to lower cholesterol either by inhibition of hepatic cholesterol synthesis or by redistribution of cholesterol from plasma to the liver (Illman, 1988). In humans, no change was seen in total cholesterol levels, although HDL (high density lipoprotein) cholesterol increased when a group of females were fed sodium propionate (Venter *et al.,* 1990). In Coronary Heart Disease (CHD) there are variable data and no firm conclusions can be drawn (Fooks *et al.,* 1999). Schaafsma and co-workers (1998) found that daily feeding of 125 mL of milk containing probiotic significantly lowered LDL (low density lipoprotein) cholesterol levels and total serum cholesterol. Fructo-oligosaccharides found to reduce total serum and LDL cholesterol (Caceres *et al.,* 2004).

The population is aging, with declining birth rates and increasing life expectancy, contributing to the growing of an aged population worldwide. Life expectancy continues to rise, as does the contribution made by older individuals to the total population (Wildman & Kelly, 2007). Thus, an aging population is accompanied by a significant rise in health costs and the need for socially acceptable care facilities is likely to impact greatly on socioeconomic parameters within Westernized countries. With aging come a reduction in overall health and an increase in morbidity and mortality due to infectious disease, many associated with the gastrointestinal tract. It is estimated that mortality due to gastrointestinal infections is up to 400 times higher in the elderly compared to younger adults (Caceres *et al.,* 2004). Therefore, stimulation of immunity and reducing harmful bacteria by probiotics is of profound importance to aging population.

5. Commercial Applications

5.1. Food Applications

Prebiotics can be used for either their nutritional advantages or technological properties, but they are often applied to offer a double benefit: an improved organoleptic quality and a better-balanced nutritional composition (Yanbo, 2009). The use of inulin and nondigestible oligosaccharides as fiber ingredients is straightforward and often leads to improved taste and texture. These specific forms of dietary fiber are readily fermentable by specific colonic bacteria, such as bifidobacteria and lactobacilli species, increasing their cell population with the concomitant production of SCFA (Whelan *et al.*, 2005). These acids, especially butyrate, acetate, and propionate, provide metabolic energy for the host and acidification of the bowel (Sghir *et al.,* 1998). Nowadays, more and more prebiotics are used in functional foods as ingredients which stimulate the growth of health-promoting gut bacteria especially probiotics and offer additional health benefits (Yanbo, 2009). Prebiotics are possible to be applied for both non-meat products and meat products.

The purpose to optimize the acceptability of two low-fat milk beverages with different types of inulin (CLR and TEX), using response surface methodology (RSM) was performed. Sixteen formulations of beverage with each inulin type were prepared, varying inulin concentration (3–8%), and sucrose concentration (0–8%). A group of 50 consumers evaluated the acceptability of the samples and tested the appropriateness of some sensory attributes intensity (color, vanilla flavor, sweetness and thickness) using just about right scales. Response surface plots showed that formulations containing 5–8% CLR and 4–6.5% sucrose and formulations containing 4–6.5% for both TEX inulin and sucrose were located in the optimum region. The sweetness and the thickness were the attributes that most affected the acceptability of the samples. The two low-fat samples (one for each inulin type) selected as the optimum showed no differences in acceptability (a = 0.05) between them neither when compared with a full fat control sample (Villegas, 2010).

Leuconostc mesenteroides B-512F and *L. mesenteroides* B-742 were cultivated in clarified cashew apple juice to produce prebiotic oligosaccharides. Yeast extract (20 g/L); K_2HPO_4 (g/L) and sucrose (50 g/L) were added to the juice to promote the microbial growth and dextransucrase production. Initial pH was adjusted to 6.5 with H_3PO_4. Fermentations were carried out at 30°C and 150 rpm for 24 h. The prebiotic effect of the fermented cashew apple juice, containing oligosaccharides, was evaluated through the *Lactobacillus johnsonii* B-2178 growth. *L. johnsonii* was incubated for 48 h using fermented cashew apple juice as substrate. Lactobacillus growth was compared to the microbial growth in non-fermented juice and in MRS broth. *L. johnsonii* growth in the fermented cashew apple juice was three folds the observed growth in the non-fermented juice (Clarice, 2010).

The effect of a short chain FOS on the sensory properties of conventional and reduced-fat cooked meat sausages has been studied in products in which a fat reduction of close to 40% was obtained. The fiber assayed was used in sufficient amounts to constitute between 2% and 12% of the final product. The energy value reduction of the final products was close to 35%. Instrumental measurements of color and texture were performed. Sensory properties were estimated by a hedonic test. A correlation principal component analysis was performed. The results showed that the sensory and textural properties and the overall acceptability were very good, which indicated that this fiber can be considered a good fat replacer in meat products. Thus, with its addition, a reduced calorie product enriched with soluble dietetic fiber is obtained (Caceres, 2004).

Low fat, dry fermented sausages were prepared with a fat content close to 50% and 25% of the original amount. The batch with the smallest proportion of fat was less tender, less springy and was gummier than the batch with the highest proportion. However, it was still considered acceptable by the panel of judges. The 25% batch was supplemented with different amounts of soluble dietetic fiber (inulin) as both a powder and in aqueous solution. Ripening was followed by physico-chemical and microbiological analysis. Sensory analysis and texture profile analysis were performed to evaluate the effect of the inulin addition. Results obtained indicated an overall improvement in the sensory properties due to a softer texture and tenderness, springiness and adhesiveness similar to the conventional high fat sausage. Thus, with the addition of inulin a low calorie product (30% of the original), enriched with soluble dietetic fiber (10% approximately) can be obtained (Mendoza, 2001).

5.2. Applications for Animals

Currently, livestock industry is growing annually and this succeed depends on a broad spectrum of economic parameters. One of important parameter is getting high production at low costs by keep feed cost and feed conversion ratio (the amount of feed needed to obtain a certain amount of weight gain) as low as possible. In addition, disease prevention and reduction of mortality are very important to achieve this target. The use of prebiotics in animal production lies in the improvement of economic results. They have also been extensively studied as possible replacements for antimicrobial growth promoters, which have been banned in the European Union since 2006. Thus prebiotics are alternative to replace antibiotic use and also for supplementing in the feed for various animals i.e. pigs, poultry, cattle, rabbits, and aquaculture.

Use prebiotic as additional feed ingredient to pig has resulted in mixed but basically nonsignificant effects regarding beneficial modulation of microbial populations in various intestinal segments and feces of swine (Mikkelsen *et al.*, 2003; Loh *et al.* 2006; Mountzouris *et al.*, 2006). However, some reports shown significantly increased bifidobacteria, lactobacilli, and enterococci (Smiricky-Tjardes *et al.*, 2003; Shim *et al.*, 2005; Tzortzis *et al.*, 2005). Prebiotics also reduce the excretion of nitrogen into the environment and suppression of boar taint, which is an off-flavor of pork (Babol and Squires, 1995).

Prebiotics in broiler diets have been shown to increase numbers of lactobacilli in the gastrointestinal tract (Yusrizal & Chen, 2003a) and inulin or fructo-oligosaccharides may improve weight gain, feed conversion and carcass weight (Yusrizal & Chen, 2003b; van Leeuwen *et al.*, 2005). Nutrient absorption was significantly improved in pigs when prebiotic was corporate in feed moreover the dense of villi distribution in chicken was also increased (Yusrizal & Chen, 2003a). Chen *et al.* (2005) demonstrated laying hens fed with inulin supplement increased egg production. van Leeuwen and Verdonk (2005) reported inulin and oligofructose were shown to increase daily weight gain and improve feed conversion in calves and improvement in fecal consistency. A positive effect of oligofructose on morbidity of rabbit was demonstrated by Morisse *et al.* (1993).

Research on prebiotics in aquaculture feeds has emerging area of application because of aquaculture is one of the fastest growing sectors of livestock production especially in fish. Mahious *et al.* (2006) showed that oligofructose increased the growth of weaning turbot, a carnivorous species. Rainbow trout, nile tilapia, catfish, and salmon are under investigation separately (Loo & Vancraeynest, 2008). To fully conclude on the effects of adding prebiotics in fish diets, more research efforts are needed to provide the aquaculture industry, the scientific community, the regulatory bodies and the general public with the necessary information and tools (Ringo *et al.*, 2010).

Conclusion

The science of prebiotics has come a long way since initiation of the concept in 1995. Prebiotics are one of the most popular food ingredients that attract significant interest by both the academic and industrial communities with health impact such as promoting beneficial bacteria, reduced colon cancer, reduced calorie, fat replacement and enriched source of soluble dietary fiber. Functional claims of prebiotics are all country allows using as source of dietary fiber. Some country approved on additional claim. Prebiotics show both important technological characteristics and interesting nutritional properties. Several are found in vegetables and fruits and can be industrially processed from renewable materials. In food formulations, they can significantly improve organoleptic characteristics, upgrading both taste and mouthfeel. For prebiotics to serve as functional food ingredients, they must be chemically stable to food processing treatments, such as heat, low pH, and Maillard reaction conditions. That is, a prebiotic would no longer provide selective stimulation of beneficial microorganisms if the prebiotic was degraded to its component mono- and disaccharides or chemically altered so that it was unavailable for bacterial metabolism. In addition to the above benefits of prebiotics, there has been considerable interest in new application areas, such as in urogenital infections and oral health, and in their applications in animal feeds, i.e. pet, poultry and aquaculture.

References

Alles, M. S., Hartemink, R., Meyboom, S., Harryvan, J. L., Van, L. K. & Nagengast, F. M. (1999). Effect of transgalactooligosaccharides on the composition of the human intestinal microflora and on putative risk markers for colon cancer. *J. Am. Clin. Nut, 69*, 980-991.

Angus, F. (2000). Suppliers. In G. R. Gibson, & F. Angus (Eds), *LFRA Ingredients Handbook Prebiotics and Probiotics* (pp. 189-195). London, UK: Food RA Leatherhead.

Babol J. & Squires E. J. (1995). Quality of meat from intact male pigs. *Food Research International, 28*, 201–212.

Ballongue, J., Schumann, C. & Quignon, P. (1997). Effects of lactulose and lactitol on colonic microbiota and enzymatic activity. *J. Scand. Gastroenterol*, *32*, 41-44.

Benno, Y., Endo, K., Shiragami, N., Sayama, K. & Mitsuoka, T. (1987). Effects of raffinose intake on human faecal microflora. *Bif. Microflora*, *6*, 59-63.

Bergman, D. N., Roe, W. E. & Kon, K. (1966). Quantitative aspects of propionate metabolism and glucogenesis in sheep. *Am. J. Physiol*, *211*, 793–799.

Bolte, G., Knauss, M., Metzdorf, I. & Stern, M. (1998). Postnatal maturation of rat small intestinal brush border membranes correlated with increase in food protein binding capacity. *Dig. Dis. Sci*, *43*, 148–155.

Bornet, F. R. (1994). Undigestible sugars in food products. *Am. J. Clin. Nutr*, *59*, 763S-769S.

Bouhnik, Y., Flourie, B., D'Agay-Abensour, L., Pochart, P., Gramet, G., Durand, M. & Rambaud, J. C. (1997). Administration of transgalactooligosaccharides increases faecal bifidobacteria and modifies colonic fermentation metabolism in healthy humans. *J. Nutr, 127*, 444-448.

Brady, L. J., Gallaher, D. D. & Busta, F. F. (2000). The role of probiotic cultures in the prevention of colon cancer. *J. Nutr*, *130*, 410S–414S.

Bucke, C. (1996). Oligosaccharides synthesis using glycosidases. *J. Chem. Technol. Biotechnol*, *67*, 217-220.

Caceres, E. C., Garcia, M. L., Toro, J. & Selgas, M. D. (2004). The effect of fructooligosaccharides on the sensory characteristics of cooked sausages. *Meat Sci, 68*, 87–96.

Casci, T., Rastall, R. A. & Gibson, G. R. (2006). Human gut microflora in health and disease: focus on prebiotics. In K. Shetty, G. Paliyath, A. Pometto, & R. E. Levin (Eds), *Food Biotechnology* (2nd ed,). New York: CRC Press.

Charalampopoulos, D. & Rastall, R. A. (2009). *Prebiotics and Probiotics Science and Technology*. New York: Springer Science.

ChenY. C., Nakthong C. & Chen T. C., (2005). Improvement of laying hen performance by dietary prebiotic chicory oligofructose and inulin. *International Journal of Poultry Science, 4*, 103–108.

Clarice, M. A., Chagas, V., Talita, L. H., Geraldo, A. M. & Sueli, R. (2010). Prebiotic effect of fermented cashew apple (*Anacardium occidentale*) juice. *Food Sci. & Technol,.43*, 141-145.

Crittenden, R. G. & Playne, M. J. (1996). Production, properties and applications of food-grade oligosaccharides. *Trends Food Sci & Technol, 7*, 353-361.

Crittenden, R. G., Morris, L. F., Harvey, M. L., Tran, L. T., Mitchell, H. L. & Playne, M. J. (2001). Selection of a Bifidobacterium strain to complement resistant starch in a synbiotic yoghurt. *J. Appl. Bacteriol*, *90*, 268-278.

De Leenheer, L. & Hoebregs, H. (1994). Progress in the elucidation of the composition of chicory inulin. *Starch*, *46*, 193-196.

De Vrese, M., Stegelmann, A., Richter, B., Fenselau, S., Laue, C. & Schrezenmeir, J. (2001). Probioticscompensation for lactase insufficiency. *Am. J. Clin. Nutr*, *73*, 421S–429S.

Fooks, L. J., Fuller, R. & Gibson, G. R. (1999). Prebiotics, probiotics and human gut microbiology. *Int.Dairy J., 9*, 53–61.

Fuller, R. (1989). Probiotics in man and animals. *J. Appl. Bacteriol, 66*, 365-378.

Garegg, P. J. (1990). Phase-transfer for selective substitutions in carbohydrates and inositols. *Abstracts of Papers of the American Chemical Society, 199*, 10-15.

Gibson, G. R. & Roberfroid, M. B. (1995). Dietary modulation of the human colonic microbiota: introducing the concept of prebiotics. *J. Nutr, 125*, 1401-1412.

Gibson, G. R. & Roberfroid, M. B. (2008). *Handbook of Prebiotics.* New York: CRC Press.

Gibson, G. R. & Rastall, R. A. (2006). *Prebiotics: Development & Application.* Chichester: John Wiley & Sons Press.

Gibson, G. R. (2004). Fibre and effects on probiotic (the prebiotic concept). *Clin. Nutr. Suppl, 1*, 25-31.

Gibson, G. R. & Wang, X. 1994. Bifidogenic properties of different types of fructo-oligosaccharides. *Food Microbiol, 11*, 491-498.

Halpern, G. M., Vruwink, K. J., Van der Water, J., Keen, C. L. & Gershwin, M. E. (1991). Influence of long-term yogurt consumption in young adults. *Int. J. Immunother, 7*, 205–210.

Hara, H., Li, S., Sasaki, M., Maruyama, T., Terada, A., Ogata, Y., Fujita, K., Ishingami, K., Ishigami, H., Hara, K., Fujimori, I. & Mitsuoka, T. (1994). Effective dose of lactosucrose on fecal flora and fecal metabolites of humans. *Bifidobacteria Microflora, 13*, 51-63.

Hayakawa, K., Mizutani, J., Wada, K., Masai, T., Yoshihara, I. & Mitsuoka, T. (1990). Effects of soybean oligosaccharides on human faecal flora. *Microb. Ecol. Health Dis, 3*, 293-303.

Hopkins, M. J., Cummings, J. H. & Macfarlane, G. T. (1998). Inter-species differences in maximum specific growth rates and cell yields of bifidobacteria cultured on oligosaccharides and other simple carbohydrate sources. *J. Appl. Microb, 85*, 381-386.

Illman, R. J., Topping, D. L., McIntosh, G. H., Trimble, R. P., Storer, G. B., Taylor, M. N. & Cheng, B. Q. (1988). Hypocholesterolaemic effects of dietary propionate studies in whole animals and perfused rat liver. *Ann. Nutr. Metab, 32*, 97–107.

Isolauri, E., Sutas, Y., Kankaanpaa, P., Arvilommi, H. & Salminen, S. (2001). Probiotics: effects on immunity. *Am. J. Clin. Nutr, 73*, 444S–50S.

Ito, M., Deguchi, Y., Miyamori, A., Matsumoto, K, Kikuchi, H., Kobayashi, Y., Yajima, T. & Kan, T. (1990). Effects of administration of galactooligosaccharides on the human fecal microflora, stool weight and abdominal sensation. *Microb. Ecol. Health and Disease, 3*, 285-292.

Ito, M., Kimura, M., Deguchi, Y., Miyamori, A., Yajima, T. & Kan, T. 1993. Effects of transgalactosylated disaccharides on the human intestinal microflora and their metabolism. *J. Nutr. Sci. Vitamin, 39*, 279-288.

Jaskari, J., Kontula, P., Siitonen, A., Jousimies-Somer, H., Mattila-Sandholm, T. & Poutanen, K. (1998). Oat β-glucan and xylan hydrolysates as selective substrates for *Bifidobacterium* and *Lactobacillus* strains. *Appl. Microbiol. Biotechnol, 49*, 175-181.

Kaila, M., Isolauri, E., Soppi, E., Virtanen, V., Laine, S. & Arvilommi, H. (1992). Enhancement of the circulating antibody secreting cell response in human diarrhea by a human *Lactobacillus* strain. *Pediatr. Res, 32*, 141–144.

Kim, H. S. & Gilliland, S. (1983). *L. acidophilus* as a dietary adjunct for milk to aid lactose digestion in humans. *J. Dairy Sci, 66*, 959–966.

Koga, Y., Shibuta, T. and O'Brien, R. 1993. Soybean oligosaccharides. In *Oligosaccharides: Production, Properties and Applications*. Nakakuki, T. (ed). pp 175-203. Japanese Technol. Rev, 3(2): 175-203. Gordon and Breach Science Publishers, UK.

Kolida, S., Tuohy, K., Gibson, G. R. (2002). Prebiotic effects of inulin and oligofructose. *British Journal of Nutrition*, *87*, S193-S197.

Kono, T. (1993). Fructooligosaccharides. In T. Nakakuki (Ed), *Oligosaccharides: Production, Properties, and Applications* (pp. 50-78). London, UK: Gordon and Breach Science Publishers.

Lin, M., Yen, C. L. & Chen, S. H. (1998). Management of lactose maldigestion by consuming milk containing lactobacilli. *Dig. Dis. Sci*, *43*, 133–137.

Loh G., Eberhard M., Brunner R. M., Hennig U., Kuhla S., Kleessen B. & Metges C. C. (2006). Inulin alters the intestinal microbiota and short-chain fatty acid concentrations in growing pigs regardless of their basal diet. *Journal of Nutrition, 136*, 1198–1202.

Loo, J. V. & Vancraeynest, D. (2008). Prebiotics and Animal Nutrition. In G. R. Gibson & M. B. Roberfroid (Eds.), *Handbook of Prebiotics* (pp. 421.431). New York: CRC Press.

Macfarlane, G. T. & Gibson, G. R. (1997). Carbohydrate fermentation, energy transduction and gas metabolism in the human large intestine. In R. I. Mackie & B. H. Placebo (Eds), *Gastrointestinal Microbiology*, (vol. 1, pp. 269-318). London: Chapman and Hall.

Maciorkowska, E., Ryszczuk, E. & Kaczmarski, M. (2010). The role of probiotics and prebiotics in apoptosis of the gastrointestinal tract. *Przeglad Gastroenterologiczny*, *5*, 88-93.

Mahious A., Gatesoupe J., Hervi M., Metailler R. & Ollevier F. (2006). Effect of dietary inulin and oligosaccharides as prebiotics for weaning turbot, *Psetta maxima* (Linnaeus, C. 1758). *Aquaculture International, 14*, 219–229.

Marteau, P. & Boutron-Ruault, M. C. (2002). Nutritional advantages of probiotics and prebiotics. *Brit. J. Nutr., 87*, S153-S157.

Marteau, P., Vaerman, J. P., Dehennin, J. P. Bord, S., Brassart, D., Pochart, P., Desjeux, J. F. & Rambaud, J. C. (1997). Effects of intrajejunal perfusion and chronic ingestion of *Lactobacillus johnsonii* strain La1 on serum concentrations and jejunal secretions of immunoglobulins and serum propteins in healthy humans. *Gastroenterol. Clin. Biol*, *21*, 293–298.

Matsubara, Y., Iwasaki, K., Nakajima, M., Nabetani, H. & Nakao, S. (1996). Recovery of oligosaccharides from steamed soybean wastewater in tofu processing by reverse osmosis and nanofiltration membranes. *Biosci. Biotech. Biochem.*, *60*, 421-428.

Mendoza, E., Garcia, M. L., Casas, C. & Selgas, M. D. (2001). Inulin as fat substitute in low fat, dry fermented sausages. *Meat Sci., 57*, 387-393.

Miettinen, M., Vuopio-Varkila, J. & Varkila. K. (1996). Production of human tumor necrosis factor alpha, interleukin-6, and interleukin-10 is induced by lactic acid bacteria. *Infect. Immun.*, *64*, 5403–5405.

Mikkelsen, L. L., Jakobsen, M. & Jensen, B. B. (2003). Effects of dietary oligosaccharides on microbial diversity and fructo-oligosaccharide degrading bacteria in faeces of piglets postweaning. *Animal Feed Science and Technology, 109*, 133–150.

Morisse, J. P., Maurice, R., Boilletot, E. & Cotte, J. P. (1993). Assessment of the activity of a fructo-oligosaccharide on different caecal parameters in rabbits experimentally infected with *E. coli* O103. *Annales Zootechniques, 42*, 81–87.

Mountzouris, K. C., Balaskas, C., Fava, F., Tuohy K. M., Gibson G. R. & Fegeros, K., (2006). Profiling of composition and metabolic activities of the colonic microflora of growing pigs fed diets supplemented with prebiotic oligosaccharides. *Anaerobe, 12*, 178–185.

Munjal, U., Glei, M., Pool-Zobel, B. L. & Scharlau, D. (2009). Fermentation products of inulin-type fructans reduce proliferation and induce apoptosis in human colon tumour cells of different stages of carcinogenesis. *British Journal of Nutrition*, *102*, 663-671.

Nakakuki, T. (1993). Oligosaccharides: production, properties, and applications. In I. Karube, & R. Kuroda (Eds), Japanese Technology Reviews (Section E: Biotechnology, pp. 1-235). UK: Gordon and Breach Science Publishers.

Oku, T. (1994). Special physiological functions of newly developed mono- and oligosaccharides. In I. Goldberg (Ed), *Functional Foods: Designer Foods, Pharmafoods and Neutraceuticals* (pp. 202-218). New York: Chapman and Hall.

Parker, R. B. (1974). Probiotics, the other half of the antibiotic story. *Ani. Nutr. Health*, *29*, 4-8.

Pelto, L., Isolauri, E., Lilius, E. M., Nuutila, J. & Salminen, S. (1998). Probiotic bacteria down-regulate the milk-induced inflammatory response in milk-hypersensitive subjects but have an immunostimulatory effect in healthy subjects. *Clin. Exp. Allergy, 28*, 1471–1479.

Perdigon, G., de Macias, M. E., Alvarez, S., Oliver, G. & Holgado, A. P. (1998). Systemic augmentation of the immune response in mice by feeding fermented milks with *Lactobacillus casei* and *Lactobacillus acidophilus*. *Immunology*, *63*, 17–23.

Praznik, W., Cie´slik, E. & Huber, A. (2003). Fructans: ocurrence and application in food. In P. Tomasik (Ed), *Chemical and Functional Properties of Food Saccharides*. London: CRC Press.

Qiang, X., YongLie, C. & QianBing, W. (2009). Health benefit application of functional oligosaccharides. *Carbohydrate Polymers*, *77*, 435-441.

Rastall, R. A. & Bucke, C. (1992). Enzymatic synthetis of oligosaccharides. *Biotechnol. Gen. Eng. Rev.*, *10*, 253-281.

Ringø, E., Olsen, R. E., Gifstad, T., Dalmo, R. A., Amlund, H., Hemre, G. I. & Bakke, A. M. (2010). Prebiotics in aquaculture: a review. *Aquaculture Nutrition*, *16*, 117-136.

Roberfroid, M. B. (1997). Health benefits of non-digestible oligosaccharides. *Adv. Exp. Med. Biol.*, *427*, 211-219.

Roberfroid, M. B. (2001). Prebiotics: preferential substrates for specific germs? *Am. J Clin. Nutr., 73*, 406S-409S.

Roberfroid, M. B., Loo, J. A. E. & Gibson, G. R. (1998). The bifidogenic nature of chicory inulin and its hydrolysis products. *J. Nutr.*, *128*, 11-19.

Roberfroid, M. B., Loo, J. A. E. & Gibson, G. R. (1998). The bifidogenic nature of chicory inulin and its hydrolysis products. *J. Nutr.*, *128*, 11-19.

Rowland, I. R. & Tanaka, R. (1993). The effects of transgalactosylated oligosaccharides on gut flora metabolism in rats associated with a human faecal microflora. *J. Appl. Bacteriol.*, *74*, 667-674.

Rowland, I. R. (1995). Toxicology of the colon-role of the intestinal microflora. In G. R. Gibson & G. T. Macfarlane (Eds.), *Human Colonic Bacteria: Role in Nutrition, Physiology, and Pathology* (pp. 115–174). New York: CRC Press.

Rycroft, C. E., Jones, M. R., Gibson, G. R. & Rastall, R. A. (2001). A comparative *in vitro* evaluation of the fermentation properties of prebiotic oligosaccharides. *J. Appl. Microbiol., 91*, 878-887.

Rycroft, C. E., Jones, M. R., Gibson, G. R. & Rastall, R. A. (2001). Fermentation properties of gentio-oligosaccharides. *Lett. Appl. Microbiol., 32*, 156-161.

Saavedra, J. M., Bauman, I., Oung, J. A. & Perman, R. H. (1994). Feeding of *Bifidobacterium bifidum* and *Streptococcus thermophilus* to infants in hospital for prevention of diarrhea and shedding of rotavirus. *Lancet, 344*, 1046–1049.

Salminen, S., Roberfroid, M., Ramos, P. & Fonden, R. (1998). Prebiotic substrates and lactic acid bacteria. In S. Salminen & A. Wright (Eds), *Lactic Acid Bacteria: Microbiology and Functional Aspects* (2nd *ed.,* pp. 343-358). New York: Marcel Dekker, Inc.

Schaafsma, G., Meuling, W. J. A., Van, D. W. & Bouley, C. (1998). Effects of a milk products by *Lactobacillus acidophilus* and with fructo-oligosaccharides added, on blood lipids in male volunteers. *J. Eur. Clin. Nutr., 52*, 436–440.

Schumann, C. (2000). Lactulose. In G. R. Gibson & F. Angus (Eds), *LFRA Ingredients Handbook: Prebiotics and Probiotics* (pp. 47-67). Surrey: Leatherhead Publishing.

Sghir, A., Chow, J. M. & Mackie, R. I. (1998). Continuous culture selection of bifidobacteria and lactobacilli from human faecal samples using fructooligosaccharide as selective substrate. *J. Appl. Microb., 85*, 769-777.

Shim, S. B., Verstegen, M. W. A., Kim, I. H., Kwon, O. S. & Verdonk, J. M. A. J. (2005). Effects of feeding antibiotic-free creep feed supplemented with oligofructose, probiotics or synbiotics to suckling piglets increases the preweaning weight gain and composition of intestinal microbiota. *Archives of Animal Nutrition, 59*, 419–427.

Smiricky-Tjardes, M. R., Grieshop, C. M., Flickinger, E. A., Bauer, L. L. & Fahey, G. C. (2003). Dietary galactooligosaccharides affect ileal and total-tract nutrient digestibility, ileal and fecal bacterial concentrations, and ileal fermentative characteristics of growing pigs. *Journal of Animal Science, 81*, 2535–2545.

Solis-Pereira, B. & Lemonnier, D. (1996). Induction of human cytokines by bacteria used in dairy foods. *Nutr. Res., 13*, 1127–1140.

Suwasono, S. & Rastall, R. A. (1998). Enzymatic synthesis of manno- and heteromanno-oligosaccharides using α-mannosidases: a comparative study of linkage-specific and non-linkage-specific enzymes. *Chem. Technol. Biotechnol., 73*, 37-42.

Tamura, Z. (1983). Nutriology of bifidobacteria. *Bifidobacteria Microflora, 2*, 3-16.

Tanaka, R., Takayama, H., Morotomi, M., Kuroshima, T., Ueyama, S., Matsumoto, K., Kuroda, A. & Mutai, M. (1983). Effects of administration of TOS and *Bifidobacterium breve* 4006 on the human fecal flora. *Bifidobacteria Microflora, 2*, 17-24.

Terada, A., Hara., H., Kataoka, M. & Mitsuoka, T. (1992). Effect of lactulose on the composition and metabolic activity of the human faecal microbiota. *Microb. Ecol. Health Dis., 5*, 43-50.

Thammarutwasik, P., Hongpattarakere, T., Chantachum, S., Kijroongrojana, K., Itharat, A. & Reanmongkol, W. (2009). Prebiotics: a review. *Songklanakarin Journal of Science and Technology, 31*, 401–408.

Timmermans, E. (1994). Lactose: its manufacture and physicochemical properties. In H. van Bekkum, H. Roper & A. G. J. Voragen (Eds), *Carbohydrates as Organic Raw Materials* (vol. 3, pp. 93-113). London: VCH-Weinheim.

Tuohy, K. M., Rouzaud, G. C. M., Brück, W. M. & Gibson, G. R. (2005). Modulation of the human gut microflora towards improved health using prebiotics – assessment of efficacy. *Curr. Pharm. Design.*, *1*, 75-90.

Tuohy, K.M., Finlay, R.K., Wynne, A.G. & Gibson, G. R. (2001). A human volunteer study on the prebiotic effects of HP-inulin-faecal bacteria enumerated using fluorescent *in situ* hybridisation (FISH). *Anaerobe*, *7*, 113-118.

Tzortzis, G., Goulas, A. K., Gee, J. M. & Gibson, G. R. (2005). A novel galactooligosaccharide mixture increases the bifidobacterial population numbers in a continuous *in vitro* fermentation system and in the proximal colonic contents of pigs *in vivo*. *Journal of Nutrition, 135*, 1726–1731.

Van der Heijden, P. J., Bianchi, A. T. J., Dol, M., Pals, J. W., Stok, W. & Bokhout, B. A. (1991). Manipulation of intestinal immune responses against ovalbumin by cholera toxin and its B subunit in mice. *Immunology, 72*, 89–93.

van Leeuwen, P., Verdonk J. M. A. J. & Kwakernaak, C. (2005). Effects of fructooligosaccharide (OF) inclusion in diets on performance of broiler chickens. *Confidential report 05/I00650 to Orafti.*

van Leeuwen, P. & Verdonk, J. M. A. M. (2005). The gastro-intestinal degradation of inulin preparations and their effects on production performance and gut microflora in calves. *Confidential report 04/I00287 to Orafti.*

Vazquez, M. J., Alonso, J. L., Dominguez, H. & Parajo, J. C. (2000). Xylooligosaccharides: manufacture and applications. *Trends Food Sci. & Technol., 11,* 387-393.

Venter, C. S., Vorster, H. H. & Cummings, J. H. (1990). Effects of dietary propionate on carbohydrate and lipid metabolism in man. *Am. J. Gastroenterol.*, *85*, 549–552.

Villegas, B., Tarrega, A., Carbonell, I. & Costell, E. (2010). Optimising acceptability of new prebiotic low-fat milk beverages. *Food Qual. & Pref.*, *21*, 234–242.

Voragen, A. G. J. (1998). Technological aspects of functional food-related carbohydrates. *Trends Food Sci. & Technol.*, *9*, 328-335.

Whelan, K., Judd, P. A., Preedy, V. R., Simmering, R., Jann, A. & Taylor, M. A. (2005). Fructooligosaccharides and fiber partially prevent the alterations in fecal microbiota and short-chain fatty acid concentrations caused by standard enteral formula in healthy humans. *Journal of Nutrition*, *135*, 1896-1902.

Wichienchot, S., Jatupornpipat, M. & Rastall, R. A. (2010). Oligosaccharides of pitaya (dragon fruit) flesh and their prebiotic properties. *Food Chemistry, 120*, 850–857.

Wichienchot, S., Prasertsan, P., Hongpattarakere, T., Gibson, G. R. & Rastall, R. A. (2006a). *In vitro* fermentation of mixed linkage gluco-oligosaccharides produced by *Gluconobacter oxydans* NCIMB 4943 by the human colonic microflora. *Current Issues in Intestinal Microbiology, 7*, 7-12.

Wichienchot, S., Prasertsan, P., Hongpattarakere, T., Rastall R. A. & Gibson, G. R. (2006b). *In vitro* Three-stage continuous fermentation of gluco-oligosaccharides, produced by *Gluconobacter oxydans* NCIMB 4943, by the human colonic microflora. *Current Issues in Intestinal Microbiology*, *7*, 13-18.

Wildman, R. E. C. & Kelly, M. (2007). Nutraceuticals and functional foods. *In* R. E. C. Wildman (Ed), *Handbook of Nutraceuticals and Funtional Foods* (2nd ed., pp. 1-22). New York: CRC Press.

Wong, C. H., Halcomb, R. L., Ichikawa, Y. & Kajimoto, T. (1995). Enzymes in organic synthesis: application to the problem of carbohydrate recognition. *Angew. Chem. Int. Ed., 34: part 1*, 412-432; *part 2*, 521-546.

Yamada, H., Itoh, K., Morishita, Y. & Taniguchi, H. (1993). Structure and properties of oligosaccharides from wheat bran. *Cereal Foods World*, *38*, 490-492.

Yanbo, W. (2009). Prebiotics: present and future in food science and technology. *Food Res. Inter.*, *42*, 8–12.

Yun, J. (1996). Fructooligosaccharides-occurrence, preparation and application. *Enz. Microb. & Technol.*, *19*, 107-117.

Yusrizal, C. & Chen, T. C. (2003a). Effect of adding chicory fructans in feed on fecal and intestinal microflora and excreta volatile ammonia. *International Journal of Poultry Science*, *2*, 188–194.

Yusrizal, C. & Chen, T. C. (2003b). Effect of adding chicory fructans in feed on broiler growth performance serum cholesterol and intestinal length. *International Journal of Poultry Science*, *2*, 214–219.

In: Oligosaccharides: Sources, Properties and Applications ISBN 978-1-61122-193-0
Editor: Nicole S. Gordon

Chapter 3

GLYCOSIDE HYDROLASES FROM HYPERTHERMOPHILES: STRUCTURE, FUNCTION AND EXPLOITATION IN OLIGOSACCHARIDE SYNTHESIS

Beatrice Cobucci-Ponzano**, *Mosè Rossi and Marco Moracci
Institute of Protein Biochemistry – CNR, Via P. Castellino 111,
80131 Naples, Italy

ABSTRACT

Hyperthermophilic microorganisms thrives at temperatures higher than 80°C and proteins and enzymes extracted from these sources are optimally stable and active in the presence of temperatures close to the boiling point of water and of other denaturants, i.e. chaotropic agents, pH, organic solvents, detergents, etc. Therefore, hyperstable enzymes are considered attractive alternatives in biocatalysis and in chemo-enzymatic synthesis. In addition, the molecular bases of the extreme stability to heat and to the ability to work optimally at high temperatures are not completely understood and intrigued biochemists, enzymologists, and biophysics in the last twenty years. In particular, hyperstable glycosidases, enzymes catalysing the hydrolysis of *O*- and *N*-glycosidic bonds, have been studied in detail as they are simple model systems promoting single-substrate reactions, and, more importantly, can be exploited for the enzymatic synthesis of oligosaccharides. The importance of these molecules increased enormously in recent years for their potential application in biomedicine. Hyperstable glycosidases, working in transglycosylation mode, can be excellent alternatives to the classical chemical methods helping in the control of regio- and stereoselectivity as conventional enzymes, but also resisting to the organics used in chemical synthesis. We will review here recent advances in the isolation and characterization of glycosidases from hyperthermophilic microorganisms and the methods used for their application in oligosaccharide synthesis.

* *To whom the correspondence should be addressed
Tel. +39-081-6132564; Fax +39-081-6132277;
E-mail b.cobucciponzano@ibp.cnr.it

INTRODUCTION

Carbohydrates serve as structural components and energy source of the cell and are involved in a variety of molecular recognition processes in intercellular communication [Varki 1993; Sears and Wong 1996]. The knowledge that these biomolecules have a key informative role in biology is becoming more obvious every day confirming their potential as therapeutic agents [Zopf and Roth 1996]. For these reasons, great interest arose in carbohydrate-based compounds and the development of techniques for the analysis and synthesis of oligosaccharides. However, despite the advances achieved with both regio- and stereoselectivity [Toshima and Tatsuta 1993], the synthesis of complex carbohydrates for large-scale production still cannot be easily performed by the classical chemical procedures, which require long protection–deprotection steps and give low final yields. Thus, the enzyme-catalyzed synthesis of oligosaccharides represents an interesting alternative to the classical chemical methods, allowing the control of both the regioselectivity and the stereochemistry of bond formation.

The enzymatic approach involves mainly two class of enzymes: glycosyl transferases and glycoside hydrolases. Glycosyl transferases have been used for oligosaccharide synthesis [Gijsen et al. 1996], but their low availability and the high cost of their substrates have limited their exploitation. Glycoside hydrolases, a widespread group of enzymes that hydrolyse glycosidic bonds, allow the use of relatively inexpensive substrates, representing an alternative choice. They play multiple roles in the cell such as the degradation of poly- and oligosaccharides to be utilized as energy sources, the hydrolysis of the cell-wall polysaccharides for the plant cell division, and the turnover of glycoconjugates involved in several biological functions. Thousands of genes encoding for glycoside hydrolases are known and they have been classified in more than 100 families and 14 clans on the basis of their sequence similarity (http://www.cazy.org) [Cantarel et al. 2008]. These enzymes follow two distinct mechanisms, which are termed *inverting* or *retaining*, depending on whether the enzymatic cleavage of the glycosidic bond liberates a sugar hemiacetal with the opposite or the same anomeric configuration compared with the glycosidic substrate, respectively. *Inverting* enzymes use a direct displacement mechanism in which the two carboxylic acid residues in the active site are positioned so that one acts as a general acid and the other provides general base catalytic assistance to the attack of water. The catalytic mechanism of *retaining* enzymes (Figure 1) proceeds via a two-step double-displacement mechanism involving the formation of a covalent glycosyl intermediate. The two carboxylic residues in the active site play different roles in this case, one acting as the nucleophile and the other as a general acid or base catalyst of the reaction. In the first step, termed glycosylation, the concerted action of the nucleophile and of the general acid residues leads to glycosidic oxygen protonation and the departure of the aglycon group with the formation of a glycosyl–ester intermediate. In the second step (deglycosylation), a water molecule partially deprotonated by the conjugate base of the catalytic acid attacks the anomeric carbon and cleaves the glycosyl–ester intermediate, leading to the overall retention of the anomeric configuration of the substrate. When acceptors other than water intercept the reactive glycosyl–enzyme intermediate, *retaining* enzymes work in transglycosylation mode. This property makes the *retaining* glycoside hydrolases interesting tools for the synthesis of carbohydrates. Despite the differences, the two mechanisms show significant similarities:

both classes of enzymes employ a pair of carboxylic acids at the active site and both mechanisms operate via transition states with substantial oxocarbenium ion character. However, strong evidence indicates that a covalent intermediate is formed in retainers [Sinnot 1990; McCarter and Withers 1994].

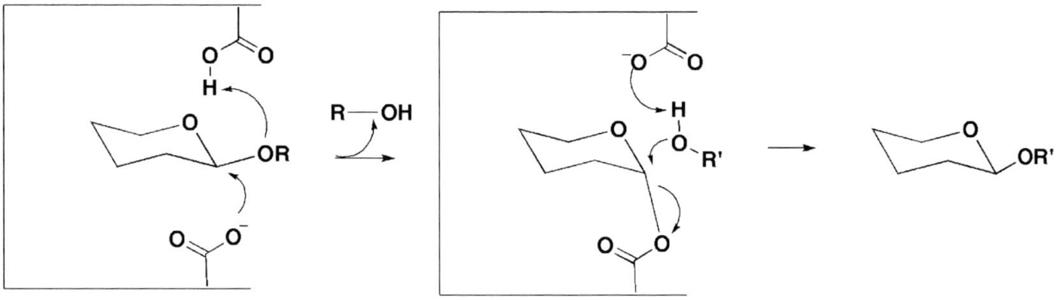

Figure 1. Reaction mechanism of *retaining* β-glycosidases. R: aglycon group. R' = H, hydrolysis. R' = alcohol, transglycosylation.

The identification of key active-site residues in glycoside hydrolase is crucial to understand the catalytic mechanism [McCarter and Withers 1994; Zechel and Withers 2000], to allow the classification of this class of enzymes [Henrissat and Bairoch 1993; Henrissat and Davies 1997], and to produce glycoside hydrolases with novel characteristics [Perugino et al. 2004]. These residues can be identified by many techniques [McCarter and Withers 1994; White and Rose 1997; Ly and Withers 1999]. Successful approaches for the identification of the nucleophile residues include specific labeling with mechanism-based inhibitors such as activated 2-deoxy-2-fluoro-glycosides, in combination with mass spectrometry, amino acid sequence alignment, and crystal structure inspection [Zechel and Withers 2000]. Among others, the use of mechanism-based inhibitors is nowadays the most powerful method for the direct identification in retainers of the carboxyl group acting as the nucleophile of the reaction, even in the absence of the amino acid sequence of the enzyme [Withers and Aebersold 1995]. By contrast, no reliable chemical methods for the identification of the acid/base catalyst are currently available. Site-directed mutagenesis followed by kinetic analysis of the mutants remains, however, the approach most used for the definition of the role played by the active site carboxylic groups. Aspartic/glutamic acid residues identified by sequence analysis and conserved in the family of interest are mutated. Mutations of the catalytic residues with non-nucleophilic amino acids lead to the strong reduction or even abolition of the enzymatic activity [Ly and Withers 1999]. However, these mutants can be reactivated in the presence of external nucleophiles such as sodium azide. Reactivated mutants in the acid/base catalyst produce glycosyl-azide with the same anomeric configuration of the substrate (Figure 2A) By contrast, the isolation of glycosyl-azide products with an anomeric configuration opposite to that of the substrate allows the identification of the catalytic nucleophile of the reaction (Figure 2B) [Ly and Withers 1999]. The classification of glycoside hydrolases into families and superfamilies on the basis of their amino acid sequence and three-dimensional (3-D) structure are commonly used to identify the active site residues and the reaction mechanism of any newly identified glycosyl hydrolase.

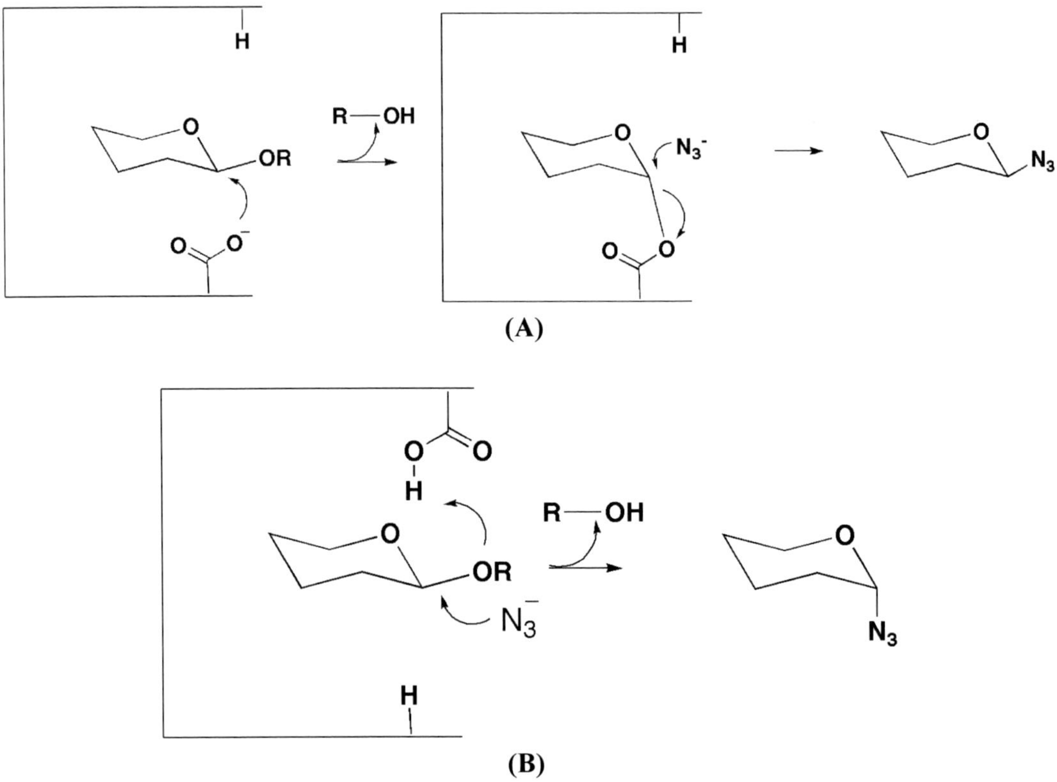

Figure 2. Azide rescue of an acid/base catalyst (A) and a nucleophile (B) mutant of a *retaining* β-glycosidase. R: aglycon group.

Glycosidases are exploited in synthetic reactions by two main approaches, reverse hydrolysis (equilibrium-controlled synthesis) or transglycosylation (kinetically controlled process), in which the glycosyl–enzyme intermediate is transferred to an acceptor other than water. The former method offers modest yields of oligosaccharide products [Withers 2001], whereas kinetically controlled synthesis, which requires a *retaining* glycosidase, provides better yields (10 to 40%), but is not generally economical for large-scale synthesis. In fact, since the product of the reaction is a new substrate for the enzyme it can be hydrolyzed to reduce the final yields of the reaction. Thus, in order to maintain high yields, the reaction conditions have to be strictly controlled. To avoid these problems, a new class of engineered glycosidase has been produced to promote the synthesis of sugars with almost quantitative yields; these novel enzymatic activities have been termed *glycosynthases* (reviewed in Perugino et al. 2004; Perugino et al. 2005, and Hancock et al. 2006). These enzymes, developed for the first time in 1998 in the laboratory of Stephen Withers, are *retaining* glycoside hydrolases mutated in the catalytic nucleophile, that synthesise complex oligosaccharides in high yields [Mackenzie *et al.* 1998]. Because of the essential nature of this residue, when it is mutated with a non-nucleophilic amino acid (typically Ala, Gly, or Ser) the mutant is almost completely inactive. Nevertheless, the rest of the active site is intact and the small cavity created on mutation can accommodate a small anion (i.e. sodium formate). Under these conditions and in the presence of a substrate with good leaving group ability, which assists the glycosylation step of the reaction, the activity of the mutant can be restored (Figure 3A). The methodology described can be usefully applied to the synthesis of carbohydrates. In fact, these modified glycoside hydrolases, lacking their catalytic nucleophile, can act only on activated substrates by synthesizing new glycosidic bonds but cannot hydrolyze the products formed. These compounds, in fact, show groups with poor leaving ability, which are resistant to the attack of the external nucleophiles and which accumulate in the reaction. In the literature, several β-glycoside hydrolases modified in glycosynthases are reported, demonstrating that this methodology could be of general applicability for β-glycoside hydrolases, involving both *exo*- and *endo*-enzymes from diverse sources and with different functions and specificities [Perugino et al. 2004].

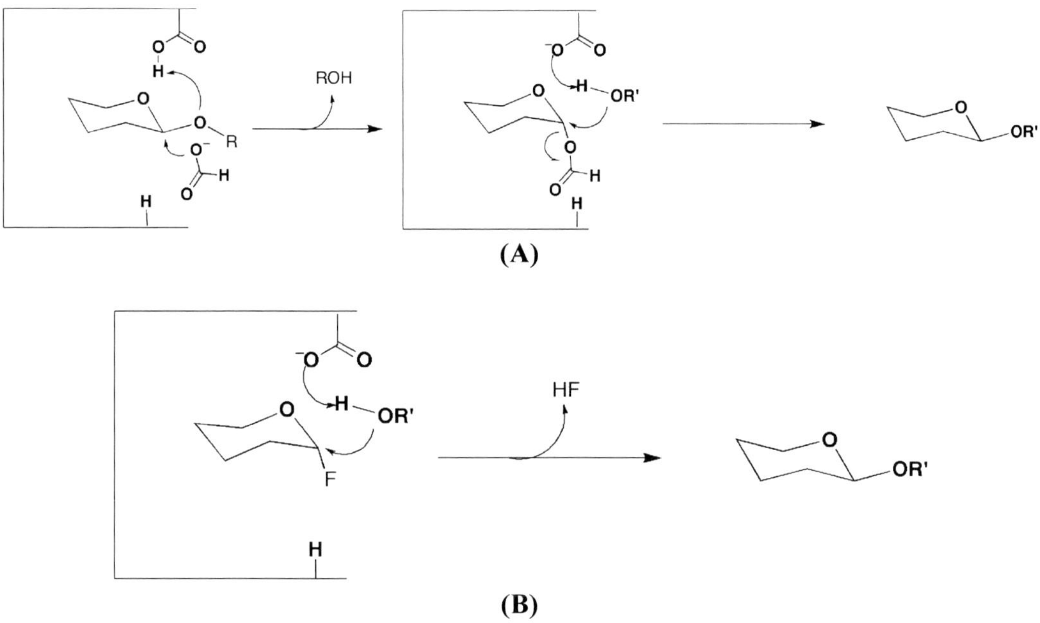

Figure 3. Reaction mechanism of *retaining* (A) and *inverting* (B) β-glycosynthases. R': alcohol.

Glycoside Hydrolases Form Hyperthermophiles

Among the glycoside hydrolases available, the enzymes from hyperthermophilic microorganisms are of particular interest for both basic and applied research. In fact, the function of the glycoconjugates identified in hyperthermophiles and of the enzymes involved in their synthesis and degradation is still largely unknown [Lower and Kennelly 2002]. On the other hand, the harsh conditions of growing of these organisms (temperatures > 80°C) have hindered *in vivo* microbiological and genetic studies; therefore, the isolation of the genes encoding for hyperthermophilic glycoside hydrolases and the detailed enzymological characterization is the only approach to define their role *in vivo*. In addition, the rise of interest in thermophilic enzymes, or thermozymes, as potential biotechnological tools is historically based on their intrinsic resistance to the harsh conditions used in several bioprocesses, such as high temperatures, high concentrations of substrate, and organic solvents. However, the interest in thermozymes is also motivated by their biodiversity, which has revealed novel enzymatic activities [Hough and Danson 1999]; in particular, thermophilic glycoside hydrolases show peculiar enzymological properties, such as unique substrate specificities or reduced substrate/product inhibition [Trincone et al. 1991], and allow the synthesis of new products that are not produced by their mesophilic counterparts [Fischer et al. 1996].

Sulfolobus solfataricus is an extremely thermoacidophilic archaeon which thrives in hot springs at 80-87 °C and pH 2.0-3.0. The genome of this microorganism has been completely sequenced [She et al. 2001] and revealed genes encoding for 22 putative glycosyl hydrolases and 33 putative glycosyl transferases. Remarkably, 36% of the glycosyl hydrolases maps in a region of the genome of about 70 kb, and are likely to be involved in the degradation of sugars for energy metabolism. However, the lack, so far, of molecular genetic tools for these microorganisms has hampered the experimental testing of the *in vivo* function of these genes. In the framework of the exploitation of glycosyl hydrolases from hyperthermophiles for the synthesis of useful oligosaccharides, the sequence of the *S. solfataricus* genome has provided access to a number of novel glycoside hydrolases.

In this chapter the study of three glycosyl hydrolases from this archaeon, focussing on their use in the synthesis of oligosaccharides, will be described.

THE β-D-GLYCOSIDASE

The β-glycosidase from *S. solfataricus* (Ssβ-gly), classified in Family 1 of glycosyl hydrolases (GH1), has been extensively studied and an increasing amount of structural and biochemical data on this enzyme is available in several reviews [Moracci et al., 1994, Moracci et al., 2001]. The enzyme is a homotetramer (240 kDa) with 56-kDa subunits; it is thermostable (half life 48 hr at 85 °C), resistant to detergents and organic solvents and shows maximal activity above 95 °C [Pouwels et al. 2000]. Recombinant Ssβ-gly has been crystallised and the 3D-structure has been solved at 2. Å [Gloster et al. 2004]. The enzyme follows a *retaining* reaction mechanism (Figure 1) with the Glu206 and Glu387 residues acting as the acid/base catalyst and the nucleophile of the reaction, respectively [Moracci et al. 1996].

Interesting properties of Ssβ-gly include wide substrate specificity, the ability to hydrolyse oligosaccharides from their non-reducing end, and to perform oligosaccharide synthesis by transglycosylation reactions with yields ranging between 10-40% [Cobucci-Ponzano et al. 2003]. The wide substrate specificity of Ssβ-gly in the reaction of hydrolysis was confirmed also in the synthesis by using, in transglycosylation reactions, alkyl- and aryl-alcohols acceptors and different aryl-glycosides as donors [Cobucci-Ponzano et al. 2003]. The study of the stereo- and regioselectivity of the enzyme revealed that the primary hydroxyl groups were always favoured if compared to the secondary hydroxyl groups, although the overall regioselectivity of the reaction depended on the complexity of the molecule. The yield of the reaction depended upon chain length and OH position but a marked improvement can be obtained increasing the molar excess of alcohol. In this respect, the increased stability of Ssβ-gly to organic solvents allowed the use of reaction mixtures in which the alcohol and/or organic cosolvent concentrations were up to 95-97%.

The glycosynthase derived from Ssβ-gly (Ssβ-glyE387G), the first hyperthermophilic enzyme of this class, promoted the oligosaccharide synthesis from activated aryl-β-glycosides in the presence of sodium formate as external nucleophile by following the mechanism described in Figure 3A [Moracci et al. 1998; Trincone et al. 2000]. In fact, Ssβ-glyE387G is almost completely inactive, but it can host a donor with the same anomeric configuration of the wild type enzyme (β-anomer). In the first step of the reaction, the external nucleophile finds room in the active site of the enzyme and attacks the anomeric centre of the substrate. In the same time, the acid/base catalyst promotes the departure of the aglycon group of the substrate whereas sodium formate and the glyconic group form a metastable intermediate. At this stage, it is essential that the aglycon group of the activate substrate has good chemical leaving ability. In fact, the mutation greatly slows the first step of the reaction and the enzyme, even in the presence of the formate ion, cannot hydrolyze stable substrates. In the second step, the attack of the intermediate by an acceptor molecule, together with the action as general base of the carboxylate, completes the reaction allowing the formation of the product. The oligosaccharides produced by the enzyme accumulate in the reaction mixture for the reasons stated above, in fact, they are not activated donors and can not be hydrolysed by the glycosynthase. These enzymes were named *retaining* glycosynthases [Perugino et al. 2004].

In addition to this mechanism it has been demonstrated that *Ss*β-glyE387G, in the presence of a substrate with an anomeric configuration opposite to that of the natural substrate, typically, α-glycosyl-fluoride (α-F-Glc) for β-glycosynthases, the enzyme catalyses the synthesis of oligosaccharides by transferring the α-F-Glc donor to sugar acceptors (Figure 3B). The oligosaccharide products, containing bonds in the β-anomeric configuration, cannot be hydrolysed by the mutant and accumulate in the reaction [Mackenzie et al. 1998; Malet et al. 1998]. On the basis of the anomeric configuration of the donor and the acceptor, the enzymes following the mechanism shown in Figure 3A were named *inverting* glycosynthases [Perugino et al. 2004].

Remarkably, Ssβ-glyE387G was able to follow both mechanisms shown in Figure 3. In the presence of 2M sodium formate as external nucleophile and activated β-glycoside donors, such as 2,4-dinitrophenyl- or 2-nitrophenyl-β-Glc and 2-nitrophenyl-β-Fuc, Ssβ-glyE387G followed the reaction mechanism shown in Figure 3B producing branched oligoglucosides (85% total efficiency, with 50% disaccharides, 40% trisaccharides and 10% tetrasaccharides using 2-Np-β-Glc) [Trincone *et al.* 2000]. In addition, in the presence of α-F-Glc substrate, Ssβ-glyE387G synthesised 2-NP-β-D-laminaribioside in 90% yields by following the mechanism in Figure 3B [Trincone *et al.* 2000].

After the first hyperthermophilic glycosynthase, three more enzymes of this class were produced by engineering the β-glycosidases from *Thermosphaera aggregans*, *Pyrococcus furiosus,* and *Pyrococcus horikoshii* [Perugino et al. 2003; Perugino et al. 2006]. Interestingly, we reported that the synthetic activity of these *retaining* glycosynthases could be greatly enhanced at acidic conditions [Perugino et al. 2003]. In fact, the reaction mechanism of *retaining* glycosynthases involves a glycosylation step in which the general acid/base catalyst and the formate ion co-operate (Figure 3A). The removal of the catalytic nucleophile in *retaining* glycosidases causes a downward shift in the pK_a of the acid/base catalyst [McIntosh et al. 1996]. Consequently, this group, which is ionised at neutral pH, could perform the first step of the reaction less efficiently. To maintain the acid/base group in the protonated catalytically efficient form, glycosynthetic reactions were performed in sodium formate buffer pH 3.0-6.0. At these conditions, the Ssβ-glyE387G enzyme, and the glycosynthases from *T. aggregans* and *P. furiosus*, showed improved efficiency in the synthetic reaction and enhanced synthetic repertoire [Perugino et al. 2003].

The branching functionalisation is a unique characteristic of the glycosynthases from *S. solfataricus* and *T. aggregans* and it has never been reported for other enzymes of this kind. *T. aggregans* and *S. solfataricus* glycosynthases differ in their regioselectivity: the former synthesises the glucose disaccharides β-1,3: β-1,4: β-1,6 as 59:28:12 ratios whereas the composition of the regioisomers synthesised by the enzyme from *S. solfataricus* was 80:2:18 [Perugino et al. 2003]. The compounds produced can be of applied interest in the pharmaceutical and food fields [Kiho et al. 1992] and can also be used as new substrates and/or inhibitors for glycosidases.

The ability of the hyperthermophilic glycosynthases to promote the synthesis of oligosaccharides is a clear example of how the unique characteristics of stability to high temperatures and acidic pH of the glycosidases from hyperthermophilic archaea allows the development of a novel strategy for the chemo-enzymatic synthesis of oligosaccharides.

The α-D-Xylosidase

Among the putative glycosyl hydrolases present in *S. solfataricus*, an α-D-xylosidase (XylS), was identified. The gene, which is actively transcribed in the archaeon, was cloned and expressed in *E. coli*, producing an enzyme that, in native conditions, is a monomer of about 85 kDa [Moracci et al. 2000]. XylS displays maximal activity at 90°C and high stability to heat (half life of 38 h at 90°C) and reveals clear selectivity for xylose-containing substrates such as 4-nitrophenyl-α-D-xyloside (4-Np-α-Xyl) and the disaccharide isoprimeverose (α-D-xylopyranosyl-(1,6)-D-glucopyranose), which is the disaccharidic unit of the hemicellulose xyloglucan, and 4-Np-β–isoprimeveroside which are hydrolysed from the non-reducing end showing that the enzyme is an exo-glycosidase (Table 1) [Moracci et al., 2000]. In contrast, the activity on maltose and different maltooligosaccharides is lower, and completely absent on isomaltose, trehalose, and sucrose. It is worth noting that the extreme specificity of the exo-xylosidase activity of XylS has been exploited to characterize complex oligosaccharides. In fact, XylS, coupled to a β-glucosidase, was used to determine unequivocally the structure of a xyloglucan oligosaccharide synthesized by a novel α-xylosyltransferase from *Arabidopsis* [Faik et al 2002].

Xyloglucan, the principal hemicellulose component in the primary cell wall and one of the most abundant storage polysaccharides in seeds, is widely distributed in plants. Xyloglucans are involved in complex biological roles. For instance, some xyloglucan oligosaccharides can promote the elongation of stem segments [Lorences and Fry 1994] and therefore play a role in the regulation of plant growth. In plant seeds, the hydrolysis of xyloglucan occurs after germination, during the mobilization of this storage polysaccharide. This polymer is composed of a β–(1,4)-glucan backbone, with α-(1,6)-D-xylose groups linked to about 75% of the glucosyl residues. Thus, the disaccharide isoprimeverose represents the building block of xyloglucan. Additional ramifications of β-D-galactosyl-(1,2)-α-xylosyl and α-L-fucosyl-(1,2)-β-D-galactosyl-(1,2)-α-xylosyl chains are α-(1,6)-linked at a lesser extent to the main backbone (Figure 4) [Crombie, et al 1998]. XylS is active on xyloglucan oligosaccharides from which it produces xylose, suggesting that the enzyme recognizes isoprimeverose units at the non-reducing end of xyloglucan fragments and promotes the release of xylose residues from these compounds [Moracci et al 2000].

Table 1. Kinetic constants of XylS

Substrates	k_{cat} (s^{-1})	K_M (mM)	k_{cat}/K_M ($s^{-1}mM^{-1}$)
4-NP-α-xylopyranoside	4.69 ± 0.27	17.0 ± 2.1	0.28
4-NP-β-isoprimeveroside	16.0 ± 1.6	1.72 ± 0.46	9.30
4-NP-α-glucopyranoside	0.05 ± 0.00	2.05 ± 0.44	0.02
Isoprimeverose	31.0 ± 1.5	28.9 ± 3.5	1.07
Maltose	1.51 ± 0.07	17.0 ± 3.2	0.09
Maltotriose	0.92 ± 0.04	3.45 ± 0.97	0.27

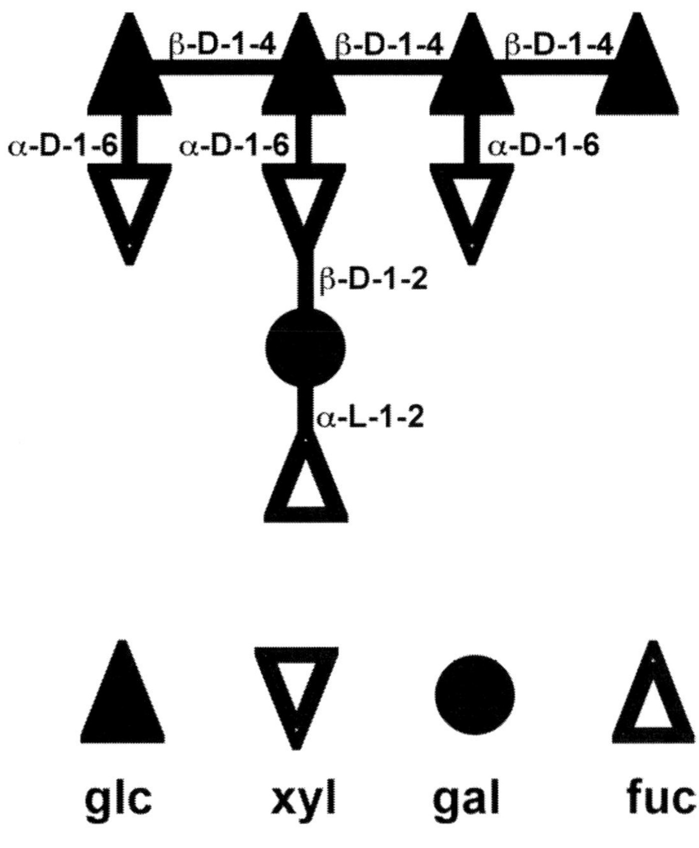

Figure 4. Schematic structure of a xyloglucan oligosaccharide.

In the xyloglucan polymer, xylose groups are α-1,6-linked to most of the glucose units forming the β-(1,4)-glucan backbone of this polysaccharide and the incubation of XylS and

*Ss*β-gly enzymes determined the complete hydrolysis of the oligosaccharides. In fact, XylS and *Ss*β-gly are both *exo*-acting enzymes, thus, they could attack alternatively the non-reducing ends of the xyloglucan oligosaccharides. The identification of *xylS* gene and the substrate specificity of its gene product strongly suggest the involvement of this enzyme in the degradation of di- and oligosaccharides containing α-1,6-linked xylose, which are the building blocks of xyloglucan. Moreover, the cooperation of XylS and *Ss*β-gly in the degradation of xyloglucan oligosaccharides *in vitro*, and the vicinity of the encoding genes on the *S. solfataricus* chromosome could suggest that the two enzymatic systems are functionally related also *in vivo*. The solfataric fields, in which *S. solfataricus* grows under aerobic conditions, are rich in plant debris containing hemicellulosic material. XylS *in vivo* could recognize isoprimeverose units at the non-reducing end of xyloglucan fragments and promote the release of the xylose that can be used as an energy source. However, the complex structure of xyloglucan would require the combined action of several enzymatic activities and protein transporters for its efficient hydrolysis and assimilation.

Table 2. Oligosaccharide synthesis by the α-xylosidase from *S. solfataricus*

Donor	Acceptor	Products
4-NP-α-Xyl	4-NP-β-Glc	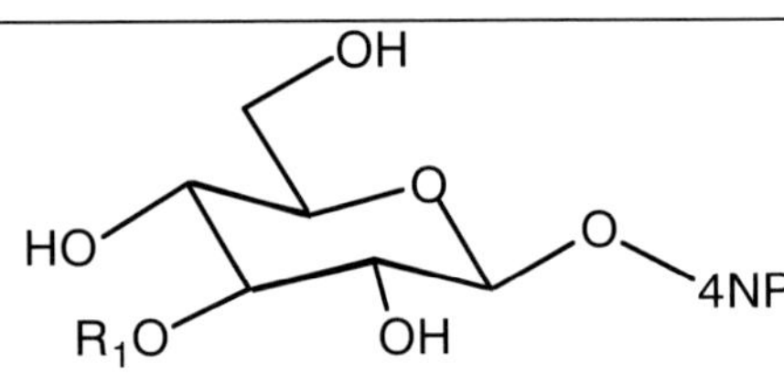 R_1=α-Xyl
α-xylosyl-F	Glucose	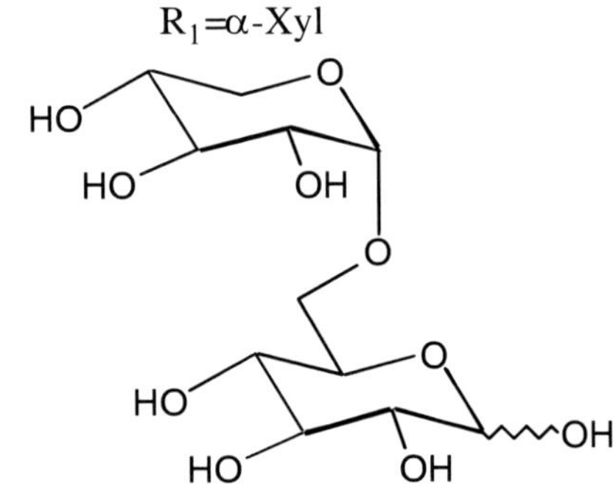
α-xylosyl-F	4-NP-β-cellobioside	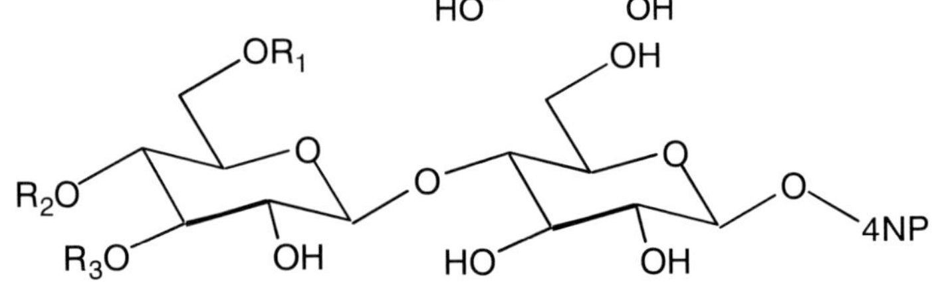 1. R_1=α-Xyl; R_2=R_3=H 2. R_1=R_2; R_3=α-Xyl 3. R_1=H; R_2=α-Xyl; R_3=H

The importance of xyloglucan oligosaccharides in several biological events determined the need of purified xyloglucan oligosaccharides and isoprimeverose for enzymological and metabolic studies [Lorences and Fry 1994; Chaillou et al. 1998]. Relatively little is known about the mechanism of xyloglucan degradation and the enzymatic systems involved in the metabolism of xyloglucan oligosaccharides and isoprimeverose. These compounds can be prepared by several steps of hydrolysis of the natural polymer xyloglucan by different glycoside hydrolases [Kato and Matsuda. 1980]. This approach is limited by the huge number of different enzymatic activities required and by the production of mixtures of oligosaccharides difficult to purify. These drawbacks can be overcome by synthesizing isoprimeverose and specific xyloglucan oligosaccharides by using the enzymes involved in the degradation of xyloglucan in transglycosylating mode. Interestingly, XylS is able to transfer xylose in a transxylosylation reaction. In fact, the enzyme, which follows the retaining reaction mechanism as do the other members of GH31, forms 4-nitrophenyl-β-isoprimeveroside (11% average yields) by transxylosylation with 4-Np-α-Xyl and 4-nitrophenyl-β-D-glucoside (4-Np-β-Glc) as donor and acceptor, respectively (Table 2) [Moracci et al. 2000]. Different regioisomers, in which xylose is transferred to different glucose positions, possibly the OH in C3 and C4, were observed in trace amounts. Free isoprimeverose was obtained, with 10-12% average yields, by using glucose and α-xylosyl fluoride as acceptor and donor, respectively [Moracci et al 2001, Trincone et al., 2001]. With the same α-xylosyl fluoride donor, and 4-Np-β–cellobioside as acceptor, XylS synthesises the trisaccharidic unit of xyloglucan XG (15% average yields) in which xylose is attached on the C6 of the external glucose of the acceptor (Table 2). These findings demonstrate that the typical exo-acting hydrolysis of this enzyme is also operative in synthetic mode [Trincone et al., 2001; Cobucci-Ponzano et al, 2003a]. For these reasons, XylS could be used for the synthesis of the building blocks of xyloglucan oligosaccharides, which, acting as growth regulators of plant cells, may find application in agro-food technology.

THE α-L-FUCOSIDASE

The interest in α-L-fucosidase and fucosyl-transferase activities is due to the central role of fucosylated oligosaccharides in a variety of biological events [Staudacher et al., 1999]. α-L-Fucosidases (3.2.1.51) are *exo*-glycosidases capable of cleaving α-linked L-fucose residues from glycoconjugates, in which the most common linkages are α-(1,2) to galactose and α-(1,3), α-(1,4), and α-(1,6) to *N*-acetylglucosamine residues. These compounds are involved in a variety of biological events as growth regulators and as the glucidic part of receptors in signal transduction, cell-cell interactions, and antigenic response [Vanhooren and Vandamme 1999]. The central role of fucose derivatives in biological processes explains the interest in α-L-fucosidase and fucosyl-transferase activities.

Family 29 of glycoside hydrolases classification (GH29) groups α-L-fucosidases from plants, vertebrates, and pathogenic microbes of plants and humans [Henrissat 1991]. The first archaeal α-L-fucosidase was identified in *S. solfataricus* (reviewed in Cobucci-Ponzano et al 2008). The analysis of the genome of *S. solfataricus* [She et al. 2001] revealed the presence of the gene fucA1, consisting of two ORFs, SSO11867 and SSO3060, encoding for polypeptides of 81 and 426 amino acid that are homologous to the *N*- and the *C*-terminal parts, respectively, of full-length bacterial and eukaryal GH29 fucosidases [Cantarel et al. 2008].

The two ORFs, actively co-transcribed *in vivo*, are separated by a -1 frameshift in a 40 bases overlap and lead to a truncated product. This suggested that the gene is expressed *in vivo* by a mechanism of *recoding*, consisiting of a localized deviation from the standard trnslational rules, called programmed -1 frameshifting [Cobucci-Ponzano et al 2003b; Cobucci-Ponzano et al 2006]. In fact, a detailed analysis of the DNA sequence of the region of overlap between the two ORFs revealed the presence of a stretch of six adenines followed by a thymine (Figure 5A) that resembles one of the heptamers involved in programmed -1 frameshifting. This is a mechanism of regulation of the expression of a minority of genes already reported in Eucarya and Bacteria, but not in Archaea, in which the modification of the translational frame takes place in a programmed way [Farabaugh 1996]. Typically, the sites cis-regulating these events consist of a 'slippery' heptameric sequence of the general form X-XXY-YYZ (codons are in the zero frame and X, Y and Z can be identical or different nucleotides) and a downstream mRNA secondary structure that promotes the pausing of the ribosome facilitating the frameshifting at the slippery sequence [Farabaugh 1996]. A similar organisation was found in the region of overlap between the ORFs SSO11867 and SSO3060, which shows the slippery sequence A-AAA-AAT immediately followed by a stem-loop (Figure 5A). In the mechanism of programmed -1 frameshifting proposed for Eukarya and Bacteria, two tRNAs, hybridised to the XXY and YYZ codons of the X-XXY-YYZ sequence, are proposed to slip simultaneously backwards on the mRNA to the -1 frame, hybridising to XXX and YYY codons. The AAT triplet, coding for Asn78 in SSO11867, corresponds to the YYZ codon, and is the last one decoded in the zero frame [Farabaugh 1996]; after this triplet, the ribosome would shift onto the TTC codon of the Phe10 (SSO3060 numbering) continuing the translation in the -1 frame (Figure 5A). To restore a single frame between the two ORFs, a T in the region following the slippery heptamer and the conservative mutation AAA → AAG (encoding for Lys77 in SSO11867), to increase the translational fidelity by disrupting the heptameric sequence, were inserted (Figure 5B). These mutations were designed on the basis of the programmed -1 frameshifting mechanism. The single ORF obtained was used to express the enzyme in *E coli.* The native recombinant enzyme, named Ssα-fuc, showed high specificity for the 4-NP-α-L-fucoside (4-NP-α-L-Fuc) substrate at 65°C (K_M and k_{cat} values of 0.028 ± 0.004 mM and 2879 ± 11 s^{-1}, respectively; k_{cat}/K_M 10,250 $s^{-1}mM^{-1}$). Moreover, Ssα-fuc is thermoactive and thermostable, as expected for an enzyme from a hyperthermophilic microorganism. The optimal temperature of the enzyme is 95°C and it displayed high stability at 75°C, showing even 40% activation after 30 min of incubation and maintaining 60% residual activity after 2 h at 80°C [Cobucci-Ponzano et al 2003b]. Small angle X-ray scattering experiments revealed that *Ssα*-fuc is a nonamer of 57 kDa molecular mass subunits, showing an unusual oligomeric assembly resulting from the association of nine subunits, endowed with 3-fold molecular symmetry [Rosano et al. 2004].

It is worth noting that the mutation inserted to obtain the recombinant Ssα-fuc was designed on the basis of the programmed -1 frameshifting mechanism [Farabaugh 1996] therefore, the functionality of the full-length enzyme gave support to the hypothesis that a translational *recoding* event, known so far only in Eukarya and Bacteria [Baranov et al. 2002] could be used to regulate the expression of this gene in *S. solfataricus* [Cobucci-Ponzano et al. 2005a]. More recently, the study of the expression of the wild-type split gene fucA1 and of its site-directed mutants in the slippery sequence demonstrated that fucA1 is expressed by programmed -1 frameshifting in both *E. coli* and *S. solfataricus*. This was the first experimental demonstration that this kind of recoding is present in the Archaea domain of life and, thus, universally conserved [Cobucci-Ponzano et al 2006].

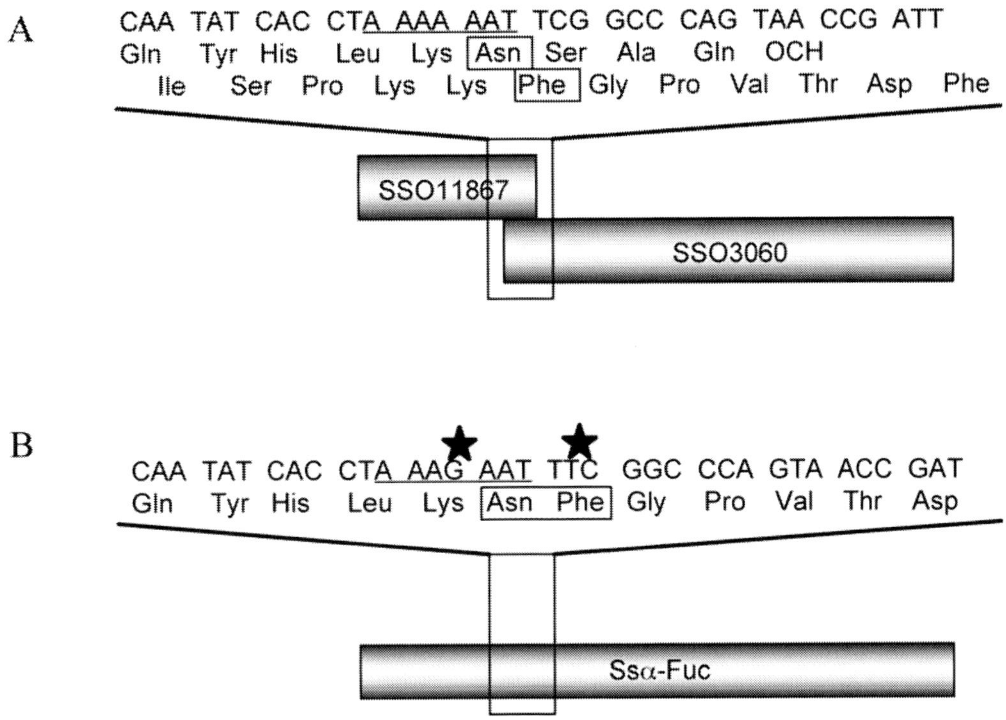

Figure 5. The α-L-fucosidase wild type gene organization (A) and preparation of the full-lenght mutant Ssα-fuc (B). The adjacent ORFs SSO11867 and SSO3060 and the sequence of overlap are shown. The slippery sequence is underlined. The mutated bases are indicated with a star.

*Ss*α-Fuc is capable of functioning in transglycosylation mode as reported for several mesophilic α-fucosidases [Murata et al 1999; Farkas et al 2000]. The synthetic ability of the thermophilic enzyme was demonstrated by using, in transfucosylation reactions, 4-NP-α-L-Fuc and 4-nitrophenyl-α-D-glucoside (4-NP-α-D-Glc) as donor and acceptor, respectively. The fucosylated products, with 14% total yield with respect to 4-NP-α-L-Fuc, were disaccharides of the acceptor in which the α-L-fucose moiety is attached at positions 2 and 3 of Glc (α-L-Fuc*p*-(1,2)-α-D-Glc-*O*-4-NP and α-L-Fuc*p*-(1,3)-α-D-Glc-*O*-4-NP, respectively) [Cobucci-Ponzano et al. 2003b]. In addition, the α-anomeric configuration of the interglycosidic linkages in the products demonstrated, for the first time by following transglycosylation reaction, that GH29 α-fucosidases follow a *retaining* reaction mechanism [Cobucci-Ponzano 2003b]. The hydrolytic activity of *Ss*α-fuc on the disaccharide α-L-Fuc-(1,3)-α-L-Fuc-*O*-4-NP revealed that the enzyme is an *exo*-glycosyl hydrolase that attacks substrates from their non-reducing end [Cobucci-Ponzano et al 2003b].

The nucleophile of GH29 α-L-fucosidases was identified, for the first time, by reactivation with sodium azide of a site-directed mutant of *Ss*α-fuc and by analyzing the anomeric configuration of the fucosyl-azide product [Cobucci-Ponzano et al. 2003c]. The residue Asp242, invariant in all members of the family, was changed into glycine by site-directed mutagenesis to completely remove the side chain of the aspartic acid. The SsαfucD242G mutant showed a k_{cat} of 0.24 s^{-1}, on 4-NP-α–L-Fuc, which is 1.0 x 10^{-3} times that of the wild-type activity (245 s^{-1}). In the presence of 2M sodium azide the mutant revealed a k_{cat} value of 11 s^{-1}, indicating a 46-fold activation by azide (Table 3) [Cobucci-Ponzano et al 2003b]. This activation falls in the range of chemically rescued activities of β-glycoside hydrolases [Wang et al 1994; Viladot et al 1998]. The reactivation experiment with sodium azide indicated that the D242G mutation affected a residue involved in catalysis in *Ss*α-fuc [Cobucci-Ponzano et al. 2003c]. As mentioned in the Introduction, one of the methods used to identify the acid/base or the nucleophile of the glycosidases is to characterise the stereochemistry of the products of mutants reactivated by azide [Ly and Withers 1999]. The isolation of glycosyl-azide products with an anomeric configuration opposite to that of the substrate would allow the identification of the catalytic nucleophile of the reaction. In this case, in fact, the azide acts as nucleophile of the reaction leading to a stable product. The fucosyl-azide product obtained by the Ssα-fucD242G mutant was found in the inverted (β-L) anomeric configuration compared with the substrate. This finding allowed, for the first time, the unambiguous assignment of Asp242 and its homologous residues as the nucleophilic catalytic residues of GH29 α-L-fucosidases [Cobucci-Ponzano et al 2003c].

Table 3. Steady-state kinetic constants of wild type and mutants α-fucosidases

	k_{cat} (s^{-1})	K_M (mM)	k_{cat}/K_M (s^{-1} mM^{-1})
Wild type	287±11	0.028±0.004	10,250
D242G			
+ sodium azide	9.66±0.28	0.19±0.02	51.55
+ sodium formate	5.91±0.20	1.03±0.11	5.76
H46G	419±99	17.0±5.7	25
H123G[a]	ND[b]	ND[b]	5.96
E58G[a]	ND[b]	ND[b]	2.63
E292G	1.860±0.093	0.056±0.013	33

[a]No saturation observed on up to 25 mM 4NP-Fuc; the specificity constants were calculated from the Lineweaver-Burk plots. [b]Not Determined.

The azide rescue method has been used so far only for *retaining* β-D-glycosidases, whereas it has never been used for *retaining* α-(D/L)-glycosidases, and it is generally based on substrates that are very reactive (showing leaving groups with p*K*a < 5) to facilitate the first step of the reaction [Shallom et al. 2002]. Therefore, it is worth noting that Ssα-fucD242G was efficiently reactivated on 4-NP-α-L-Fuc, which shows leaving group with p*K*a = 7.18. By comparison, β-glycosidases from GH1 and GH52, mutated in the nucleophile, could not be reactivated by azide on 4-NP-glycosides, but require substrates with excellent leaving groups for an efficient chemical rescue of their activity [Moracci et al 1998; Shallom et al 2002]. The identification, for the first time, of the nucleophile of a α-L-glycosidase by following this approach showed that it could be of general applicability for *retaining* enzymes. By following a different approach, the corresponding residue of the α-L-fucosidase from the hyperthermophilic bacterium *Thermotoga maritima* (Tmα-fuc) was identified

[Tarling et al. 2003]. In addition, the same authors obtained the crystal structure of complexes of Tmα-fuc with α-L-fucose and with a mechanism-based inhibitor. This study allowed the identification of Glu266 as the general acid/base catalyst of the enzyme through the inspection of the 3D-structure and the detailed kinetic characterization of active-site mutants [Sulzenbacher et al. 2004]. Remarkably, although the residues Glu266 and Glu66, with the latter being involved in the substrate binding, were both essential for activity, they were not conserved in GH29 and could be identified only through inspection of the 3D structure.

To identify other amino acids involved in the mechanism of reaction as acid/base catalyst or have different funcion in Ssα-fuc, several residues that, from sequence alignments of GH29 and from the 3D-structure of Tmα-fuc, were predicted to be involved in catalysis, were modified by site-directed mutagenesis. The kinetic characterization of the mutants Ssα-fucH46G and Ssα-fucH123G in Ssα-fuc revealed that these residues are involved in substrate binding [Cobucci-Ponzano et al. 2005b] as predicted by the crystal structure of Tmα-fuc complexed to α-L-fucose [Sulzenbacher et al. 2004]. Thus, the reduced affinity for the substrate of Ssα-fucH46G and Ssα-fucH123G mutants (Table 3) and the absence of reactivation by external ions clearly support the hypothesis that these residues interact with the hydroxyl groups bound to C4 and C2 of the substrate. In contrast, as mentioned above, the acid/base catalyst of Tmα-fuc, Glu266, and another residue, Glu66, which is also located in the active site of the enzyme, are conserved to a lesser extent in GH29. However, a pairwise alignment of Ssα-fuc and Tmα-fuc showed that these residues corresponded to Ssα-fuc Glu58 and Glu292. To determine whether Glu58 or Glu292 was the acid/base catalyst, the classical approach was followed in which the glutamic acids are replaced by glycine and the obtained mutants are kinetically characterized. The mutation of the residues Glu58 and Glu292 almost inactivates the enzymes [Cobucci-Ponzano et al 2005b]. The first diagnostic tool to test if the general acid/base has been removed by mutation is the analysis of the steady state kinetic constants; in fact, it has been pointed out that, if the acid/base of a retaining glycosidase is removed, this affects both steps of the reaction, but the effect depends on the substrate used [Ly et al 1999]. In the case of poor substrates, showing groups with worse leaving ability (aglycons with p*K*a > 8), the first step of the reaction (glycosylation) requires the participation of the general acid/base catalyst. Instead, for good substrates (aglycons with p*K*a < 8), the glycosylation step needs less assistance and the second step of the reaction (deglycosylation) is dependent on the assistance of the acid/base catalyst that functions as a general base at this stage [MacLeod et al 1996; Vallmitjana et al 2001; Li at al 2002]. The 4-NP-α-L-Fuc substrate has a p*K*a value of the leaving group 4-nitrophenol of 7.18 suggesting that the limiting step in the hydrolysis is at the borderline between glycosylation and deglycosylation and, thus, hampering to establish in advance which is the limiting step of the reaction of Ssα-fuc with the 4NP-Fuc substrate. The kinetic characterization of the Ssα-fucE58G and Ssα-fucE292G mutants gave unexpected results. In fact, the lack of saturation on the 4-NP-α-L-Fuc substrate of Ssα-fucE58G mirrors that of the acid/base mutant in *T. maritima* enzyme (Tmα-fucE266A) [Sulzenbacher et al 2004]. Instead, Ssα-fucE292G showed the classical behavior of a mutated acid/base catalyst with a 150-fold reduction of k_{cat} and a K_M almost unchanged (Table 3) [Cobucci-Ponzano et al. 2005b]. In fact, the leaving ability of 4-NP-α-L-Fuc could be good enough to complete the glycosylation step; therefore, when the acid/base catalyst is removed, the k_{cat} value, assessing the deglycosylation step, is significantly reduced. The reduction in the catalytic rate results in accumulation of the fucosyl-enzyme intermediate and consequently determines a decrease of the K_M value [Ly et al 1999]. Presumably, the K_M did not decrease in Ssα-fucE292G because the K_M of the wild-type is already very low. It is worth noting that the removal of the residue acting as the acid/base catalyst in Tmα-fuc produced a mutant that could not be saturated with 4-NP-α-L-Fuc [Sulzenbacher et. al 2004]. This behavior has never been reported in retaining glycosidases and may suggest that the catalytic machinery of α-L-fucosidase may differ from those known so far. A second diagnostic tool for the assessment of the acid/base catalyst is the pH dependence. Typically, glycosidases have a bell-shaped pH dependence, and when the acid/base is removed, the basic limb is severely affected [Ly et al 1999]. Wild-type Ssα-fuc has a peculiar pH dependence showing a reproducible increase of activity at pH 8.6 (Figure 6). This behavior suggests that more than two ionizable groups are involved in catalysis [Debeche et al 2002]. The pH profile resulted in a large change in the case of the Ssα-fucE58G mutant, producing a typical bell-shaped curve with a pH optimum at 4.6 sharper than that of the wild-type (3.0-5.0) (Figure 6), suggesting that the removal of Glu58 unmasked the ionization of a group responsible for the basic limb (p*K*a 5.3) and possibly increased the p*K*a of the nucleophile of the reaction mainly determining the acidic limb [Cobucci-Ponzano et al. 2005b].

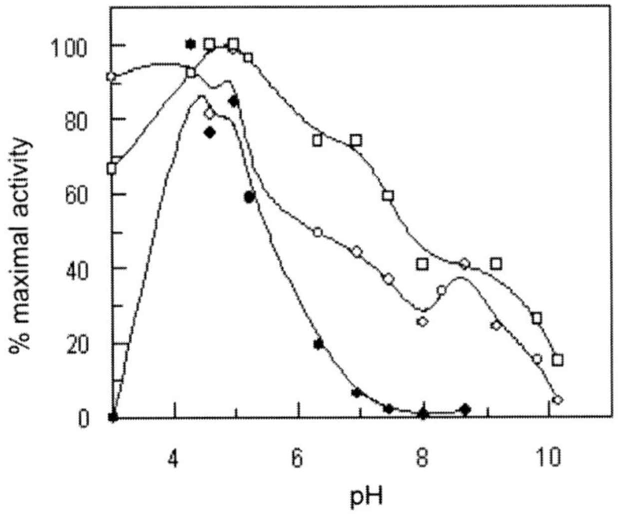

Figure 6. Dependence on pH of Ssa-Fuc and mutants The wild-type is reported as ❍, and E58G and E292G are reported as ● and ❑, respectively.

The most definite tool to determine the acid/base catalyst is the chemical rescue of the activity of the mutants by nucleophilic anions. Replacing the acid/base catalyst with the small non-ionizable glycine residue generally reduces dramatically the activity of the mutant. In addition, the presence of the glycine results in sufficient space in the active site so that a small nucleophile can be accommodated in this cavity after the formation of the glycosyl-enzyme intermediate. In these cases, rate enhancement and the isolation of a glycosyl-azide product with the same anomeric configuration of the substrate are expected [Ly et al 1999]. Activity rate enhancements by means of external nucleophiles is the most effective method to unequivocally identify the acid/base catalyst. Remarkably, Ssα-fucE58G was activated by more than 70-fold in the presence of sodium azide, formate, acetate, and chloride (Table 4). However, no α-fucosyl-azide product could be identified. In striking contrast, the activity of Ssα-fucE292G could not be rescued by any of the nucleophiles used. From these results, the role of the acid/base catalyst of Ssα-fuc was assigned to Glu58 and suggested that in Ssα-fuc a catalytic triad, namely, Glu58, Glu292, and Asp242, is involved in catalysis [Cobucci-Ponzano et al. 2005b].

Table 4. Steady-state kinetic constants of wild type and E58G in different buffers

	k_{cat} (s^{-1})	K_M (mM)	k_{cat}/K_M ($s^{-1}M^{-1}$)	Reaction conditions	pH
wt	287±11	0.028±0.004	10,250	Sodium phosphate	6.3
	430±49	0.26±0.09	1,624	Sodium citrate	5.0
E58G	ND[a]	ND[a]	2.63	Sodium phosphate	6.3
	143±8	1.6±0.3	89	Sodium citrate	4.6
	586±43	0.62±0.18	950	Sodium acetate	4.6
	846±46	1.07±0.18	790	Sodium formate	4.6
	679±32	2.96±0.43	229	Sodium phosphate + NaN_3	6.3

[a]Not Determined; no saturation was observed with up to 25 mM 4NP-Fuc. The specificity constant was calculated from the Lineweaver-Burk plots.

Fucose derivatives play a central role in biological processes including the regulation of plant growth [de La Torre et al. 2002] and several physiological and pathological events in mammals [Listinsky et al. 1998; Mori et al. 1998; Noda et al. 1998]. Therefore, the efficient synthesis of fucosylated oligosaccharides has potential applications in biomedicine [Vanhooren and Vandamme 1999]. Although the preparation of a glycosynthase from a specific glycosidase is, in principle, a simple strategy, this has not led to a large variety of different enzymes and the glycosynthases available so far are limited to glycoside families GH1, 2, 5, 7, 8, 10, 16, 17, 26 and 31 out of the more than 100 glycosidase families described. In fact, there are reports of several enzymes which are recalcitrant to becoming glycosynthases: for instance, only one α-glycosynthase is known so far [Okuyama et al. 2002], and enzymes from families 29, 35 and 39 modified to this end, did not yield glycosynthases (Cobucci-Ponzano et al. 2003c; Perugino et al. 2005). Attempts to produce a fucosynthase from Ssα-fuc were not successful. The mutant Ssα-fucD242G in the presence of sodium formate acting as external nucleophile and the substrate 4-NP-α-L-Fuc produced a 16-fold reactivation, but no fucosylated oligosaccharide could be observed [Cobucci-Ponzano et al. 2003b]. The absence of synthetic products could be due to the nature of the ion used to the acidic pH exploited or to the insufficient leaving ability of the aglycon group in 4-nitrophenol to perform glycosynthesis. The analysis of the activity of the Ssα–fucD242G mutant under a variety of conditions, including sodium formate or sodium acetate buffers at different pHs, sodium formate, sodium acetate or sodium chloride as different nucleophiles in sodium phosphate buffer, in the presence of glycoside acceptors, and at different concentrations of 4-NP-α-L-Fuc substrate donor showed that, under most of the reactivation conditions tested, the hydrolytic activity of Ssα–fucD242G was increased [Cobucci-Ponzano et al. 2008]. In particular, sodium formate was by far the best nucleophile tested while higher concentrations of 4-NP-α-L-Fuc did not improve the activity. Moreover, addition of either α- or β-D-gluco, or -galactosides to the reaction mixture did not increase the activity; instead, it hampered the chemically rescued activity, suggesting that they were poor acceptors and probably compete with 4-NP-α-L-Fuc in the donor site [Cobucci-Ponzano et al. 2008]. However, the analysis of the reaction mixture did not reveal any product of synthesis, confirming that the mutant did not work as a glycosynthase. Hyperthermophilic β-glycosynthases efficiently produced oligosaccharides when using 2-nitrophenyl-β-D-glycoside substrates, in which 2-nitrophenol, although having a pKa similar to that of 4-nitrophenol (pKa 7.22 and 7.18, respectively), can form a chelate ring by hydrogen bonding, thereby increasing the leaving ability upon protonation [Perugino et al. 2003]. However, since no α-L-fucosides containing aglycons with a leaving ability better than 4-nitrophenol are commercially available, the substrate 2-chloro-4-nitrophenyl α-L-fucopyranoside (2-Cl-4-NP-α-L-Fuc), was synthesized [Cobucci-Ponzano et al. 2008]. The aglycon 2-chloro-4-nitrophenol has a pKa of 5.45 [Tehan et al. 2002], which is noticeably lower than that of 4-nitrophenol (7.18); this difference making the 2-chloro-4-nitrophenol a much better leaving group. The steady-state kinetic constants of the mutant Ssα–fucD242G on 4-NP- and 2Cl-4-NP-α-L-Fuc in the presence of 2M sodium azide were compared with those of the wild type Ssα–fuc without external ions (Table 5). The specificity constant k_{cat}/K_M of the mutant on 2-Cl-4-NP-α-L-Fuc was more than 7-fold higher than that found on 4-NP-α-L-Fuc [Cobucci-Ponzano et al. 2008]. This activation reflected mainly the increment in k_{cat}, while the K_M values remained almost unaltered, indicating that the affinity of Ssα-fucD242G was the same for 4-NP- and 2-Cl-4-NP-α-L-Fuc. Interestingly, the wild type on the latter substrate produced kinetic constants similar to those obtained with 4-NP-α-L-Fuc (Table 5), indicating that the different leaving abilities of the aglycons in the two substrates did not change the reaction rates [Cobucci-Ponzano et al, 2008]. In contrast, the presence of the activated leaving group greatly enhanced the activity of the mutant showing that the D242G mutation affects the first step of the reaction.

Table 5. Steady-state kinetic constants of wild type and D242G mutant Ssαfuc

	k_{cat} (sec^{-1})	K_M (mM)	k_{cat}/K_M (sec^{-1} mM^{-1})
D242G+ 2M NaN_3			
4Np-Fuc	9.66±0.28	0.19±0.02	51.55
2-Cl-4-NP-α-L-Fuc	55.3±2.7	0.14±0.02	384.29
wild type			
4Np-Fuc	287±11	0.028±0.004	10,250
2-Cl-4-NP-α-L-Fuc	157±9	0.013±0.004	11,602

However, also the impaired catalytic activity of the α-L-fucosidase nucleophile mutant was significantly improved in the presence of external ions and of a synthetic aryl-α-L-fucoside substrate, the enzyme did not synthesize products and, hence, was not a glycosynthase, suggesting that a successful glycosynthetic reaction is the result of a delicate balance between the most suitable characteristics of the mutant enzyme, the substrate, and external nucleophiles [Cobucci-Ponzano et al, 2008]. The work on this enzyme is currently in progress to define the reaction conditions to obtain an α-L-fucosintase.

CONCLUSION

Traditionally, the enzymes from thermophilic sources find most of their applications in biotransformations performed under harsh conditions. However, these applications, especially in the field of carbohydrate modifications, are of real interest in only a limited number of cases. The exploration of the biodiversity of glycosidases from hyperthermophiles can widen their applicative perspective. We have shown in this chapter that glycosidases from hyperthermophiles can be useful tools for the synthesis of target oligosaccharides not only because of their intrinsic stability to heat, high molar excess of organics and nucleophilic compounds, and acid conditions, but also because of their intrinsic catalytic properties, such as efficiency in performing transglycosylation reactions, and their regio-, and stereospecificity. Our studies indicate that the expression of functional genes obtained from genomic data will give access to a vast repertoire of these enzymes. These studies confirmed the importance of the biodiversity of hyperthermophilic enzymes for the isolation of new enzymatic activities and for increasing the development of oligosaccharides to be used for both applied and basic research.

REFERENCES

Baranov, P.V., Gurvich, O.L., Fayet, O., Prere, M.F., Miller, W.A., Gesteland, R.F., Atkins, J.F. & Giddings, M.C. (2001). RECODE: a database of frameshifting, bypassing and codon redefinition utilized for gene expression. *Nucleic Acids Res, 29*, 264–267.

Cantarel, B.L., Coutinho, P.M., Rancurel, C., Bernard, T., Lombard, V., & Henrissat, B. (2008) The Carbohydrate-Active EnZymes database (CAZy): an expert resource for Glycogenomics. *Nucleic Acids Res*. doi:10.1093/nar/gkn663

Chaillou, S., Lokman, B.C., Leer, R.J,. Posthuma, C., Postma, P.W. & Pouwels, P.H. (1998). Cloning, sequence analysis, and characterization of the genes involved in isoprimeverose metabolism in *Lactobacillus pentosus*, *J Bacteriol, 180*, 2312–2320

Cobucci-Ponzano, B., Perugino, G., Trincone, A., Mazzone, M., Di Lauro, B., Giordano, A., Rossi, M. & Moracci, M. (2003a). Applications in biocatalysis of glycosyl hydrolases from the hyperthermophilic archaeon *Sulfolobus solfataricus*. *Biocat. Biotrans, 21*, 215-221

Cobucci-Ponzano, B., Trincone, A., Giordano, A., Rossi, M. & Moracci, M., (2003b). Identification of an archaeal alpha-L-fucosidase encoded by an interrupted gene. Production of a functional enzyme by mutations mimicking programmed -1 frameshifting, *J Biol Chem, 278*, 14622–14631.

Cobucci-Ponzano, B., Trincone, A., Giordano, A., Rossi, M. & Moracci, M. (2003c). Identification of the catalytic nucleophile of the family 29 alpha-L-fucosidase from *Sulfolobus solfataricus* via chemical rescue of an inactive mutant. *Biochemistry, 42*, 9525-9531.

Cobucci-Ponzano, B., Rossi, M. & Moracci, M. (2005a) Recoding in archaea. *Mol Microbiol. 55*, 339-348.

Cobucci-Ponzano, B., Mazzone, M., Rossi, M. & Moracci, M. (2005b) Probing the catalytically essential residues of the alpha-L-fucosidase from the hyperthermophilic archaeon *Sulfolobus solfataricus*. *Biochemistry, 44,* 6331-6342.

Cobucci-Ponzano, B., Conte, F., Benelli, D., Londei, P., Flagiello, A., Monti, M., Pucci, P., Rossi, M. & Moracci, M. (2006). The gene of an archaeal alpha-L-fucosidase is expressed by translational frameshifting. *Nucleic Acids Res, 34*, 4258-4268.

Cobucci-Ponzano, B., Conte, F., Mazzone, M., Bedini, E., Corsaro, M. M., Rossi, M. & Moracci, M. (2008). Design of new reaction conditions for characterization of a mutant thermophilic α-l-fucosidase. *Biocatalysis and Biotransformation*, *26*, 18-24.

Crombie, H.J., Chengappa, S., Hellyer, A. & Reid, J.S.G. (1998). A xyloglucan oligosaccharide-active, transglycosylating beta-D-glucosidase from the cotyledons of nasturtium (*Tropaeolum majus* L) seedlings — purification, properties and characterization of a cDNA clone. *Plant J.,* 15, 27–38.

Debeche, T., Bliard, C., Debeire, P. & O'Donohue, M. J. (2002). Probing the catalytically essential residues of the a-L-arabinofuranosidase from *Thermobacillus xylanilyticus*. *Protein Eng. 1*, 21-28.

de La Torre, F., Sampedro, J., Zarra, I. & Revilla, G. (2002). AtFXG1, an Arabidopsis gene encoding alpha-L-fucosidase active against fucosylated xyloglucan oligosaccharides. *Plant Physiol. 128*, 247–255

Faik, A., Price, N.J., Raikhel, N.V. & Keegstra, K. (2002). An Arabidopsis gene encoding an α-xylosyltransferase involved in xyloglucan biosynthesis. *Proc. Natl. Acad. Sci. USA,* 99, 7797–7802.

Farabaugh, P.J. (1996) Programmed translational frameshifting. *Annu. Rev. Genet.*, *30*, 507-528.

Farkas, E., Thiem, J. & Ajisaka, K., (2000). Enzymatic synthesis of fucose-containing disaccharides employing the partially purified alpha-L-fucosidase from *Penicillium multicolor. Carbohydr. Res., 328*, 293–299.

Fischer, L., Bromann, R., Kengen, S.W.M., de Vos, W.M. & Wagner, F. (1996) Catalytical potency of b-glucosidase from the extremophile *Pyrococcus furiosus* in glucoconjugate synthesis. *Biotechnology 14*, 88–91.

Gijsen, H.J.M., Qiao, L., Fitz, W. & Wong, C.H. (1996) Recent advances in the chemoenzymatic synthesis of carbohydrates and carbohydrate mimetics. *Chem Rev, 96*, 443–473.

Gloster, T.M., Roberts, S., Ducros, V.M., Perugino, G., Rossi, M., Hoos, R., Moracci, M., Vasella, A. & Davies, G.J. (2004) Structural studies of the beta-glycosidase from *Sulfolobus solfataricus* in complex with covalently and noncovalently bound inhibitors. *Biochemistry*, *43*, 6101-6109.

Hancock, S.M., Vaughan, M.D. & Withers, S.G. (2006) Engineering of glycosidases and glycosyltransferases. *Curr Opin Chem Biol. 10*, 509-519.

Henrissat, B. & Bairoch, A. (1993) New families in the classification of glycosyl hydrolases based on amino acid sequence similarities, *Biochem. J. 293*, 781-788.

Henrissat, B. & Davies, G. (1997) Structural and sequence-based classification of glycoside hydrolases, *Curr. Opin. Struct. Biol. 7*, 637-644.

Henrissat, B. (1991) A classification of glycosyl hydrolases based on amino acid sequence similarities, *Biochem. J.,* 280, 309–316.

Hough, D.W. & Danson, M.J. Extremozymes. (1999) *Curr Opin Chem Biol, 3,* 39-46.

Kato, Y. & Matsuda, K. (1980) Structure of oligosaccharides obtained by controlled degradation of mung bean xyloglucan with acid and *Aspergillus oryzae* enzyme preparation. *Agric Biol Chem 44* , 1751–1758

Kiho, T., Katsuragawa, M., Nagai, K., Ugai, S. & Haga, M. (1992) Structure and antitumor activity of a branched (1-3)-□-D-glucan from the alkaline extract of *Amanita muscaria. Carbohydr. Res. 224*, 237-243.

Li, Y. K., Chir, J., Tanaka, S. & Chen, F. Y. (2002). Identification of the general acid/base catalyst of a family 3 *b*-glucosidase from *FlaVobacterium meningosepticum, Biochemistry 41*, 2751-2759.

Listinsky, J.J., Siegal, G.P. & Listinsky, C.M. (1998). Alpha-L-fucose: a potentially critical molecule in pathologic processes including neoplasia. *Am J Clin Pathol, 110,* 425–440.

Lorences, E.P. & Fry, S.C. (1994). Sequencing of xyloglucan oligosaccharides by partial Driselase digestion: the preparation and quantitative and qualitative analysis of two new tetrasaccharides. *Carbohydr Res. 263*, 285-293.

Lower, B.H. & Kennelly, P.J. (2002) The membrane-associated protein-serine/threonine kinase from Sulfolobus solfataricus is a glycoprotein. *J Bacteriol, 184,* 2614-2619.

Ly, H. D. & and Withers, S. G. (1999) Mutagenesis of glycosidases. *Annu. ReV. Biochem. 68*, 487-522.

Malet, C., & Planas, A. (1998) From beta-glucanase to beta-glucansynthase: glycosyl transfer to alpha-glycosyl fluorides catalyzed by a mutant endoglucanase lacking its catalytic nucleophile. *FEBS Lett. 440*, 208-212.

Moracci, M., Capalbo, L., Ciaramella, M., & Rossi, M. (1996) Identification of two glutamic acid residues essential for catalysis in the □-glycosidase from the thermoacidophilic archaeon *Sulfolobus solfataricus*. *Protein Eng. 12*, 1191-1195.

Moracci, M., Ciaramella, M. & Rossi, M. (2001) □-Glycosidase from *Sulfolobus solfataricus. Methods Enzymol. 330*, 201-215.

McCarter, J.D. & Withers, S.G. (1994) Mechanisms of enzymatic glycoside hydrolysis. *Curr Opin Struct Biol, 4*, 885–892.

Mackenzie, L.F., Wang, Q., Warren, R.A.J., & Withers, S.G. (1998) Glycosynthases: mutant glycosidases for oligosaccharide synthesis. *J. Am. Chem. Soc. 120*, 5883-5884.

MacLeod, A. M., Tull, D., Rupitz, K., Warren, R. A. & Withers, S. G. (1996) Mechanistic consequences of mutation of active site carboxylates in a retaining *b*-1,4-glycanase from *Cellulomonas fimi*, *Biochemistry 35*, 13165-13172.

McIntosh, L.P., Hand, G., Johnson, P.E., Joshi, M.D., Korner, M., Plesniak, L.A., Ziser, L., Wakarchuk, W.W. & Withers, S.G. (1996) The pKa of the general acid/base carboxyl group of a glycosidase cycles during catalysis: a 13C-NMR study of *Bacillus circulans* xylanase. *Biochemistry 35*, 9958-9966.

Moracci, M., Ciaramella, M., Nucci, R., Pearl, L.H., Sanderson, I., Trincone, A. & Rossi, M. (1994) Thermostable □-glycosidase from *Sulfolobus solfataricus. Biocatalysis, 11*, 89-103.

Moracci, M., Trincone, A., Perugino, G., Ciaramella, M. & Rossi, M. (1998) Restoration of the activity of active-site mutants of the hyperthermophilic beta-glycosidase from

Sulfolobus solfataricus: dependence of the mechanism on the action of external nucleophiles. *Biochemistry, 37*-17262 17270.

Moracci, M., Cobucci-Ponzano, B., Trincone, A., Fusco, S., De Rosa, M., van der Oost, J., Sensen, C.W., Charlebois, R.L. & Rossi M (2000) Identification and molecular characterization of the first a-xylosidase from an archaeon. *J Biol Chem 275*, 22082–22089

Moracci, M., Trincone, A., Cobucci-Ponzano, B., Perugino, G., Ciaramella, M., & Rossi, M. (2001) Enzymatic synthesis of oligosaccharides by two glycosyl hydrolases of *Sulfolobus solfataricus*. *Extremophiles*. 5:145-152.

Mori, E., Hedrick, J.L., Wardrip, N.J., Mori, T. & Takasaki, S. (1998) Occurrence of reducing terminal N-acetylglucosamine 3-sulfate and fucosylated outer chains in acidic N-glycans of porcine zona pellucida glycoproteins. *Glycoconj J, 15*, 447–456.

Noda, K., Miyoshi, E., Uozumi, N., Gao, C.X., Suzuki, K., Hayashi, N., Hori, M. & Taniguchi, N. (1998) High expression of alpha-1-6 fucosyltransferase during rat hepatocarcinogenesis. *Int J Cancer, 75*, 444–450.

Murata, T., Morimoto, S., Zeng, X., Watanabe, S. & Usui, T. (1999) Enzymatic synthesis of alpha-L-fucosyl-N-acetyllactosamines and 3'-O-alpha-L-fucosyllactose utilizing alpha-L-fucosidases. *Carbohydr Res, 320*, 192-199.

Okuyama, H. Mori, K. Watanabe, A. & Kimura, S. Chiba. (2002). glucosidase mutant catalyzes 'a-glycosynthase'-type reaction. *Biosci Biotechnol Biochem, 66,* 928-933.

Perugino, G., Trincone, A., Giordan,o A., van der Oost, J., Kaper, T., Rossi, M., & Moracci, M., (2003) Activity of hyperthermophilic glycosynthases is significantly enhanced at acidic pH. *Biochemistry, 42*, 8484-8493.

Perugino, G., Trincone, A., Rossi, M., & Moracci, M., (2004) Oligosaccharide synthesis by glycosynthases. *Trends Biotechnol, 1*, 31-37.

Perugino, G., Cobucci-Ponzano, B., Rossi, M., & Moracci, M, (2005) Recent advances in the oligosaccharide synthesis promoted by catalytically engineered glycosidases. *Ad. Synth. Catal., 347*, 941-950.

Perugino, G., Falcicchio, P., Corsaro, M.M., Matsui, I., Parrilli, M., Rossi, M., & Moracci, M., (2006) Preparation of a glycosynthase from the b-glycosidase of the hyperthermophilic Archaeon *Pyrococcus horikoshii*. *Biocat, Biotrans., 24*, 23-29

Pouwels, J., Moracci, M., Cobucci-Ponzano, B., Perugino, G., van der Oost, J., Kaper, T., Lebbink, J.H., de Vos, W.M., Ciaramella, M., & Rossi, M. (2000) Activity and stability of hyperthermophilic enzymes: a comparative study on two archaeal -glycosidases. *Extremophiles, 3*, 157-164.

Sears, P. & Wong, C.H. (1996) Intervention of carbohydrate recognition by proteins and nucleic acids, *Proc. Natl. Acad. Sci. USA, 93*, 12086–12093.

Shallom, D., Belakhov, V., Solomon, D., Shoham, G., Baasov, T., and Shoham, Y. (2002) Detailed kinetic analysis and identification of the nucleophile in alpha-L-arabinofuranosidase from Geobacillus stearothermophilus T-6, a family 51 glycoside hydrolase. *J. Biol. Chem. 277*, 43667-43673.

She, Q., Singh, R. K., Confalonieri, F., Zivanovic, Y., Allard, G., Awayez, M. J., Chan-Weiher, C. C., Clausen, I. G., Curtis, B. A., De Moors, A., Erauso, G., Fletcher, C., Gordon, P. M., Heikamp-de Jong, I., Jeffries, A. C., Kozera, C. J., Medina, N., Peng, X., Thi-Ngoc, H. P., Redder, P., Schenk, M. E., Theriault, C., Tolstrup, N., Charlebois, R. L., Doolittle, W. F., Duguet, M., Gaasterland, T., Garrett, R. A., Ragan, M. A., Sensen, C.

W. & Van der Oost, J. (2001) The complete genome of the crenarchaeon Sulfolobus solfataricus P2. *Proc. Natl. Acad. Sci. U. S. A., 98,* 7835–7840

Sinnot, M.L. (1990) Catalytic mechanisms of enzymic glycosyl transfer. *Chem Rev, 90,* 1171–1202

Staudacher, E., Altmann, F., Wilson, I.B. & März, L. (1999) Fucose in N-glycans: from plant to man. *Biochim Biophys Acta. 147,* :216-236.

Sulzenbacher, G., Bignon, C., Nishimura, T., Tarling, C. A., Withers, S. G., Henrissat, B. & Bourne, Y. (2004) Crystal structure of *Thermotoga maritima* a-L-fucosidase. Insights into the catalytic mechanism and the molecular basis for fucosidosis, *J. Biol. Chem. 279*, 13119-13128.

Tarling, C. A., He, S., Sulzenbacher, G., Bignon, C., Bourne, Y., Henrissat, B. & Withers, S. G. (2003) Identification of the catalytic nucleophile of the family 29 a-L-fucosidase from *Thermotoga maritima* through trapping of a covalent glycosylenzyme intermediate and mutagenesis, *J. Biol. Chem. 278*, 47394-47399.

Tehan, B.G., Lloyd, E.J., Wong, M.G., Pitt, W.R., Montana, J.G., Manallack, D.T. & Gancia, E. (2002). Estimation of pKa using semiempirical molecular orbital methods. part 1: application to phenols and carboxylic acids. *Quant Struct-Act Relat, 21,* 457-472.

Toshima, K., Tatsuta, K. (1993) Recent progress in *O*-glycosylation methods and its application to natural products synthesis. *Chem Rev* , *93*, 1503–1531

Trincone, A., Nicolaus, B., Lama, L. & Gambacorta, A. (1991) Stereochemical studies of enzymatic transglycosylation using *Sulfolobus solfataricus*. *J Chem Soc Perkin Trans, 1,* 2841–2844

Trincone, A., Cobucci-Ponzano, B., Di Lauro, B., Rossi, M., Mitsuishi, Y. & Moracci, M. (2001) Enzymatic synthesis and hydrolysis of xylogluco-oligosaccharides using the first archaeal alpha-xylosidase from *Sulfolobus solfataricus*. *Extremophiles*, *5,* 277-282.

Trincone, A., Perugino, G., Rossi, M., & Moracci, M. (2000) A novel thermophilic glycosynthase that effects branching glycosylation. *Bioorg. Med. Chem. Lett.*, *10*, 365-368.

Vallmitjana, M., Ferrer-Navarro, M., Planell, R., Abel, M., Ausin, C., Querol, E., Planas, A. & Perez-Pons, J. A. (2001) Mechanism of the family 1 *b*-glucosidase from *Streptomyces* sp: Catalytic residues and kinetic studies, *Biochemistry 40*, 5975-5982.

Vanhooren, P.T. & Vandamme, E.J. (1999) L-Fucose: occurrence, physiological role, chemical, enzymatic and microbial synthesis. *J. Chem. Technol. Biotechnol., 74*, 479–497.

Varki, A. (1993) Biological roles of oligosaccharides: all of the theories are correct, *Glycobiology, 2*, 97–130.

Viladot, J.L., de Ramon, E., Durany, O. & Planas, A. (1998) Probing the mechanism of Bacillus 1,3-1,4-beta-D-glucan 4-glucanohydrolases by chemical rescue of inactive mutants at catalytically essential residues. *Biochemistry*, *37*, 11332-11342

Wang, Q., Graham, R.W., Trimbur, D., Warren, R.A.J. & Withers, S.G. (1994). Changing enzymatic reaction mechanisms by mutagenesis: conversion of a retaining glucosidase to an inverting enzyme. *J Am Chem Soc, 116,* 11594 11595.

White, A. & Rose, D.R. (1997) Mechanism of catalysis by retaining b-glycosyl hydrolases. *Curr Opin Struct Biol, 7*, 645–651

Withers, S.G. & Aebersold, R. (1995) Approaches to labeling and identification of active site residues in glycosidases. *Protein Sci.* , *4*, 361-372

Withers, S.G. (2001) Mechanisms of glycosyl transferases and hydrolases. *Carb. Polymers.*, *44*, 325–337.

Zechel, D.L. & Withers, S.G. (2000) Glycosidase mechanisms: anatomy of a finely tuned catalyst. *Acc Chem Res, 33*, 11–18

Zopf, D. & Roth, S. (1996) Oligosaccharide anti-infective agents. *Lancet*, *347*, 1017–1021.

In: Oligosaccharides: Sources, Properties and Applications ISBN 978-1-61122-193-0
Editor: Nicole S. Gordon

Chapter 4

ENZYMATIC SYNTHESIS OF LINEAR, CYCLIC AND COMPLEX TYPE OLIGOSACCHARIDES

***Piamsook Pongsawasdi*[1] *and Kazuo Ito*[2]**
[1] Department of Biochemistry, Faculty of Science, Chulalongkorn University, Bangkok, 10330, Thailand
[2] Laboratory of Enzyme Chemistry, Graduate School of Science, Osaka City University, Osaka 558-8585, Japan

ABSTRACT

For oligosaccharide synthesis, an enzymatic process is preferred over the complicated multi-step reactions of chemical synthesis. The advantages of the enzymatic process are the regio- and stereo-selectivity, potential for large-scale synthesis, and also the environmental-friendly synthesis conditions. Two main classes of enzymes, the glycosidases and the glycosyltransferases, are exploited in the common approach for oligosaccharide synthesis via reverse hydrolysis and transglycosylation reactions. The glycosynthase class has recently been developed from specific mutations of glycosidases through site-directed mutagenesis to minimize the natural hydrolysis activity and thereby boost up the synthetic activity of the enzyme. In this chapter, oligosaccharides synthesized by enzymes are classified into three main types, the linear, cyclic and complex oligosaccharides. Our work on three major enzymes, comprised of two glycosyltransferases and one glycosidase that are capable of synthesizing oligosaccharides is outlined. The use of bacterial cyclodextrin glycosyltransferase and amylomaltase in intermolecular transglucosylation to produce functional linear oligosaccharides is presented. In addition, these two enzymes, through the intramolecular transglucosylation reaction, are used for the synthesis of small and large cycloamyloses (cyclodextrins) that are widely used as stabilizers and solubilizers. For the complex-type oligosaccharides, the transglycosylation activity of endo-β-*N*-acetylglucosaminidase to yield the high mannose-GlcNAc-Glc oligosaccharides with potential applications in biological functions, such as therapeutic agents, is discussed. The chapter ends with concluding remarks on the commercially available oligosaccharides, with the emphasis on the significance of oligosaccharides and oligosaccharide-producing enzymes and their promising future.

INTRODUCTION

Oligosaccharides : Definition and Classification

Oligosaccharides are defined as a group of carbohydrate oligomers with 2 - 10 monosaccharide units linked by a covalent glycosidic bond. The three main types of oligosaccharides are the linear, branching and cyclic. Branching oligosaccharides with different types of sugar constituents (hetero-oligosaccharides) are called the complex type. A disaccharide is the simplest oligosaccharide and can be referred to as a monosaccharide glycoside. The occurrence of di- and tri-saccharides is relatively common in nature in the free form where small oligosaccharides may function as an energy source or cellular metabolites [1]. Sucrose, a non-reducing and the most abundant disaccharide, occurs throughout the plant kingdom and is familiar as common table sugar, whilst the reducing disaccharide lactose (milk sugar) occurs naturally only in milk. However, tri- and larger oligosaccharides are far less frequent in the free form in nature and do not occur in significant amounts [2]. Rather, they are often associated with proteins or lipids as glycoconjugates in which they have both structural and regulatory functions. These oligosaccharides form two types of direct attachments to proteins, the *N*-linked (e.g. in immunoglobulin M or G) and the *O*-linked (e.g. in mucin). Most are branching or complex type oligosaccharides that confer a specific recognition role [1]. The cyclic type of oligosaccharides have a unique structure consisting of glucose oligomers linked by an α-1,4 bond with no reducing end, and are known as cyclodextrins (CDs) or cycloamyloses (CAs) [3].

Oligosaccharides are highly diverse, due to the possibility of forming multiple linkages (e.g. 1,2 or 1,3 or 1,4 or 1,6 for hexoses) of different monomers (e.g. glucose, galactose, mannose or fructose) and anomers (α and β). In addition, the branching structure makes oligosaccharides even more complex. Among all the biological molecules, carbohydrates can display the largest number of structures in a short sequence [4]. Thus, a high level of information-storing potential is inherent in complex carbohydrate ligands, which are recognized by specific protein receptors.

Oligosaccharides are widely known as functional food ingredients that have a great potential to improve the quality of many foods. The functional properties of oligosaccharides are divided into the physicochemical properties (e.g. sweetness, bitterness, hygroscopicity, reinforcement, preservation and stability), biological properties (e.g. digestibility, non- or anti-cariogenicity, bacteriostatic action and proliferation of probiotics) and other properties (e.g. as specific substrate or inhibitor for enzymes, as elicitors) [2].

Prebiotic oligosaccharides, a group of non-digestible oligosaccharides (NDOs), have recently become of interest as nutrachemicals because they can be beneficial to human health. Structurally, these molecules are mostly linear oligosaccharides with linkages other than (1,4), so they cannot be digested by the α1-4 glycosidase enzymes in the human digestive system allowing them to reach the large intestine in an intact form where they can function as nutrients for the probiotic microorganisms [5]. As a result, the population of probiotic organisms is increased resulting in a healthy intestinal environment and several beneficial effects, such as increased calcium adsorption and a possible decline in colon cancer and diseases related with dyslipidemias [6]. Therfore, there is currently a great demand for this class of oligosaccharides for industrial use.

Oligosaccharide Synthesis by Chemical vs. Enzymatic Methods

At present, oligosaccharides have been widely utilized in foods and beverages due to their various useful properties. In addition, oligosaccharides in the form of glycoconjugates have key informative roles in biological systems and so these molecules have potential applications as therapeutics. Oligosaccharide synthesis is therefore becoming increasingly important and two distinct strategies for their synthesis, chemical and enzymatic, are known [7-8]. Chemical synthesis requires complicated multi-step reactions, where many protecting-group manipulations are needed to control the stereo- and regio-specificity of the products. Thus, the production is inefficient with a typically low yield being obtained. Classical chemical synthesis is especially difficult in the synthesis of complex oligosaccharides. In contrast, larger - scale synthesis can be achieved by enzymatic means due to the specificity characteristics of enzymes. Other advantages of using enzymes are the rate enhancement and the mild and more environmentally friendly synthesis conditions [9]. However, the cost effectiveness of using enzymes depends on the availability and the stability of the enzymes and the cost of substrates. Therefore, only certain enzymes are suitable for application in the industrial synthesis of oligosaccharides. This then drives the need to isolate other natural enzymes that are suitable and / or to genetically modify existing known enzymes to make them more suitable for specific purposes.

In addition to chemical and enzymatic synthesis, a few oligosaccharides can be obtained from the direct extraction from natural sources, such as soybean oligosaccharides and beet raffinose [10]. There are also a few examples of oligosaccharides obtained from chemical synthesis. For example, lactulose (4-β-D-galactosylfructose), a disaccharide that is used for the treatment of constipation and also as a nutritional supplement in food and animal feed, is obtained by the isomerization reaction from lactose. In addition, lactitol (4-α-D-galactosylglucitol), a sugar alcohol with a low glycemic index and is used as food additive, is produced by the catalytic hydrogenation of lactose. However, today, most oligosaccharides are manufactured by an enzymatic process and are either built up from simple sugars, such as sucrose or lactose, by enzymatic transglycosylation, or they are formed by the controlled enzymatic hydrolysis of polysaccharides, such as starch or xylan.

ENZYMES THAT ARE USED IN OLIGOSACCHARIDE SYNTHESIS

Originally, two main classes of enzymes were used in oligosaccharide synthesis, the glycosidases (glycosyl hydrolase, GH, EC 3.2.1) and the glycosyltransferases (EC 2.4) [11]. The **glycosidases** catalyze the reaction:

$$(\text{Glycosyl})_n - \text{OR} + H_2O \rightleftharpoons (\text{Glycosyl})_n - \text{OH} + \text{H-O-R}$$

where R is the oligosaccharide at the reducing end of a glycosyl chain.

These enzymes, although they are most efficient at the hydrolysis of glycans, can also synthesize oligosaccharides via the reverse hydrolysis reaction and so are referred to as catalyzing an equilibrium-controlled synthesis mode. Many glycosidases can also transfer the glycosyl moiety to an acceptor sugar rather than to water, the so-called transglycosylation reaction, which is a kinetically-controlled mode of synthesis [12]. Some glycosidases are retaining while others are inverting enzymes, where the glycosyl transfer proceeds through a nucleophilic displacement (SN2 reaction: double- or a single-displacement mechanism, respectively) or an oxocarbonium ion intermediate (SN1 reaction) [13]. Among nearly a hundred different families of glycosyl hydrolases that have been grouped together based on amino acid sequence similarity, approximately 48 families are retaining enzymes, such as α-amylase, α-glucosidase and pullulanase, whilst 36 families belong to inverting enzymes, such as β-amylase and glucoamylase [8]. Not so many glycosyl hydrolases are used in industry, but examples include α-amylase, β-amylase, glucoamylase, pullulanase, cellulase and hemicellulase, pectinase, lactase and invertase. These enzymes are mostly used for hydrolysis of carbohydrate polymers or oligomers [9]. Only a few enzymes are used for oligosaccharide synthesis by the reverse or controlled hydrolysis reactions and include the malto-oligosaccharide-forming amylase, pullulanase and xylanase. These enzymes and their substrates are relatively cheap, have a high stability and are mostly used for the synthesis of short-chain oligosaccharides [11]. However, a high concentration of enzyme and substrate are required and a low product yield is usually obtained since the hydrolysis direction is more favored for glycosidases. The transglycosylation reaction is more efficient (in terms of yields obtained) than the reverse hydrolysis reaction, but it still has only a low to moderate (in the range of 10-40%) product yield since the products can become substrates and undergo hydrolysis [7]. To improve the oligosaccharide synthetic yield, the strategy of modifying the glycosidases to reduce their hydrolysis activity has been adopted. A novel **"glycosynthase"** class, whose members have an inactive or low hydrolytic reaction have been produced by protein engineering of the glycosidases [8].

Glycosynthase was first named when β-glucosidase from *Agrobacterium* sp., which is a retaining enzyme and a member of the glycoside hydrolase 1 (GH1) family, was mutated at its nucleophile residue (Glu358), giving it the ability to synthesize various β-glucosides [7]. The Glu358Ala mutant led to the production of oligosaccharide products with a yield of > 60 % [14]. The success in the production of β-1, 4-linked cello-oligosaccharides, a group of potential cellulose inhibitors, was also reported [7,15]. The Glu358Ser mutation was superior to the Glu358Ala in having a higher enzyme activity with a 24-fold improvement in the enzyme synthesis rate. This mutant could synthesize 4-nitrophenyl-β-*N*-acetyllactosamine, a precursor of various cell-surface antigens, with a 63% yield [16]. Besides β-glucosynthase, other exoenzymes, such as β-galactosynthase, β-mannosynthase and α-glucosynthase, have been successfully mutated from their wild type glycosidase counterparts resulting in the ability to synthesize high yields of oligosaccharide products. Endo-glycosynthases, such as the β-glucansynthase mutated from β-glucanase and cellulase, are valuable tools for the synthesis of longer or complex oligosaccharides and even polysaccharides. These enzymes are all expressed in recombinant form and can yield concentrations as high as 150 - 250 mg of pure protein / liter of culture [7].

For large-scale oligosaccharide synthesis using glycosylsynthases, recent developments in solid-phase synthesis by immobilization of the sugar have led to significant improvements in the recovery of the product. In addition, the solid-phase technique offers easy workup procedures and the prospect of automation. The Glu358Gly mutant of β-glucosidase from

Agrobacterium sp. can catalyze the glycosyl transfer onto resin-bound acceptor glycosides with a yield of > 83% [17].

Glycosyltransferases are IUB Class II enzymes that catalyze the transfer of the glycosyl group. The specific glycosyltransferases of vertebrates are endomembrane proteins that are involved in the synthesis of the oligosaccharide moiety of glycoproteins. The enzyme requires the activation of the sugar donor substrate by a nucleotide or dolichol phosphate molecule to push the reaction to proceed [1]. Typical examples of the class include β-1,4-UDP-galactosyltransferase EC 2.4.1.38 and α-2,6-CMP-sialyltransferase EC 2.4.99.3. So far, the use of these enzymes in oligosaccharide synthesis is not practical due to the low availability of these enzymes and the high cost of the substrates. One unique class of glycosyltransferases, the bacterial CD glycosyltransferase (CGTase) is able to transfer the glucosyl group intramolecularly and intermolecularly resulting in the formation of cyclic and linear oligosaccharides with no requirement for a nucleotide activator [18]. At present, the CGTase is the only glycosyltransferase used as industrial enzyme for the synthesis of CDs, a group of cyclic oligosaccharides with numerous applications in industry [9].

Oligosaccharide Synthesis by CGTase

CGTase (EC 2.4.1.19) is a bacterial transferase enzyme that belongs to the α-amylase family and is characterized by an α-1,4-specificity, a retaining mechanism and four conserved regions in the amino acid sequence within the acidic residues active site [19-20]. Within the α-amylase family, CGTase is a member of the 4α-Glycosyltransferase (4αGTase) group. This group comprises of four types of enzymes; (i) CGTase (Type I), (ii) amylomaltase and plant D-enzyme (Type II, EC 2.4.1.25), (iii) glycogen debranching enzyme (Type III, EC 3.2.1.33 and 2.4.1.25) and (iv) other 4αGTases (Type IV, found in hyperthermophilic bacteria/ archaea) [21]. This group of enzymes catalyzes the transfer of a part of the α-1,4-D-glucan from a donor to an acceptor molecule, such as glucose or another α-1,4-D-glucan with a free 4-hydroxyl group as shown in the following equation :

$$(\alpha\text{-1,4-glucan})_m + (\alpha\text{-1,4-glucan})_n \longleftrightarrow (\alpha\text{-1,4-glucan})_{m-x} + (\alpha\text{-1,4-glucan})_{n+x}$$

This is an intermolecular transglycosylation reaction, and is often referred to as a disproportionation reaction. CGTase and amylomaltase can also catalyze an intramolecular glucan transfer reaction within a single linear glucan molecule (cyclization reaction) to produce a cyclic α-1,4-glucan (CA or CD), as follows:

$$(\alpha\text{-1,4-glucan})_n \longleftrightarrow \text{cyclic } (\alpha\text{-1,4-glucan})_x + (\alpha\text{-1,4-glucan})_{n-x}$$

This reaction is reversible, with the reverse reaction being the coupling reaction. In addition to these three reactions above, CGTase and amylomaltase also possess a weak hydrolytic activity [21].

Although CGTase and amylomaltase catalyze similar reactions, the two enzymes are different in both their structure and function. Based on amino acid sequence similarities, CGTase and α-amylase are classified as members of the GH13 family, while amylomaltase is a part of the GH77 family. Amylomaltase exhibits a remarkably low hydrolytic activity that is at least one order of magnitude lower than that of the CGTase's from GH13 [22]. Although CGTase and amylomaltase share a characteristic (βα8) barrel in domain A, they show some different subdomain structures, with the presence of domain C being found only in the CGTases [23]. In addition, the two enzymes differ in cycloamylose production where the main products of CGTase are small-ring CAs or CDs (SR-CDs) composed of glucose with a degree of polymerization (DP) of 6 - 8 units (named α, β, γ-CD, respectively), while the main products of amylomaltase are large-ring CDs (LR-CDs) that consist of glucose with a DP of 16 up [21, 24].

Several CGTases have been reported in bacteria, especially within the genus *Bacillus*. It is an extracellular enzyme and can be classified as an α-, β-, α/β-, γ- or β/γ- CGTase, depending on the main type of CD produced (Table 1). However, so far, only a few strains producing γ-CGTase have been found.

Table 1. Examples of CGTases from different sources [2,19,25]

Bacteria	**CD product**	**Opt. pH**	**Opt. temp. °C**
CGTases used in industrial production of CD			
B. macerans	α	5.0 - 5.7	55
B. stearothermophilus	α, β	5.0 - 5.5	75
B. coagulans	α, β	6.0 - 6.5	65
Alk. Bacillus 38-2 (*B. circulans*)	β	4.5 - 9.5	65
K. pneumoniae	α	5.0 - 6.0	65
GMO* from *B. circulans* 251	β	6.5	50
Other CGTases			
B. megaterium	β	5.0 - 5.7	55
Paenibacillus sp. A11	β	5.0 - 6.0	40
B. firmus	γ, β	6.0 - 8.0	50
B. clarkii	γ	10.5 -11.0	60
Thermoanaerobacter sp.	α, β	4.5 - 7.0	80 - 85

*GMO = Genetically modified organism; *B* = *Bacillus*; Opt. = Optimal

Significance of Oligosaccharide Products

As mentioned, the oligosaccharide products formed from the action of CGTase may be linear (formed by a disproportionation reaction) or cyclic (formed by a cyclization reaction). For linear oligosaccharide products, the only striking example is the use of CGTase from *Thermoanaerobacter* sp. for the industrial production of glucosylsucrose or coupling sugar (GnF) from starch as the glucosyl donor and sucrose as the acceptor (Figure 1) [26]. This functional oligosaccharide is 60% as sweet as sucrose and is used in food products to replace sucrose due to its anticariogenicity [27].

Starch + Sucrose (G-F) $\xrightarrow{\text{CGTase}}$ G-F
$G^{1}\text{-}{}^{4}G\text{-}F$ (G2F)*
$G^{1}\text{-}{}^{4}G^{1}\text{-}{}^{4}G\text{-}F$ (G3F)*
$G^{1}\text{-}{}^{4}G^{1}\text{-}{}^{4}G^{1}\text{-}{}^{4}G\text{-}F$ (G4F)
$G^{1}\text{-}{}^{4}G^{1}\text{-}{}^{4}G^{1}\text{-}{}^{4}G^{1}\text{-}{}^{4}G\text{-}F$ (G5F)

*main components

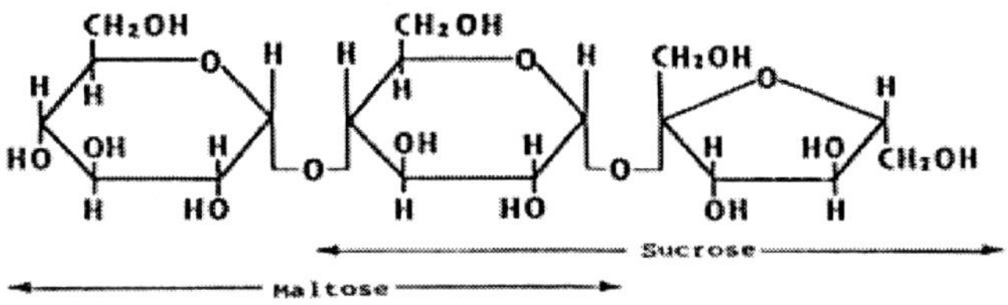

Figure 1. Synthetic reaction of glucosylsucrose by CGTase [26].

CGTase can transfer a glucosyl group to other acceptors besides carbohydrates, such as flavonoid compounds. Most flavonoids have useful biological activities and especially an antioxidant activity. Stevioside, a natural glucoside with a steviol flavonoid moiety that is obtained from the plant *Stevia rebaudiana* Bertoni, has a therapeutic application as a sugar substitute for diabetics. The relative sweetness is 143-fold higher than that of sucrose at 0.025% (w/v) concentration, and so it is beneficial as a sweetener in foods, but its drawback in the natural form is a bitter aftertaste. However, the glucosylated stevioside products (with mono-, di-, tri-, and more glucosyl units), produced from intermolecular transglucosylation reactions catalyzed by CGTase [28], have a reduced bitter taste and an increased sweetness. Several other flavonoid glucosides with improved properties have been produced by the action of CGTase, such as hesperidine mono- and di-glucosides, naringin monoglucoside, 2-O-D-α-ascorbic glucoside and O-α-glucosylthiamin [29-32], So far, these products have not been commercialized.

For cyclic oligosaccharide products, CGTase has been used industrially for the synthesis of SR- CDs (Figure 2). The unique characteristic of CDs is their molecular structures with hydrophobic inner cavities, each of a different size depending on the number of glucose residues in the ring. This structure allows host-guest interactions or molecular encapsulation that can lead to inclusion complex formation between the CD host and the suitable guests. CDs have been widely used as stabilizers and solubilizers in the food, pharmaceutical, agricultural chemistry and cosmetic fields [33-34]. Commercial products, such as fruit powder, spice powder (garlic, wasabi and ginger), health-supplement drinks, peppermint chewing gum, cholesterol-reduced egg/butter, slow-released fragrance/drug, active agent-stabilizer complex (CD complexed with vitamin A/E, squalene moisturizer and antibactericide) and steroid drug-CD solubilizers, have been marketed, especially in Japan, Europe and US [35].

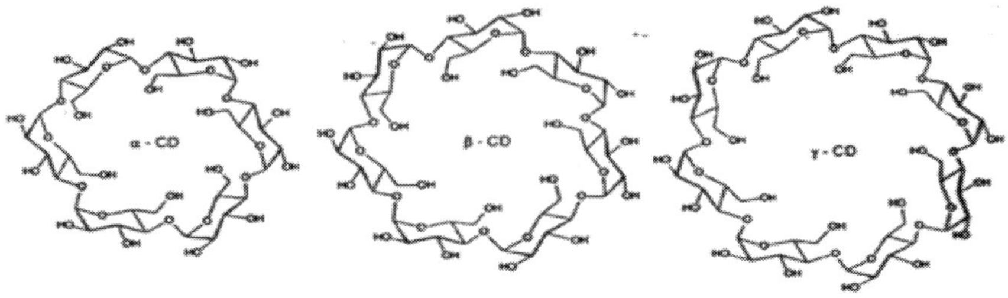

Figure 2. Structure of small-ring cyclodextrins (SR-CDs) [33].

In the industrial production of CDs, the CGTase used are mostly derived from the genus *Bacillus*. For example, for α-CGTase: *B. macerans* and *K. pneumoniae*; for β-CGTase: alkaline *B. circulans* 38-2 (ATCC 21783); for α/β-CGTase: *B. coagulans* and *B. stearothermophilus* (Table 1) [2]. At present, nearly half of the CGTase used is commercially produced from the genetically modified strain of *B. circulans* 251, which produces a β-CGTase [36]. The starting materials are generally potato or corn starch and two main processes of CD production, the solvent and non-solvent processes, are employed [37]. In the solvent process, the addition of an organic solvent that can form a complexing agent with the CD makes the separation of the CD product(s) from the reaction mixture much easier. The limitation of the process is that the solvent must have a low toxicity and be easily removed by distillation or evaporation. Toluene and cyclohexane have been used to precipitate β-CD, and bromobenzene likewise to precipitate γ-CD. The development of lower toxic solvents, such as linear alcohols, ethers, esters and ketones, such as 1-decanol, to precipitate α-CD has been successful [38]. The FDA of the US and the European Health Council do not allow more than 5 ppm of residual 1-decanol in the products. The solvent process is mainly used in western countries. On the contrary, no solvent is added in the non-solvent process, and so the CD products obtained are mixtures of the three CDs depending on the enzyme used. This process is mainly used in Japan which markets the CD products either as a mixture form or as separate CDs after enrichment by cation-exchange chromatography, hydrophobic adsorption, ultrafiltration and crystallization [39].

A β-CGTase from a Thermotolerant *Paenibacillus*

A thermotolerant high CGTase-producing bacteria was screened from samples taken at the Hot Spring area in the Ratchaburi Province of Thailand [40]. Bacterial identification through physiological and biochemical properties, and supported by phylogenetic analysis based upon the 16S rRNA gene sequence, led to the strain classification as *Paenibacillus* sp. RB01. This organism has a maximal growth rate at 37 °C, whilst it produced the highest CGTase activity when cultured in modified Horikoshi's medium (pH 10) with 1.0% (w/v) soluble starch (potato) at 40 °C for 60 h. The enzyme was purified to apparent homogeneity by starch adsorption, DEAE-cellulose and BioGel P-100 chromatography (Table 2) [41], and found to have a molecular weight of 65 kDa and was characterized as a β-CGTase. By measuring the cyclization activity, amylose was the best substrate, while branching in the molecule, such as that seen in amylopectin, reduced the activity (Table 3). Longer oligosaccharides were better substrates than shorter ones, with maltotriose being the smallest acceptable substrate. The enzyme was established as a glycoprotein by the Periodic Acid-Schiff staining of the SDS-PAGE resolved protein band [42].

In order to overexpress the enzyme, the CGTase gene was cloned in the pET-19b vector and expressed in *E. coli* [43]. The recombinant CGTase also displayed three molecular isoforms with identical masses but different net charges (Figure 3), the same as that seen with the wild-type enzyme. The two main isoforms (isoforms I and II) were separated from each other by FPLC on a mono P chromatofocusing column. Each isoform showed a similar optimum pH (pH 6 - 7 for cyclization activity) and temperature (65 - 70 °C) as well as the for product ratio (0.2:1.0:0.6 for α-: β-: γ-CD), while the kinetic parameter k_{cat} / K_m for the coupling reaction with β-CD as the donor and cellobiose as the acceptor was different. However, in terms of the maximal yields of CDs, the two isoforms differed in their optimal reaction temperature and time required, being optimal at 40 °C for 6 h for isoform I and at 60 °C for 24 h for isoform II. The isoform formation was caused by the non-enzymatic deamidation of some labile asparagines at the surface of the enzyme structure. Currently, the significance of the formation of the different isoforms is unclear but it might be involved in the regulation of catalytic activity of this CGTase.

Table 2. Purification of CGTase from *Paenibacillus* sp. RB01 [41]

Step	Volume (ml)	Total activity* (U x 10^3)	Total protein (mg)	Specific activity* (U / mg)	Purification (fold)	Yield (%)
Crude	1,000	59.5	388.9	153	1	100
Starch adsorption	85	42.8	9.66	4,437	29	72
DEAE-cellulose	362	30.9	5.59	5,539	36	52
Bio-Gel P-100	34	20.8	2.86	7,268	48	35

*Dextrinizing activity

Table 3. Specificity of substrates in cyclization reaction of CGTase [42]

Substrate	Relative cyclization activity (%)*
Amylose	100
Amylopectin	78.8
Soluble starch (potato)	95.0
Pullulan	0
Maltoheptaose	93.5
Maltohexaose	88.5
Maltopentaose	87.4
Maltotetraose	84.1
Maltotriose	36.2
Maltose	0
Glucose	0

* The assay was performed at pH 6.0 and 70 °C for 30 min.

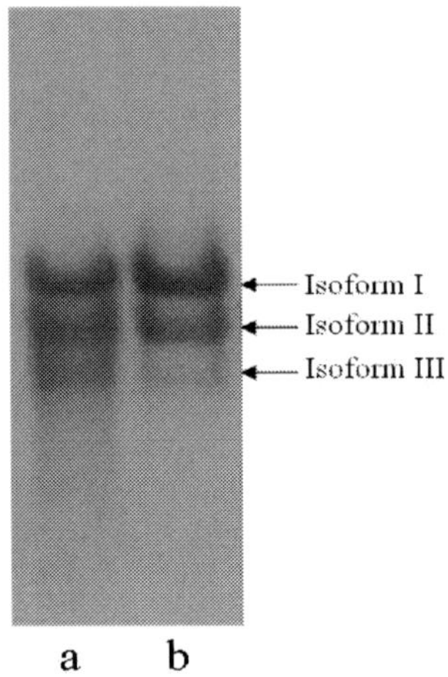

Figure 3. Native-PAGE CGTase. The native (lane a) and recombinant (lane b) *Paenibacillus* sp. RB01 enzyme.

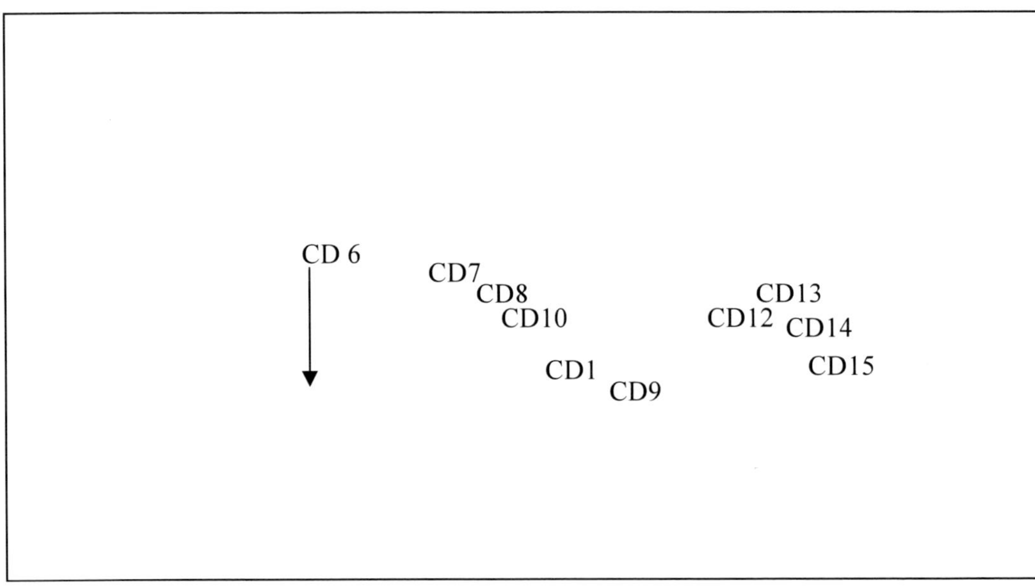

Figure 4. HPAEC Chromatogram of the CD products of CGTase. Samples were loaded onto a CarboPac PA-100 column (250 mm x 4 mm) and eluted at a flow rate of 1 ml/min with a linear gradient of 0 - 100% (increasing at 1%/min) of 200 mM $NaNO_3$ in 150 mM NaOH.

Oligosaccharide Synthesis by Recombinant CGTase from *Paenibacillus* sp. RB01

We examined the synthesis of cyclic oligosaccharides by the recombinant CGTase. When the enzyme was incubated with 4% (w/v) soluble starch (potato) at pH 6.0 and 60 °C for 24 h, the ratio of the α-: β-: γ-CD products were 0.18:1.00:0.36, as determined by HPAEC-PAD

(High Performance Anion Exchange Chromatography with Amperometric Detection). In the initial phase of the reaction (6 h), the products were also composed of LR-CDs from CD9 to CD25 (Figure 4), but as the reaction time was increased to 24 h, the proportion of LR-CDs decreased whilst the proportion of the SR-CDs significantly increased.

Oligosaccharide Synthesis by Amylomaltase

Amylomaltase (EC 2.4.1.25) is a glucosyltransferase that catalyzes the formation of LR-CDs with a DP of 16 up [21, 24]. The enzyme is a member of the α-amylase family and belongs to the type II 4αGTase group. The mode of action and general knowledge on the enzyme are mentioned in Section 3, as compared to the CGTase.

Amylomaltase, in contrast to CGTase, is an intracellular enzyme. In plants, such as potato, barley and pea, the enzyme is known as a disproportionation or D-enzyme and is believed to be involved in starch metabolism [21]. In microorganisms, amylomaltase was first found in *E. coli* as a maltose-inducible enzyme which is essential for the metabolism of maltose [44]. The amylomaltase gene has subsequentially been cloned from a number of microorganisms, including *E. coli*, *Clostridium butyricum*, *Thermococcus litoralis*, *Thermus aquaticus*, *T. thermophilus*, *T. brockianus* and *Aquifex aeolicus* [24, 45-49]. However, the purification of amylomaltase and the detection of CD-producing activity have been reported in only a few bacterial strains, such as *T. aquaticus* ATCC 33923 and *A. aeolicus*.

Significance of Oligosaccharide Products

Similar to CGTase, the oligosaccharide products from amylomaltase can be linear or cyclic depending on the substrate and reaction conditions. Until now, no linear oligosaccharides have been produced by amylomaltase at the industrial scale. However, the use of amylomaltase has recently been explored for its potential applications. The enzyme from *Thermotoga maritime* was used in combination with a maltogenic amylase from *Bacillus stearothermophilus* to produce isomalto-oligosaccharides (IMOs) from starch [50]. IMOs are NDO that can be applied as a substitute sugar to support the growth of the beneficial intestinal microflora, or to prevent dental caries.

The use of amylomaltase in the production of a thermoreversible starch gel has been reported [51]. The gel is of commercial interest since it can be used as a substitute for gelatin that is not acceptable by vegetarians and certain religious groups due to its animal origin. The amylomaltase-treated starch could also be used in the improvement of food products, such as in the improvement of the creaminess of low-fat yoghurt [52] and as a fat substitute in reduced-fat mayonnaise when used in combination with xanthan gum [53].

For the synthesis of cyclic oligosaccharides, LR-CDs are formed from the action of amylomaltase on starch. LR-CDs are highly water-soluble compounds that are assumed to form a single helical V-amylose conformation and a toroidal shape, with an anhydrophilic chanel-like cavity [54] (Figure 5). As mentioned above, those produced by amylomaltase show a DP of 16 up [49]. Similar to SR-CDs, they can form an inclusion complex with suitable hydrophobic guests [21]. Recently, LR-CD has been used as an artificial chaperone for protein refolding and is added as an ingredient in commercial protein refolding kits [55].

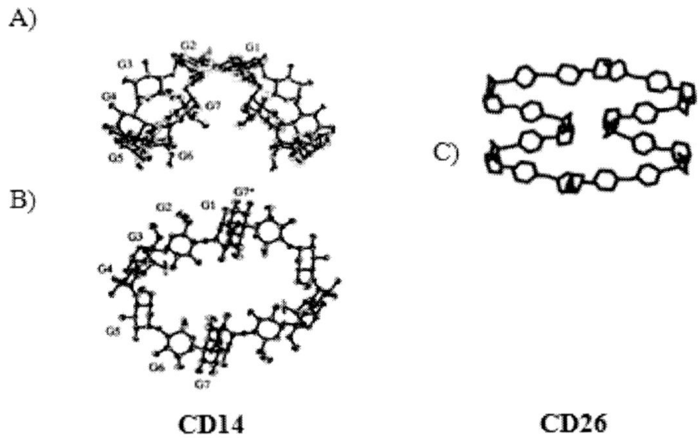

Figure 5. Structures of LR-CDs. A) Side view of CD 14, B) Top view of CD14, C) Top view of CD26 [54].

A Novel Amylomaltase from *Corynebacterium glutamicum*

The amylomaltase gene from *C. glutamicum* ATCC 13032 was PCR amplified using *Nde*-I and *Xho*-I tagged specific primers, as described [41]. Then the *Nde*-I and *Xho*-I digested DNA fragment was directionally inserted into the pET-19b expression vector preceding the His-tag sequence, transformed into *E. coli* BL21 (DE3) cells and the transformants were selected on ampicillin-LB agar plates. Plasmids were extracted and the inserted gene fragment was sequenced. The nucleotide sequence of 2121 bp obtained matched the sequence in the GenBank library reported for *C. glutamicum* 4αGTase [56]. However, the deduced amino acid sequence of 706 residues showed only a very low sequence identity (20 - 25%) to the well-characterized amylomaltases from *Thermus* and *Aquifex* [57].

The recombinant cells were able to express maximum amylomaltase activity when cultured at 37 °C for 2 h upon induction by 0.4 mM IPTG. The $(His)_6$-amylomaltase recombinant enzyme was highly enriched with HiTrap affinity column chromatography to apparent homogeneity, resolving as a single band on coomassie stained SDS-PAGE gels with a molecular weight of 84 kDa.

Oligosaccharide Synthesis by Recombinant Amylomaltase from *C. glutamicum*

The ability of this enzyme to synthesize both linear and cyclic oligosaccharides was evaluated, with the TLC analysis of the products revealing that the enzyme was able to catalyze the intermolecular transglucosylation reaction when incubated with all tested malto-oligosaccharide substrates (G2 to G7). However, no reaction was observed with glucose (G1), and so the enzyme requires at least a maltose unit for the disproportionation reaction, an observation that is similar to that for the amylomaltase of *T. aquaticus* but which differs from those enzymes of *E. coli* and *T. maritima* [21,55].

For the cyclization activity, a unique LR-CDs product profile was obtained (Figure 6). When incubated with 0.2% (w/v) of pea starch at pH 6.0 and 30 °C for 4 h, LR-CDs from CD19 to CD50 were synthesized, with CD19 as the smallest product and CD27 and CD28 as the two principal products. In term of the smallest product produced, this amylomaltase is different from that of *T. aquaticus* ATCC 33923 and the potato D-enzyme which produced CD22 and CD17, respectively [48]. While the chimeric enzyme containing amylomaltase from *T. aquaticus* YT-1 and the starch binding domain of *B. stearothermophilus* ET1 CGTase gave products in the range of CD19 to CD35, with CD25 as the principal one [55].

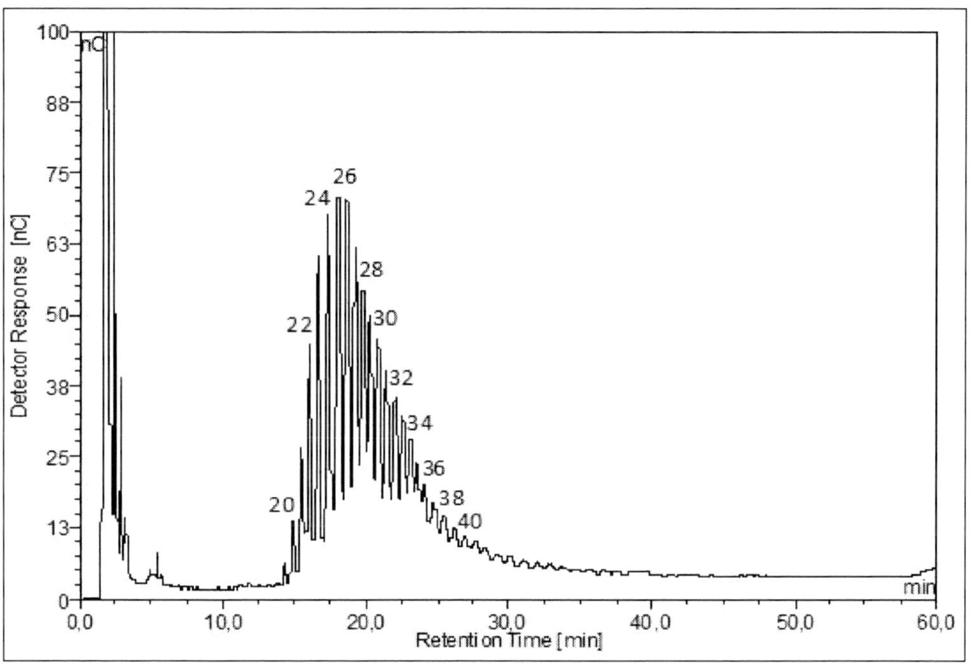

Figure 6. HPAEC chromatogram of LR-CD products of amylomaltase. The column and the elution conditions were as in Figure 4, except that a gradient of 0 - 63% sodium nitrate within 60 min was used. Peak numbers indicate the DP of LR-CDs. CD20 and CD21, confirmed by mass spectrometry, were used as references.

Oligosaccharide Synthesis by Endo-β-*N*-acetylglucosaminidase

Significance of Oligosaccharide Products

Structure and Function of N-Linked Oligosaccharides

Many proteins expressed in eukaryotes are glycoproteins bearing *N*-linked oligosaccharides [58]. In *N*-linked oligosaccharides, the *N*-acetylglucosamine (GlcNAc) of the reducing terminal is linked to asparagine (found in the Asn-Xaa-Ser/Thr sequence), forming an *N*-aspartylglycosylamine linkage. *N*-linked oligosaccharides are classified into three types, namely the high mannose, hybrid and complex types (Figure 7). High mannose-type oligosaccharides contain only mannose residues attached to the core portion, but the number of mannose residues varies. In hybrid-type oligosaccharides, mannose residues are attached to the 1-6 arm, and an *N*-acetyllactosamine side chain is attached to the 1-3 arm (Figure 7). Complex-type oligosaccharides have *N*-acetyllactosamine residues at the both arms. The side chains are modified by sialic acid residues or additional *N*-acetyllactosamines. Complex-type oligosaccharides can have variable numbers of side chains, such as bi-, tri- and tetra-antennary.

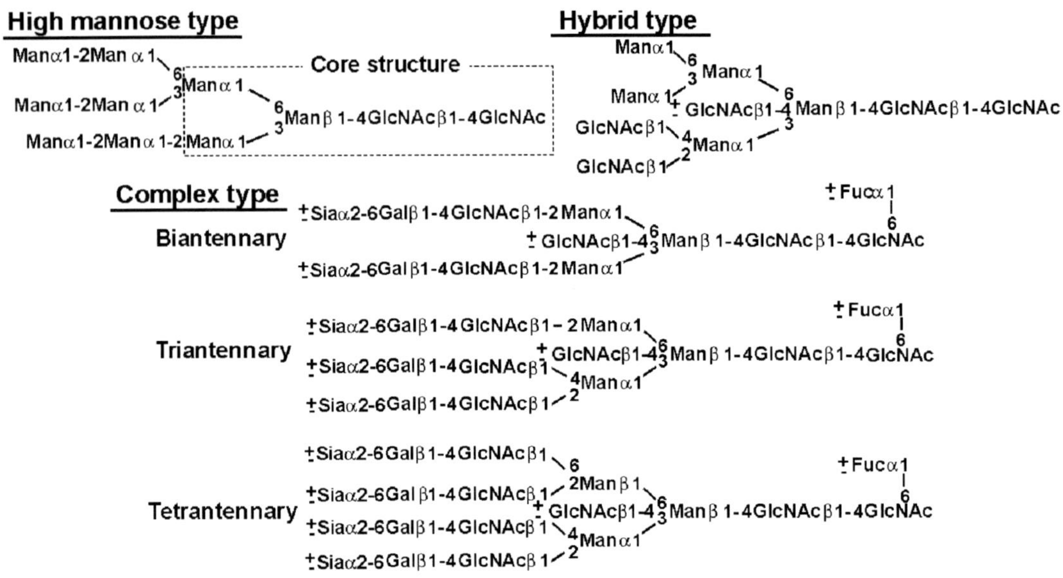

Figure 7. Representative structures of *N*-linked oligosaccharides.

N-glycosylation gives many new or altered functions to proteins and indeed is thought to be important for their functions [59]. *N*-linked oligosaccharides on glycoproteins interact with oligosaccharide-binding proteins (lectins), and this oligosaccharide-lectin interaction mediates important biological recognition, adhesion and signaling events [60]. In addition, *N*-linked oligosaccharides affect the folding, oligomerization and stability of proteins [61]. *N*-linked glycosylation has been reported as a process that when faulty can lead to some diseases, such as cancer and arthritis [62].

General Knowledge on Endo-β-*N*-Acetylglucosaminidase

Endo-β-*N*-acetylglucosaminidase (EC 3.2.1.96) hydrolyzes the *N*, *N'*-diacetylchitobiose moiety in *N*-linked oligosaccharides. One GlcNAc residue still remains attached to a protein or a peptide through an asparagine residue (Figure 8).

The enzyme was found first in the culture medium of *Streptococcus pneumoniae* by Muramatsu in 1971 [63], and was named Endo D. Endo-β-*N*-acetylglucosaminidase was expected to be useful in order to analyze the structure and function of *N*-linked oligosaccharides. Similar enzymes were subsequently found in other microorganisms, such as *Streptomyces plicatus* (formerly *Streptomyces griseus*; Endo H) [64], *Clostridium perfringens* (Endo C-I and Endo C-II) and *Elizabethkingia meningoseptica* (former *Flavobacterium meningosepticum*; Endo F_1, Endo F_2 and Endo F_3). Besides these microorganisms, the enzyme has been found in the culture medium of many other bacteria and fungi. Almost all of them can preferentially act on high mannose and hybrid-type oligosaccharides. Only a few enzymes can act on complex-type oligosaccharides of bi- or tri-antennary.

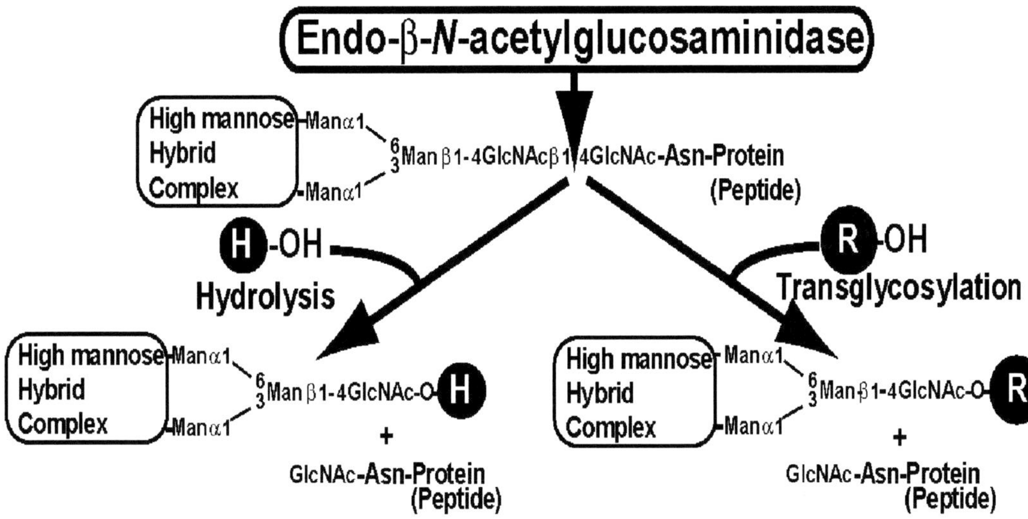

Figure 8. Action of endo-β-*N*-acetylglucosaminidase on *N*-linked oligosaccharides.

(H)-OH, water molecule; (R)-OH, acceptor molecule.

Endo-β-*N*-acetylglucosaminidase has also been found in mammalian tissues [65], including rat liver and kidney [66] and human skin fibroblasts [67], and also in hen oviducts [68] and the nematode, *Caenorhabditis elegans* [69]. Animal enzymes are located in the cytoplasm in a soluble form and preferentially act on the high mannose-type oligosaccharides. However, the Fuc residue attached to the proximal GlcNAc is an obstacle to the action of animal enzymes. The enzyme has also been found in plants [70], where they are also thought to be located in the cytoplasm and to be involved in the formation of the free *N*-linked oligosaccharides present in the cytoplasm.

Endo-β-*N*-acetylglucosaminidases are structurally classified into two GH families, GH18 and GH85. The GH18 family enzymes commonly have a chitinase active site motif and are found only in prokaryotes, whereas the GH85 family enzymes have no chitinase active site motif and have been found in both eukaryotes and prokaryotes.

Preparation and Biochemical Properties of Endo-β-*N*-Acetylglucosaminidases

Endo HS

The endo-β-*N*-acetylglucosaminidases described above preferentially act on *N*-linked oligosaccharides of the high mannose and hybrid type. Our group had been working on an endo-β-*N*-acetylglucosaminidase (Endo HS) involved in the deglycosylation of human salivary α-amylase family A [71- 72]. This endo HS is a membrane-bound enzyme found on the epithelial cells in the human oral cavity. It was purified by a series of column chromatography steps after detergent solubilization [72], and shows an optimum pH of around 6 and can easily release *N*-linked oligosaccharides of bi-, tri-, tetra- and penta-antennary complex-types, regardless of the existence of the Fuc residue on the proximal GlcNAc. In contrast, it hardly releases any high mannose or hybrid type *N*-linked oligosaccharides from glycoproteins [72]. Endo-β-*N*-acetylglucosaminidases are often less active at releasing oligosaccharides from native mature glycoproteins, but rather require prior denaturation or digestion of the protein moiety first for effective deglycosylation. However, Endo HS can easily release complex-type oligosaccharides from native glycoproteins as well as glycopeptides and glycoasparagines. Especially, the enzyme can release all of the complex-type oligosaccharides from highly glycosylated glycoprotein's, such as the human α_1-acid glycoprotein [73]. Thus, Endo HS is specific for the complex-type *N*-linked oligosaccharides and is apparently distinct from any other known endo-β-*N*-acetylglucosaminidases. Endo HS can be expected to be an effective deglycosylation tool for the analysis of structure and function of *N*-linked oligosaccharides on glycoproteins.

Endo FV

Recently, an endo-β-*N*-acetylglucosaminidase, named Endo FV, has been purified from the basidiomycete fungi, *Flammulina velutipes* [74]. The enzyme was purified from the fruiting body of *F. velutipes* and separated into two multiple forms by ion exchange chromatography. They show an optimum pH of around 6 and are stable over a pH range of 5.5 - 9.5 and at temperatures of up to 40 °C. The multiple forms of Endo FV have the same molecular weight of 34 or 31 kDa as estimated by SDS-PAGE and by gel filtration, respectively. Their substrate specificity for *N*-linked oligosaccharides is also the same. The action of Endo FV on fluorescence-labeled glycoasparagines indicates that Endo FV can preferentially release high mannose-type oligosaccharides and is less active on hybrid type oligosaccharides. However, the enzyme cannot act on any complex-type oligosaccharides. Endo FV is less active for *N*-linked oligosaccharides attached to glycoproteins.

The Endo FV gene has no genomic introns and is composed of a 996-bp open reading frame encoding for a predicted 331 amino acid residue protein. Little nucleotide sequence identity is found between the Endo FV gene and the genes of other endo-β-*N*-acetylglucosaminidases. Endo FV has a chitinase active site motif, suggesting on current classification systems that Endo FV belongs to the GH18 family and if so extends the GH18 family range into eukaryotes in addition to prokaryotes. The recombinant Endo FV expressed in *E. coli* is similar to the original enzyme in *F. volutipes* in terms of its *in vitro* substrate specificity.

Endo FV is similar to fungal class III chitinases in primary structure but clearly different from other GH18 family members, such as Endo H and Endo F_1. However, Endo FV has no detectable chitinase activity even though it has the proposed active site motif. On the other hand, Endo FV is predicted to be homologous in tertiary structure with Endo H and Endo F_1, with a similar specificity for *N*-linked oligosaccharides. It is suggested that Endo FV may have acquired similar specificity as that of Endo H or Endo F_1 as a result of convergent evolution.

Oligosaccharide Synthesis through Hydrolysis and Transglycosylation Activities of Endo-β-*N*-Acetylglucosaminidases

Endo-β-*N*-acetylglucosaminidase can also cleave the *N, N'*-diacetylchitobiose linkage in the presence of hydroxyl groups of suitable acceptor molecules and then transfer the released oligosaccharides to form neoglycoconjugates (Figure 8). The transglycosylation activity of endo-β-*N*-acetylglucosaminidase was first reported with Endo FV [75], and has subsequently been found in other endo-β-*N*-acetylglucosaminidases where the *N*-linked oligosaccharides can be transferred to monosaccharide and oligosaccharide acceptor molecules [76-77]. Endo-β-*N*-acetylglucosaminidases available for the synthesis of neoglycoconjugates bearing *N*-linked oligosaccharides are described as follows.

Endo A

Endo A from *Arthrobacter protoformiae* preferentially hydrolyzes high mannose type oligosaccharides and also transfers the oligosaccharide from glycopeptides to monosaccharides, such as mannose, glucose and GlcNAc [76]. The enzyme shows a high transglycosylation activity and can transfer a high mannose-type oligosaccharide from a glycoasparagine to GlcNAc-RNase B [78], resulting in the remodeling of the oligosaccharide in RNase B. Endo A can be used for transferring high mannose-type oligosaccharides to some functionalized GlcNAc glycosides to form glycopolymers [79], and can also transfer high-mannose type oligosaccharides to GlcNAc-induced calcitonin [80].

Endo M

Endo M from *Mucor hiemalis* shows hydrolysis activity for *N*-linked oligosaccharides of the biantennary complex type as well as for the high mannose type. The enzyme can be used for transferring biantennary complex-type oligosaccharides to various monosaccharides [77], and bioactive peptides, such as calcitonin [81]. The glycosylated peptides are biologically active and acquire resistance for proteolysis.

Endo HS

The structure of the transferred *N*-linked oligosaccharide is likely to depend on the specificity of each endo-β-*N*-acetylglucosaminidase for the hydrolysis reaction. Many enzymes preferentially hydrolyze high mannose and hybrid types but Endo HS can also specifically release various complex-type oligosaccharides [73]. As for the transglycosylation reaction, Endo HS can transfer a biantennary complex type oligosaccharide from the native human transferrin to *p*-nitrophenyl-β-D-Glc (*p*NP-β-D-Glc) [82], as confirmed by MALDI-

TOF analysis. The amount of the transglycosylation product depends on the concentration of the acceptors. Endo HS can also directly transfer the *N*-linked oligosaccharide from human transferrin to *p*NP-α-D-Glc, *p*NP-α-D-Gal, *p*NP-β-D-Gal, *p*NP-β-D-Man, *p*NP-β-D-Xyl, *p*NP-β-D-GlcNAc as well as to glucose, mannose, galactose and GlcNAc. In addition to sugars, Endo HS can transfer the oligosaccharide from human transferrin to *p*NP-glycerol. Apparent differences in the K_m values for the oligosaccharide donor (human transferrin) are not observed when using different acceptors, such as *p*NP-β-D-Glc and *p*NP-glycerol. Moreover, Endo HS can transfer triantennary complex-type oligosaccharides from native calf fetuin to *p*NP-β-D-Glc, as well as *N*-linked oligosaccharides, such as bi-, tri- and tetra-antennary complex types from human α_1-acid glycoprotein to *p*NP-β-D-Glc [82]. Besides glycoproteins, Endo HS is also able to transfer a bi-antennary complex-type oligosaccharide from glycopeptides to *p*NP-β-D-Glc.

Endo HS can also transfer bi- and tri-antennary complex-type oligosaccharides from glycoasparagines prepared from human transferrin and calf fetuin, to various monosaccharides, such as glucose, mannose, galactose, xylose, fucose and fructose, and to bioactive glycosides, such as arbutin, salicin and linamarin [83]. The enzyme also transfers *N*-linked oligosaccharides to acceptors, such as maltose, sucrose, lactose, malto-oligosaccharides and the sugar alcohols, such as mannitol, sorbitol, xylitol, maltitol, maltotriitol, glycerol and 3-chloro-1, 2-propanediol.

Endo A and Endo M cannot transfer oligosaccharides to galactose, fucose and GlcUA, suggesting that they strictly require both the 3'-OH and 4'-OH of a sugar molecule to be equatorial [76-77]. However, Endo HS can transfer *N*-linked oligosaccharides to galactose, indicating that Endo HS does not recognize the configuration of the hydroxyl groups [82]. The transglycosylation products formed by Endo HS hardly decrease while those formed by Endo A or Endo M are rehydrolyzed and disappear, indicating that Endo HS is quite distinct from the other enzymes in transglycosylation reaction [83].

Endo HS is a useful enzyme for enzymatic synthesis of neoglycoconjugates bearing *N*-linked oligosaccharides of a wide variety of complex types. The synthesized neoglycoconjugates can be expected to show hybrid functions of acceptor molecules and *N*-linked oligosaccharides of a complex type.

Endo FV

Endo FV from the basidiomycete, *Flammulina velutipes,* can also transfer *N*-linked oligosaccharides of the high mannose type from a glycoasparagine prepared from ovalbumin to glucose [74]. The rate of the transglycosylation reaction and the hydrolysis reaction are similar. The transglycosylation product is rehydrolyzed and disappears during the reaction procedure. On the other hand, no transglycosylation product is formed using galactose as the oligosaccharide acceptor, suggesting that Endo FV is similar to Endo M [77] and Endo A [76] in acceptor specificity and rehydrolysis of transglycosylation products. Endo FV may also strictly recognize the configuration of the hydroxyl groups of an acceptor sugar molecule for the transglycosylation reaction, as do the Endo A [79] and Endo M [84] enzymes. The motif containing the essential amino acids for transglycosylation is commonly conserved in enzymes belonging to the GH85 family, such as Endo A and Endo M, but not in the GH18 family enzymes [84-85]. Transglycosylation is characteristic of the GH85, but not the GH18 family enzymes [78]. However, Endo FV, a GH18 family member also shows transglycosylation activity, suggesting that the transglycosylation activity of Endo FV has a distinct mechanism from that of GH family 85 enzymes.

CONCLUDING REMARKS

Research and development of novel oligosaccharides with physicochemical and biological functional properties has been continually growing since the onset of the industrial synthesis of oligosaccharides from the early 1980s. The market for oligosaccharides is substantial and still gradually increasing. Japanese biotech companies dominate the worldwide oligosaccharide production, as well as the research and development of oligosaccharides. Since 1970, several novel microbial enzymes producing specific oligosaccharides have been discovered and developed for use in a commercial scale production of oligosaccharides. Numerous oligosaccharides with the required characteristics, such as low and non digestibility, low sweetness, low calorie and low or no anticariogenicity have been evaluated. They are classified into starch-, sucrose- and lactose-related, and other types of, oligosaccharides (Table 4) [2]. Among those presently on the market, the demand and production for starch oligosaccharides as new sweeteners is the largest. The recent development in industrial enzymology has made possible a series of new starch oligosaccharides such as gentio-oligosaccharide (β-1,6 linked), nigero-oligosacchride (α-1,3 linked) and trehalose (α-1,1 linked).

With increasing consumer health consciousness and awareness of physiologically beneficial and functional foods, the future for oligosaccharides is greatly promising. In oligosaccharide production, the need to achieve a continuous and efficient production of oligosaccharides, and to maintain a low cost of high-purity oligosaccharides, is required [2]. Significant progress has been made in the synthesis of oligosaccharides in recent years, increasing the potential of these molecules as therapeutics. Solid-phase oligosaccharide synthesis has gained considerable attention as a promising step to improve product recovery [7]. The use of whole microorganisms as cell factories in large-scale oligosaccharide production has also been reported [86]. In addition, oligosaccharides with highly functionalized properties must be developed, either from the discovery of novel enzymes or the successful altering of the enzyme specificity or catalysis of existing known enzymes by means of genetic engineering. Furthermore, when the large genomic and structural database on glycosidases is available, more glycosynthases will be efficiently designed [8]. The significance of functional oligosaccharides in the near future is foreseen in its important role in the reduction of lifestyle-related diseases, as well as the improvement of human health [2].

ACKNOWLEDGMENTS

The authors would like to thank all the supports from the Asian Core Program of JSPS-NRCT and from Chulalongkorn University and Osaka City University. We also thank Dr. Robert Butcher of the Publication Counseling Unit of the Faculty of Science, Chulalongkorn University for editing the chapter.

Table 4. Examples of industrial oligosaccharide products and the oligosaccharide-producing enzymes [2]

Raw material	Oligosaccharide product	Enzyme	Microbial origin
Starch-related	Trehalose	MTSase and MTHase*	*Arthrobacter* sp. & *Sulfolobus* sp.
	Malto-oligosaccharides (MOS) (mixture of G2-G8, with G4 as main component)	Maltotetraohydrolase coimmobilized with Pullulanase	*Pseudomonas stutzeri* *K. pneumoniae*
	Isomalto-oligosaccharides (IMO)	α-Glucosidase	Fungal enzyme
	Cyclodextrins	CGTase	*B. stearothermophilus*, etc. (see Table 1)
Sucrose-related	Fructo-oligosaccharides (FOS)	β-Fructofuranosidase	*Aspergillus niger*
	Glucosylsucrose (Coupling sugar)	CGTase	*Thermoanaerobacter* sp.
	Lactosucrose	β-Fructofuranosidase	*A. niger*
	Palatinose (isomaltulose)	α-Glucosyltransferase	*Protaminobacter rubrum*
Galactose-related	Galacto-oligosaccharides	Lactase	*Aspergillus oryzae* or *Streptococcus thermophilus*
Xylan, agar, mannan, chitin/chitosan	Xylo-oligosaccharides	Xylanase	
	Agaro-oligosaccharides	Agarase	
	Manno-oligosaccharides	Mannanase	
	Chitin/Chitosan-oligosaccharides	Chitinase	

* Malto-oligosyltrehalose synthase and Malto-oligosyltrehalose trehalohydrolase.

REFERENCES

[1] Voet, D. & Voet, J.G. (2004) *Biochemistry* (Third Edition, 356-381). USA: John Wiley & Sons, Inc.

[2] Nakakuki, T. (2002) Present status and future of functional oligosaccharides development in Japan. *Pure Appl. Chem., 74*, 1245-1251.

[3] Szejtli, J. (1998) Introduction and general overview of cyclodextrin chemistry. *Chem. Rev., 98*, 1743-1753.

[4] Berg, J.M., Tymoczko, J.L. & Stryer, L. (2002) *Biochemistry* (Fifth Edition, 295-318). New York: W.H. Freeman and Company.

[5] Roberfroid, M.B. (1997) Health benefits of non-digestible oligosaccharides. In Kritchecsky, D. & Bonfield, C. (eds.), *Dietary Fiber in Health and Disease*, (211-219). New York: Plenum Press.

[6] Ouwehand, A.C., Derrien, M., Vos, W., Tiihonen, K. & Rautonen, N. (2005) Prebiotics and other microbial substrates for gut functionality. *Curr. Opin. Biotechnol., 16*, 212-217.

[7] Perugino, G., Trincone, A., Rossi, M. & Moracci, M. (2004) Oligosaccharide synthesis by glycosynthases. *Trends Biotechnol., 22(1),* 31-37.

[8] Takanori, N. (2007) Evolution of the oligosaccharide synthesis by glycosynthases. *Trends Glycosci. Glycotechnol., 19 (105),* 49-50.

[9] Liese, A., Seelbach, K. & Wandrey, C. (2000) *Industrial Biotransformation,* Weinheim : Wiley-VCH.

[10] Sako, T., Matsumoto, K. & Tanaka, R. (1999) Recent progress on research and applications of non-digestible galacto-oligosaccharides. *Int. Dairy J., 9*, 69-80.

[11] Nilsson, K.G.I. (1988) Enzymatic synthesis of oligosaccharides. *TIBTECH, 6,* 256-264.

[12] Crout, D.H. & Vic, G. (1998) Glycosidases and glycosyltransferases in glycoside and oligosaccharide synthesis. *Curr. Opin. Chem. Biol., 2*, 98-111.

[13] Kuriki, T. (2000) Catalytic mechanism of Glycoenzymes. In Ohnishi, M. (ed.), *Glycoenzymes*, (3-13). Tokyo: Japan Scientific Societies Press.

[14] Mackenzie, L.F., Wang, Q., Warren, R.A.J. & Withers, S.G. (1998) Glycosynthases: mutant glycosidases for oligosaccharide synthesis. *J. Am. Chem. Soc., 120*, 5583-5584.

[15] Withers, S.G., Mackenzie, L. & Wang, Q. (2001) Methods and compositions for synthesis of oligosaccharides, and the products formed thereby. *US Patent* 6284494 B1.

[16] Mayer, C., Zechel, D.L., Reid, S.P., Warren, R.A.J. & Withers, S.G. (2000) The E358S mutant of *Agrobacterium* sp. β-glucosidase is a greatly improved glycosynthase. *FEBS Lett., 466*, 40-44.

[17] Tolborg, J.F., Petersons, L., Jensen, K.J., Mayer, C., Jakeman, D.L., Warren, R.A.J. & Withers, S.G. (2002) Solid-phase oligosaccharide and glycopeptides synthesis using glycosynthases. *J. Org. Chem., 67*, 4143-4149.

[18] Maitin, V., Athanasopoulos, V. & Rastall, R.A. (2004) Synthesis of FimH receptor-active manno-oligosaccharides by reverse hydrolysis using α-mannosidase from *Penicillium citrinum, Aspergillus phoenicis* and almond. *Appl. Microbiol Biotechnol. 63*, 666-671.

[19] Tonkova, A. (1998) Bacterial cyclodextrin glucanotransferase. *Enzyme Microb. Technol., 22*, 678-686.

[20] Janecek, S. (2000) Structural features and evolutionary relationships in the α-amylase family. In Ohnishi, M. (ed.) *Glycoenzymes,* (13-54) Tokyo: Japan Scientific Societies Press.

[21] Takaha, T. & Smith S.M. (1999) The function of 4-α-glucanotransferase and their use for the production of cyclic glucans. *Biotechnol. Genet. Eng. Rev.*, *16*, 257-280.

[22] Kaper, T., Leemhuis, H., Uitdehaag, J.C.M., van der Veen, B.A., Dijkstra, B.W., van der Maarel, M.J.E.C., & Dijkhuizen, L. (2007) Identification of acceptor substrate binding subsites +2 and +3 in the amylomaltase from *Thermus thermophilus* HB8. *Biochemistry, 46,* 5261-5269.

[23] Sträter, N., Przylas, I., Saenger, W., Terada, Y., Fujii, K. & Takaha, T. (2002) Structural basis of the synthesis of large cycloamyloses by amylomaltase. *Biologia Bratislava, 57 (Suppl. 11),* 93-99.

[24] Bhuiyan, S.H., Kitaoka, M. & Hayashi, K. (2003). A cycloamylose-forming hyperthermostable 4-α-glucanotransferase of *Aquifex aeolicus* expressed in *Escherichia coli. J. Mol. Catal., B Enzym., 22*, 45-53.

[25] Qi, Q. & Zimmermann, W. (2005) Cyclodextrin glucanotransferase : from gene to applications. *Appl. Microbiol. Biotechnol., 66,* 475 - 485.

[26] Martin, M.T., Cruces, M.A., Alcade, M., Plou, F.J., Bernabe, M. & Ballesteros, A. (2004) Synthesis of maltooligosyl fructofuranosides catalyzed by immobilized cyclodextrin glucosyltransferase using starch as donor. *Tetrahedron, 60*, 529-534.

[27] Crittenden, R.G. & Playne, M.J. (1996) Production, properties and applications of food-grade oligosaccharides. *Trends Food Sci. Technol., 7,* 353-361.

[28] Tanaka, O. (1997) Improvement of taste of natural sweeteners. *Pure Appl. Chem., 69(4),* 675-683.

[29] Kometani, T., Terada, Y., Nishimura, T., Takii, H. & Okada, S. (1994) Transglycosylation to hesperidin by cyclodextrin glucanotransferase from alkalophilic *Bacillus* species in alkaline pH and properties of hesperidin glycosides. *Biosci. Biotech. Biochem., 58,* 1990 - 1994.

[30] Kometani, T., Nishimura, T., Nakae, T., Takii, H. & Okada, S. (1996) Synthesis of neohesperidin glycosides and naringin glycosides by cyclodextrin glucanotransferase from an alkalophilic *Bacillus* species. *Biosci. Biotech. Biochem., 60(4),* 645 - 649.

[31] Aga, H., Yoneyama, M., Sakai, S. & Yamamoto, I. (1991) Synthesis of 2-O-α-D glucopyranosyl L-ascorbic acid by cyclomaltodextrin glucanotransferase from *B. stearothermophilus. Agric. Biol. Chem., 55,* 1751 - 1756.

[32] Uchida, K. & Suzuki, Y. (1998) Enzymatic synthesis of a new derivative of thiamin, O-alpha-glucosylthiamin. *Biosci. Biotech. Biochem., 62,* 221 - 224.

[33] Szejtli, J. (1988) Cyclodextrins In *Cyclodextrin Technology*, (chapter 1, pp. 1-78). Dordrecht: Kluwer Academic Publishers,

[34] Hedges, R. (1998) Industrial applications of cyclodextrins. *Chem. Rev., 98,* 2035 - 2044.

[35] Hashimoto, H. (2002) Present Status of industrial application of cyclodextrins in japan. *J. Incl. Phenom. Macroc. Chem., 44,* 57 - 62.

[36] Pongsawasdi, P. (2008) *Carbohydrate-modifying Enzymes in Industry* (in Thai). (Chapter 10). Bangkok: ChulaPress.

[37] Bender, H. (1986) Production, characterization, and application of cyclodextrins. *Adv. Biotech. Proc., 6,* 31 – 71.

[38] Flaschel, E., Landert, J., & Spiesser, D. (1984) The Production of α-cyclodextrin by enzymatic degradation of starch. *Ann. NY. Acad. Sci., 434,* 70 - 77.
[39] Horikoshi, K., Yanamoto, M., Nakamura, N., & Kawano, M. (1981) Process for recovering cyclodextrins. *US. Patent* 4, 303, 787.
[40] Tesana, S. (2001) *M.Sc. Thesis (Biotechnology),* Faculty of Science, Chulalongkorn University.
[41] Yenpetch, W. .M ,Iizuka ,.J ,Kaulpiboon &Thermostable (2004) .P ,Pongsawasdi and its 01RB .sp *Paenibacillus* CGTase fromc hemicalm odification withglucomannan. Association de *,Symposium Proceedings of the Twelfth International Cyclodextrin* .112-107 ,France ,Pharmacie Galenique Industrielle
[42] Yenpetch, W. (2002) *M.Sc. Thesis (Biochemistry).* Faculty of Science, Chulalongkorn University.
[43] Yenpetch, W. (2009) *Ph.D. Dissertation (Biochemistry),* Faculty of Science, Chulalongkorn University.
[44] Monod, J. & Torriani, A.M. (1950) Amylomaltase of *Escherichia coli. Ann. Inst. Pasteur. (Paris), 78(1),* 65-77.
[45] Pugsley, A.P. & Dubreuil, C. (1988) Molecular characterization of malQ, the structural gene for the *Escherichia coli* enzyme amylomaltase. *Mol. Microbiol., 2,* 473-479.
[46] Goda, S.K., Eissa, O., Akhtar, M. & Minton, N.P. (1997) Molecular analysis of a *Clostridium butyricum* NCIMB 7423 gene encoding 4-alpha-glucanotransferase and characterization of the recombinant enzyme produced in *Escherichia coli. Microbiology, 143,* 3287-3294.
[47] Jeon, B.S., Taguchi, H., Sakai, H., Ohshima, T., Wakagi, T. & Matsuzawa, H. (1997) 4-alpha-glucanotransferase from the hyperthermophilic archaeon *Thermococcus litoralis*: enzyme purification and characterization, and gene cloning, sequencing and expression in *Escherichia coli. Eur. J. Biochem., 248,* 171-178.
[48] Terada, Y., Fujii, K., Takaha, T. & Okada, S. (1999*) Thermus aquaticus* ATCC 33923 amylomaltase gene cloning and expression and enzyme characterization: Production of cycloamylose. *Appl. Environ. Microbiol., 65,* 910-915.
[49] Young, B.B., Kim, H.J., Kim, H.Y., Baik, M.Y., Ahn, S.C., Kim, C.H. & Park, C.S. (2006) Cloning and overexpression of 4α-glucanotransferase from *Thermus brockianus* (TBGT) in *E. coli. J. Microbiol. Biotechnol. 16,* 1809-1813.
[50] Lee, H.S., Auh, J.H., Yoon, H.G., Kim, M.J., Park, J.H., Hong, S.S., Kang, M.H., Kim, T.J., Moon, T.W., Kim, J.W. & Park, K.H. (2002) Cooperative action of alpha-glucanotransferase and maltogenic amylase for an improved process of isomaltooligosaccharide (IMO) production. *J. Agric. Food Chem., 50,* 2812-2827.
[51] Kaper, T., van der Maarel, M.J., Euverink, G.J. & Dijkhuizen, L. (2004) Exploring and exploiting starch-modifying amylomaltases from thermophiles. *Biochem. Soc. Trans., 32,* 279-282.
[52] Alting, A.C., van de Velde, F., Kanning, M.W., Burgering, M., Mulleners, L., Sein, A. & Buwalda, P. (2009) Improved creaminess of low-fat yoghurt: The impact of amylomaltase-treated starch domains. *Food Hydrocolloids, 23,* 980–987.
[53] Mun, S., Kim, Y., Kang, C., Park, K., Shim, J. & Kim, Y. (2009) Development of reduced-fat mayonnaise using 4αGTase-modified rice starch and xanthan gum. *Int. J. Biol. Macromol., 44,* 400-407.

[54] Endo, T., Zheng, M. & Zimmermann, W. (2002). Enzymatic Synthesis and Analysis of Large-Ring Cyclodextrins. *Aust. J. Chem., 55,* 39-48.

[55] Park, J., Kim, H., Kim, Y., Cha, H., Kim, Y., Kim, T., Kim, Y. & Park, K. 2007. The action mode of *Thermus aquaticus* YT-1 4-α-glucanotransferase and its chimeric enzymes introduced with starch-binding domain on amylose and amylopectin. *Carbohydr. Polym., 67,* 164-173.

[56] Ikeda, M. & Nakagawa, S. (2003) The *Corynebacterium glutamicum* genome: features and impact on biotechnological processes. *Appl. Microbiol. Biotechnol. 62*, 99-109.

[57] GenBank accession numbers BAA33728.1, YP_144527.1, AAC06897.1

[58] Sharon, N. & Lis, H. (1982) Glycoproteins In Neurath, H. & Hill, R. L. (Eds.) *The Proteins*, Volume V, pp. 1-144, New York: Academic Press.

[59] Rademacher, T.W., Parekh, R.B. & Dwek, R.A. (1988) Glycobiology. *Ann. Rev. Biochem., 57*, 785-838.

[60] Taylor, M.F. & Drickamer, K. (2006) Carbohydrate recognition in cell adhesion and signaling In Taylor, M. E. & Direckamer, K. *Introduction to Glycobiology*, (chapter 8, pp. 122-149). New York: Oxford University Press.

[61] Mitra, N., Sinha, S., Ramya, T.N.C. & Surolia, A. (2006) *N*-linked oligosaccharides as outfitters for glycoprotein folding, form and function. *Trends Biochem. Sci., 31*, 156-163.

[62] Freeze, H.H. & Balancing, W.V. (2001) *N*-linked glycosylation to avoid diseases. *Biochimie*, *83*, 791-799.

[63] Muramatsu, T. (1971) Demonstration of an endo-glycosidase acting on a glycoprotein. *J. Biol. Chem., 246*, 5535-5537.

[64] Tarentino, A.L., Plummer, T.H. & Maley, F. (1972) A re-evaluation of the oligosaccharide sequence associated with ovalbumin. *J. Biol. Chem., 247*, 2629-2631.

[65] Nishigaki, M., Muramatsu, T. & Kobata, A. (1974) Endoglycosidases acting on carbohydrate moieties of glycoproteins: demonstration in mammalian tissue. *Biochem., Biophys., Res., Commun., 59*, 638-645.

[66] Pierce, R. J., Spik, G. & Montreuil, J. (1979) Cytosolic location of an endo-*N*-acetyl-β-D-glucosaminidase activity in rat liver and kidney. *Biochem. J., 180,* 673-676.

[67] Tachibana, Y., Yamashita, K., Kawaguchi, M., Arashima, S., & Kobata, A. (1981) Digestion of asparagine-linked oligosaccharides by endo-β-*N*-acetylglucosaminidase in the human skin fibroblasts obtained from fucosidosis patients. *J. Biochem., 90*, 1291-1296.

[68] Kato, T., Hatanaka, K., Mega, T. & Hase, S. (1997). Purification and characterization of endo-β-*N*-acetylglucosaminidase from hen oviduct. *J. Biochem., 122*, 1167-1173.

[69] Kato, T., Fujita, K., Takeuchi, M., Kobayashi, K., Natsuka, S., Ikura, K., Kumagai, H. & Yamamoto, K. (2002) Identification of an endo-β-*N*-acetylglucosaminidase gene in *Caenorhabditis elegans* and its expression in *Escherichia coli. Glycobiology, 12*, 581-587.

[70] Kimura, Y., Tokuda, T., Ohno, A., Tanaka, H. & Ishiguro, Y. (1998) Enzymatic properties of endo-β-*N*-acetylglucosaminidases from developing tomato fruits and soybean seeds: substrate specificity of plant origin endoglycosidase. *Biochim. Biophys. Acta, 1381*, 27-36.

[71] Ito, K., Li, L.-F., Nishiwaki, M., Okada, Y., & Minamiura, N. (1992) Evidence for conversion of human salivary α-amylase family A to family B by an enzyme action. *J. Biochem. (Tokyo), 112*, 88-92.

[72] Ito, K., Okada, Y., Ishida, K. & Minamiura, N. (1993) Human salivary endo-β-*N*-acetylglucosaminidase HS specific for complex type sugar chains of glycoproteins. *J. Biol. Chem., 268*, 16074-16081.

[73] Ito, K. (2006) Formation of glycoisoforms of human salivary α-amylase and endo-β-*N*-acetylglucosaminidase. *Trends Glycosci. Glycotechnol., 18*, 73-84.

[74] Hamaguchi, T., Ito, T., Inoue, Y., Limpaseni T., Pongsawasdi, P. & Ito, K. (2010) Purification, characterization and molecular cloning of a novel endo-β-*N*-acetylglucosaminidase from the basidiomycete, *Flammulina velutipes*. *Glycobiology, 20*, 420-432.

[75] Trimble, R.B., Atkinson, P.H., Tarentino, A.L., Plummer Jr., T.H., Maley F. & Tomer, K.B. (1986) Transfer of glycerol by endo-β-*N*-acetylglucosaminidase F to oligosaccharides during chitobiose core cleavage. *J. Biol. Chem., 261,* 12000-12005.

[76] Takegawa, K., Yamaguchi, S., Kondo, A., Iwamoto, H., Nakoshi, M., Kato, I., & Iwahara, S. (1991) Transglycosylation activity of endo-β-*N*-acetylglucosaminidase from *Arthrobacter protophormiae*. *Biochem. Int., 24*, 849-855.

[77] Yamamoto, K., Kadowaki, S., Watanabe, J. & Kumagai, H. (1994) Transglycosylation activity of *Mucor hiemalis* endo-β-*N*-acetyl-glucosaminidase which transfers complex oligosaccharides to the *N*-acetylglucosamine moieties of peptides. *Biochem. Biophys. Res. Commun., 203*, 244-252.

[78] Takegawa, K., Tabuchi, M., Yamaguchi, S., Kondo, A., Kato, I., & Iwahara, S. (1995) Synthesis of neoglycoproteins using oligosaccharide-transfer activity with endo-β-*N*-acetylglucosaminidase. *J. Biol. Chem., 270*, 3094-3099.

[79] Fan, J.Q., Quesenberry, M.S., Takegawa, K., Iwahara, S., Kondo, A., Kato, & Lee, Y.C. (1995) Synthesis of neoglycoconjugates by transglycosylation with *Arthrobacter protophormiae* endo-β-*N*-acetylglucosaminidase Demonstration of a micro-cluster effect for mannose-bindings. *J. Biol. Chem., 270*, 17730-17735.

[80] Yamamoto, K., Haneda, K., Iguchi, R., Inazu, T., Mizuno, M., Takegawa, K., Kondo, A. & Kato, I. (1999) Chemo-enzymatic synthesis of a calcitonin derivative containing a high-mannose type oligosaccharide by endo-β-*N*-acetylglucosaminidase from *Arthrobacter protophormiae*. *J. Biosci. Bioeng., 87*, 175-179.

[81] Haneda, K., Tagashira, M., Yoshino, E., Takeuchi, M., Inazu, T., Toma, K., Iijima, H., Isogai, Y., Hori, M., Takamatsu, S., Hujibayashi, Y., Kobayashi, K., Takecushi, M., & Yamamoto, K. (2004) Chemo-enzymatic synthesis and structure-activity study of artificially *N*-glycosylated eel calcitonin derivatives with a complex type oligosaccharide. *Glycoconj. J., 21*, 377-386.

[82] Ito, K., Miyasdgawa, K., Matsumoto, M., Yabuno, S., Kawakami, N., Hamaguchi, T., Iizuka, M. & Minamura, N. (2006) Evidence for the transglycosylation of complex type oligosaccharides of glycoproteins by endo-β-*N*-acetylglucosaminidase HS. *Arch. Biochem. Biophys. 454*, 89-99.

[83] Miyagawa, K., Matsumoto, M., Yabuno, S., Kawakami, N., Hamaguchi, T., Iizuka, M., Minamiura, N. & Ito, K. (2007) Transglycosylation of asparagines-linked complex-type

oligosaccharides from glycoproteins by endo-β-*N*-acetylglucosaminidase HS. *J. Appli. Glycosci.*, *54*, 139-146.

[84] Fujita, K., Takami, H., Yamamoto, K. & Takegawa, K. (2004) Characterization of endo-β-*N*-acetylglucosaminidase from alkaliphilic *Bacillus halodurans* C-125. *Biosci. Biotechno. Biochem.*, *68*, 1059-1066.

[85] Fujita K. & Takegawa K. (2001). Tryptophan-216 is essential for the transglycosylation activity of endo-β-*N*-acetylglucosaminidase A. *Biochem. Biophys. Res. Commun.*, *283*, 680-686.

[86] Endo, T. & Koizumi, S. (2000) Large-scale production of oligosaccharides using engineered bacteria. *Curr. Opin. Struct, Biol.*, *10*, 536-541.

In: Oligosaccharides: Sources, Properties and Applications ISBN 978-1-61122-193-0
Editor: Nicole S. Gordon

Chapter 5

PRECURSOR N-LINKED OLIGOSACCHARIDES AS CODES FOR GLYCOPROTEIN FOLDING STATUS

Ron Benyair and Gerardo Z. Lederkremer*

Department of Cell Research and Immunology, George Wise Faculty of Life Sciences, Tel Aviv University, Tel Aviv 69978, Israel

ABSTRACT

The majority of eukaryotes share a common N-linked oligosaccharide precursor, $Glc_3Man_9GlcNAc_2$, which is transferred to proteins during glycoprotein biosynthesis. This precursor is then sequentially processed in the endoplasmic reticulum (ER), creating a series of oligosaccharide structures that are recognized as codes by specific lectins and that inform of the folding status of the glycoprotein. For example, the chaperones/ lectins calnexin and calreticulin bind to monoglucosylated oligosaccharides after the excision of the two terminal glucose residues by glucosidases I and II. The last glucose and sometimes one or even two mannose residues are excised, and the resulting structures $Man_{7\text{-}9}GlcNAc_2$ are recognized by the lectins ERGIC53, VIP36 and others that are involved in transport of glycoproteins to the Golgi. These lectins associate with properly folded glycoproteins that can exit the ER. In contrast, N-linked oligosaccharides on misfolded glycoproteins are more extensively trimmed to $Man_{5\text{-}6}GlcNAc_2$. A high local concentration of ER mannosidase I in an ER-derived quality control compartment is mainly responsible for this trimming, with the possible participation of other mannosidases. $Man_{5\text{-}6}GlcNAc_2$ oligosaccharides are then recognized by the lectins OS9 and XTP3-B that target the misfolded glycoprotein for ER-associated degradation.

LIST OF ABBREVIATIONS

CNX	calnexin

---

* Email: gerardo@post.tau.ac.il
Phone: 972-3-640-9239. FAX: 972-3-642-2046

CRT	calreticulin
ERAD	endoplasmic reticulum associated degradation
EDEM	ER degradation-enhancing α-mannosidase–like protein
ERGIC53	53 kDa ER to Golgi intermediate compartment protein
ERManI	ER α1,2 mannosidase I
ERQC	ER-derived quality control compartment
GI	glucosidase I
GII	glucosidase II
Golgi Man IA, B, C	Golgi mannosidases IA, IB, IC
GT	UDPGlc:glycoprotein glucosyltransferase
UPR	unfolded protein response
VIP36	36 kDa vesicular-integral membrane protein
VIPL	VIP36-like protein
OST	oligosaccharyltransferase

Note: For abbreviations of N-linked oligosaccharides see Figure 1, the generic form is GxMy = $Glc_xMan_yGlcNAc_2$.

INTRODUCTION

Most of the proteins that follow the secretory pathway are N-linked glycoproteins. These are proteins which have been modified by the addition of an oligosaccharide precursor, consisting of 2 N-acetylglucosamine, 9 mannose and 3 glucose residues in a branched structure ($Glc_3Man_9GlcNAc_2$ or G3M9, Figure 1), at an aspargine residue in the context of an N-X-S/T sequence (where X may be any amino acid except proline). The addition of the N-linked oligosaccharide is usually co-translational. The oligosaccharide precursor is assembled in two main stages, the first occurring on the cytoplasmic side of the ER membrane and the second on the luminal side of the membrane, following a flipping step. The process of constructing the $Glc_3Man_9GlcNAc_2$ oligosaccharide and its transfer to a nascent polypeptide involves many different enzymes, consuming much energy and many building blocks. During the process of protein folding and maturation, this oligosaccharide will be extensively trimmed. Thus the question arises, why such a large oligosaccharide precursor is needed for protein N-glycosylation, if it will later be extensively pruned. Knowledge has been gained on the importance of oligosaccharide structures, products of stepwise processing of this precursor in the ER. These different oligosaccharide structures form a sort of "glyco-code" on glycoproteins which can be read by different lectins, revealing the folding state of the glycoprotein and allowing for discrimination between partially folded, terminally misfolded and correctly folded glycoproteins. The different lectins, which identify the different oligosaccharide structures, deliver the glycoproteins to their intended fate: Partially folded glycoproteins will be returned to chaperones for further folding attempts, terminally misfolded glycoproteins will be targeted for ER-associated degradation (ERAD) and properly folded glycoproteins will exit to the Golgi apparatus for further processing (Hebert and Molinari 2007; Molinari 2007).

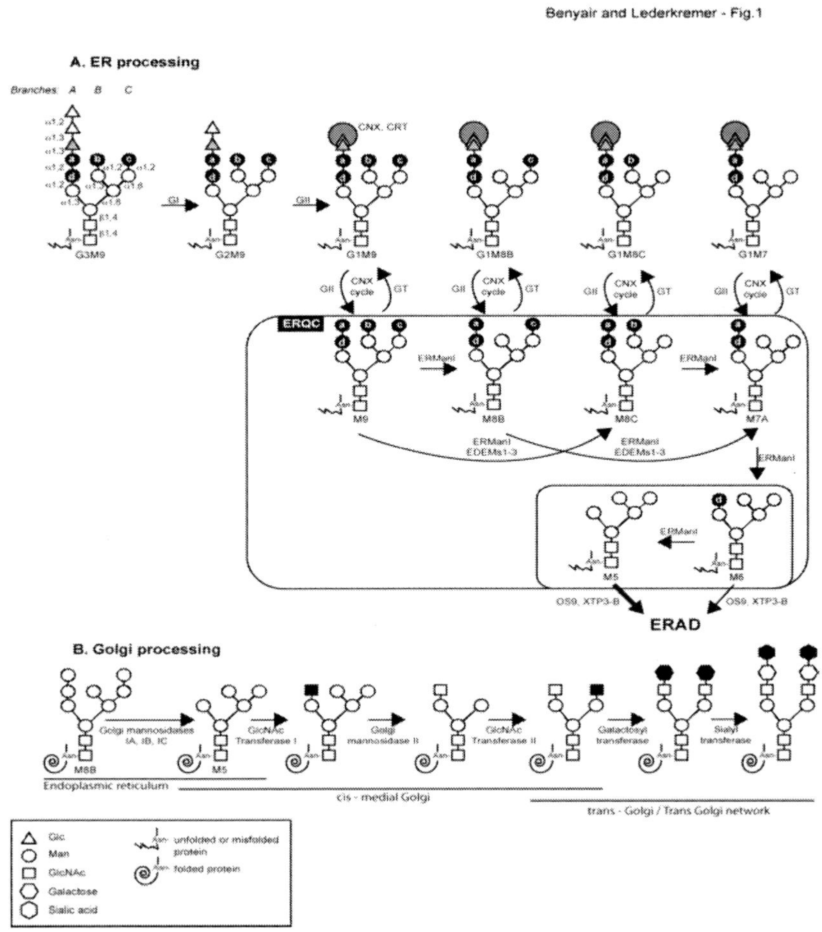

Figure1. Processing of N-linked oligosaccharides. A) Following transfer of the $Glc_3Man_9GlcNAc_2$ oligosaccharide precursor to a nascent polypeptide, on an Aspargine residue, the non-reducing terminal glucose is removed by GI to yield $Glc_2Man_9GlcNAc_2$. This oligosaccharide is in turn trimmed by GII, which removes a second glucose residue to form $Glc_1Man_9GlcNAc_2$, which is recognized by the lectin chaperone CNX or its soluble homolog CRT. In the CNX folding cycle, the remaining glucose residue is removed by GII, which precludes binding of the glycoprotein to CNX. The folding sensor GT can re-glucosylate unfolded glycoproteins, allowing renewed binding to CNX and further folding attempts. During folding attempts, glycoproteins are exposed to ERManI and EDEMs1-3 which elicit mannose residue trimming. Reglucosylation by GT can occur as long as the acceptor mannose (mannose-a) is still present. Extensive mannose trimming in the ERQC yields $Man_{6\text{-}5}GlcNAc_2$ which cannot be reglucosylated nor bind CNX. Glycoproteins bearing these oligosaccharides are recognized by OS9 and XTP3B in the ERQC and are targeted for degradation by ERAD. The oligosaccharides M8C and M7A can also be identified by OS9 and XTP3-B. However, glycoproteins bearing these oligosaccharides can still be reglucosylated, bind to CNX and fold correctly, thus preventing their sequestration by OS9 and XTP3-B. B) Properly folded glycoproteins can reach the Golgi, bearing 9-7 mannose residues. In the Golgi, the mannosidases ManIA, IB and IC trim the oligosaccharide, leaving it with 5 mannose residues. GlcNAc transferase I identifies the M5 structure and adds a GlcNAc residue on the A branch of the oligosaccharide (black square). The oligosaccharide is trimmed further by Golgi mannosidase II which removes 2 mannose residues. GlcNAc transferase II now adds another GlcNAc residue (black square) to the oligosaccharide. Two galactoses residues are added to the recently added GlcNAcs by a galactosyl transferase and 2 sialic acid residues are added upon these by a sialyl transferase. Residues added at each step are in black.

It was once thought that following sequential removal of the three terminal glucose residues, removal of one mannose from the deglucosylated precursor M9 to yield M8B (Figure 1) targets a terminally misfolded glycoprotein to ERAD. However, it is now recognized that a glycoprotein carrying M8B can still be identified as a partially folded protein and it can be reglucosylated, causing it to remain in the calnexin (CNX) folding cycle, achieve proper folding and later exit to the Golgi. Recent studies indicate that glycoprotein targeting to ERAD involves extensive mannose trimming from M9 to M6 and M5 (Figure 1) (Lederkremer and Glickman 2005; Molinari 2007; Lederkremer 2009). This could potentially pose a problem, as N-glycans of properly folded glycoproteins are also later trimmed to M5, raising the question of how the cell discriminates correctly folded M5 bearing glycoproteins from those who are terminally misfolded. The solution lies in compartmentalization of glycoproteins in different stages of maturation between the ER and the Golgi apparatus.

The oligosaccharide moieties of properly folded glycoproteins are trimmed to M5 in the Golgi, after the quality control checkpoint and are thus exposed to a different milieu of lectins than that which is present in the ER. M5 bearing glycoproteins in the Golgi may proceed to further processing and productive maturation. In contrast, recent evidence shows that terminally misfolded glycoproteins undergo extensive mannose trimming in the ER-derived quality control compartment (ERQC), which allows their identification by the lectins OS9 and XTP3-B. These lectins associate with the retrotranslocation machinery and enable delivery of misfolded glycoproteins to ERAD. Thus, the fate of an M5-bearing glycoprotein is decided solely by its location: in the Golgi, these glycoproteins will mature but in the ERQC, they will be targeted to ERAD.

CONSTRUCTION OF THE $GLC_3MAN_9GLCNAC_2$ OLIGOSACCHARIDE PRECURSOR

The oligosaccharide precursor, the structure of which is conserved in plants, fungi and animals, is gradually constructed on a lipid anchor, dolichol-pyrophosphate. At first, phospho-GlcNAc is added to dolichol-monophosphate, facing the cytoplasmic side of the ER membrane, from an activated donor, UDP-GlcNAc, resulting in a GlcNAc-dolichol-diphosphate structure and a UMP residue. A second GlcNAc is added, again using UDP-GlcNAc as a donor, but now with the release of UDP. The second GlcNAc will now receive 5 mannose residues, contributed by activated GDP-mannose found in the cytosol, in a reaction catalyzed by mannosyltransferases 1-5.

This structure, $Man_5GlcNAc_2$, and the dolichol-diphosphate lipid, undergo a flipping reaction in the ER membrane, catalyzed by a flippase, bringing the sugar residues into the ER lumen (Sanyal and Menon 2009). There, 4 more mannose residues are transferred to the oligosaccharide from dolichol-phosphomannose donors, in a reaction catalyzed by mannosyltransferases 6-9, which is followed by addition of 3 glucose residues from dolichol-phosphoglucose donors. Addition of the first two glucose residues is catalyzed by an α1,3 glucosyltransferase enzyme, while the third, terminal glucose is transferred by an α1,2 glucosyltransferase. Every step in the gradual construction of the oligosaccharide is catalyzed by a different glycosyltransferase. Following construction of this oligosaccharide and recognition of a proper N-glycosylation signal, the entire branched structure is transferred *en bloc* to an aspargine residue on a nascent polypeptide in a reaction catalyzed by oligosaccharyltransferase (OST).

Processing of Glucose Residues - Early Stages of Glycoprotein Quality Control

Immediately following transfer of the oligosaccharide $Glc_3Man_9GlcNAc_2$ to a nascent protein, the terminal α1,2 bound glucose is removed by glucosidase I (GI), producing an oligosaccharide bearing 2 glucose residues. The fast removal of the terminal glucose may serve to shift the equilibrium of the reaction catalyzed by OST in favor of the product, allowing for more efficient N-glycosylation reactions (Aebi, Bernasconi et al. 2010). This short lived di-glucosylated oligosaccharide form was once thought to be no more than a mere intermediate between G3M9 and the more stable G1M9 but recent evidence has shown that a novel ER lectin, malectin, binds specifically to this oligosaccharide form and may serve to modulate the activity of glucosidase II (GII) (Schallus, Jaeckh et al. 2008). The G2M9 oligosaccharide is identified by GII which removes the terminal α1,3 bound glucose residue to yield a monoglucosylated oligosaccharide which is now recognized by CNX.

The type 1 transmembrane protein CNX and its soluble luminal homolog, CRT, are lectin chaperones that associate with nascent glycoproteins after the trimming of the two terminal glucose residues from the precursor N-linked oligosaccharide G3M9 by GI and GII. CNX allows for the ER-retention of these unfolded glycoproteins, giving them time to achieve proper folding and preventing their aggregation (Reviewed in (Caramelo and Parodi 2008)). In order to test the folding state of a glycoprotein, it must first dissociate from CNX in order to allow access to its oligosaccharide structure. The dissociation of CNX from the glycoproteins is elicited by removal by GII of the third, α1,3 bound, mannose-linked glucose, producing a non-glucosylated $Man_9GlcNAc_2$ (M9) oligosaccharide. Recent reports show similar kinetics for removal of glucose from G1M9, G1M8 or G1M7 by glucosidase II, suggesting that mannose-trimmed oligosaccharides can still associate with CNX (Totani, Ihara et al. 2006; Bosis, Nachliel et al. 2008) (Figure 1). Following the dissociation from CNX, the glycoprotein is examined in order to ascertain its folding state. If the glycoprotein is found to be correctly folded, it will be free to exit the ER to the Golgi. However, if it is found to be still incompletely folded, the terminal glucose residue can be re-added to the oligosaccharide by the folding sensor UDPGlc:glycoprotein glucosyltransferase (GT) thereby allowing renewed CNX binding. GT contains an REEL, ER retention signal (Arnold, Fessler et al. 2000) and is thus found mainly in the ER and ER to Golgi intermediate compartment (ERGIC), along with GII (Zuber, Fan et al. 2001), where it comes in contact with glycoproteins that are still involved in folding attempts. GT recognizes both the core pentasaccharide $Man_3GlcNAc_2$ in high mannose N-glycans and exposed hydrophobic regions of target glycoproteins, allowing it to examine folding states of non-glucosylated glycoproteins (Sousa and Parodi 1995; Totani, Ihara et al. 2009). This is consistent with the wide specificity of GT as long as the mannose residue that acts as an acceptor for reglucosylation (mannose a, see Figure 1) is present. GT is required for certain slow-folding or misfolded domains in glycoproteins (Pearse, Gabriel et al. 2008) but not for normal folding of many glycoproteins, implying that they can complete their folding properly in a single

Table 1. N-linked oligosaccharides are recognized by a variety of intracellular proteins

Function	Localization	Protein		Oligo-saccharide
		S.cerevisiae	Mammalian	
Glycosyltransferase, Transfer of G3M9 to proteins	ER	OTase	OST	G3M9
Glycosidase, glucose trimming	ER	Gls1p	GI	
Lectin, unknown	ER	-	Malectin	
Glycosidase, glucose trimming, quality control	ER	Gls2p	GII	G2M9
Glycosidase, glucose trimming, quality control	ER	Gls2p	GII	
Lectin, folding, quality control	ER	CNE1	Calnexin	G1M9-7
Lectin, folding, quality control	ER	-	Calreticulin	
Lectin, folding, quality control	ER	-	Calmegin[1]	
Lectin, ER to Golgi transport (?)	ER, ER exit sites	-	Intracellular MBP	
Lectin, ER to Golgi transport	ERGIC	-	ERGIC-53	
Lectin, ubiquitination	Cytosol	-	Fbs1	
Lectin, ubiquitination	Cytosol	-	Fbs2	High mannose
Lectin, glycosidase, ERAD	ER, ERQC	Htm1p	EDEM1	
Lectin, glycosidase, ERAD	ER(?)		EDEM2	
Lectin, glycosidase, ERAD	ER(?)		EDEM3	
Lectin, ER to Golgi transport	ERGIC, cis-Golgi	-	VIP36	
Lectin, ER to Golgi transport	ER		VIPL	M9-M7
Glycosyltransferase, transfer of glucose, quality control	ER	-	GT	
Glycosidase, ERAD	ERQC	Mns1p	ERManI	
Glycosidase, maturation and ERAD(?)	ER(?), ERGIC, cis-Golgi(?)	-	Golgi ManIA	
Glycosidase, maturation and ERAD(?)	Cis-Golgi	-	Golgi ManIB	M9-M6
Glycosidase, maturation and ERAD(?)	Cis-Golgi	-	Golgi ManIC	
Glycosyltransferase, targeting to lysosomes	Cis-Golgi	-	GlcNAc1-phospho-	M8-M7

			transferase	
Lectin, ERAD	ER, ERQC	YOS9[2]	OS9	M8C-M5
Lectin, ERAD			XTP3-B	
Glycosidase, removal of N linked glycans from proteins	Cytosol	PNG1	PNGase	Any N-linked
Lectin, ER to Golgi transport	Unknown	-	ERGL[3]	Unknown

Notes: The oligosaccharide structures are recognized by different proteins, with varying functions along the quality control and maturation pathways.

1) Calmegin is a calnexin homologue which is found solely in testis.

2) The affinity of the yeast Yos9 is higher for M5.

3) The specificity and localization of ERGL are not yet clear.

Question marks are shown where function or localization are hypothesized. In yeast, "-" symbolizes no known homolog.

CNX binding event (Solda, Galli et al. 2007). Consistent with this, GT was found to be essential in lower eukaryotes only under conditions of high stress (Izquierdo, Atrih et al. 2009) and is completely lacking in lower organisms which utilize low mannose versions of the oligosaccharide precursor (Banerjee, Vishwanath et al. 2007).

The deglucosylation and reglucosylation of glycoproteins allows for intermittent cycles of folding and quality control to take place. These processes are known as the “CNX cycle” and they can continue until proper folding of the glycoprotein is achieved, which prevents further recognition by the folding sensor GT, allowing exit to the Golgi (Figure 1, Table 1). It is important to note that the CNX cycle does not exist in *S.cerevisiae* (Caramelo and Parodi 2008), which does not possess a functional homolog of GT. Besides their lectin activity, CNX and CRT can bind non-glycosylated proteins and prevent their aggregation *in vitro* (Sandhu, Duus et al. 2007; Brockmeier, Brockmeier et al. 2009; Michalak, Groenendyk et al. 2009), also shown recently *in vivo* (Ireland, Brockmeier et al. 2008). Nevertheless, most CNX and CRT binding is carbohydrate-dependent (Caramelo and Parodi 2008). A reasonable model is that the initial binding is generally carbohydrate-dependent, leading then to direct peptide binding.

ERp57 and Other Oxidoreductases

Of the several oxidoreductases and disulfide isomerases that exist in the ER, ERp57 acts specifically on glycoproteins by forming a complex with CNX or CRT (Jessop, Tavender et al. 2009). CNX and ERp57 associate with nascent glycoproteins in a complex but can dissociate independently from the bound glycoprotein. ERp57 dissociates rapidly from a misfolded glycoprotein, whereas its association with a substrate undergoing productive folding persists, the opposite being true for CNX (Frenkel, Shenkman et al. 2004). Furthermore, ERp57 does not need to interact with CNX and CRT to act as a chaperone and e.g. promote assembly of MHC class I (Zhang, Kozlov et al. 2009).

Another recently described selenocysteine-containing oxidoreductase, Sep15, is suggested to participate in glycoprotein folding in a complex with GT (Labunskyy, Hatfield et al. 2007).

MANNOSE TRIMMING – THE PATH TO ERAD LEADS FROM M9 TO M6-5

If left undisturbed, the CNX cycle could potentially continue forever for a glycoprotein that has problems in its folding. Thus, a crucial decision is when to abandon refolding attempts through the CNX cycle and target a misfolded glycoprotein to ERAD. During the time a glycoprotein is engaged in the CNX cycle, mannose residues may be trimmed from its N-linked oligosaccharide. This trimming of α1,2 linked mannose residues from the oligosaccharide of a glycoprotein is essential for the decision making process and recently the extent of trimming and the required oligosaccharide structures have become more clear (Lederkremer and Glickman 2005; Lederkremer 2009). It was initially proposed that removal of a single specific mannose residue (mannose-b in Figure 1) from M9 to generate M8B would mark a glycoprotein for ERAD. However, M8B-bearing glycoproteins can remain in the CNX cycle (Totani, Ihara et al. 2006) and can later be transported to the Golgi for further processing and maturation. Furthermore, in *S.cerevisiae*, lack of the middle mannose-b (as in M8B, Figure 1) has been shown not to be sufficient for ERAD targeting (Quan, Kamiya et al. 2008; Clerc, Hirsch et al. 2009). The same is true in mammalian cells, as in mutants that transfer a particular oligosaccharide isomer to nascent proteins (M5-2, also devoid of mannose-b), removal of mannose-a residue is still required for ERAD (Ermonval, Kitzmuller et al. 2001). Thus, it is known that removal of a single mannose residue is insufficient for ERAD targeting but the question remained, how mannose trimming targets glycoproteins to ERAD. In wild type mammalian cells it was found that the oligosaccharides of ERAD substrates are processed from M9 to M6 and M5, which means that most or all α1,2 linked mannose residues are removed from the 9-mannose oligosaccharide precursor and that this extensive trimming is essential for targeting misfolded glycoproteins to ERAD (Frenkel, Gregory et al. 2003; Avezov, Frenkel et al. 2008). This extensive trimming does not take place on the bulk of the glycoproteins traversing the secretory pathway as most are correctly folded (Avezov, Ron et al. 2010). The same extensive mannose trimming which is required for ERAD in mammalian cells may also exist in lower eukaryotes. During ERAD in *S. cerevisiae*, trimming to an M7 isomer lacking mannose-c (Figure 1) and only a small amount of M5 were observed, raising the possibility that processing of M9 to M7 was sufficient for ERAD targeting (Clerc, Hirsch et al. 2009). This was observed however, without inhibition of the proteasomes, which are the final destination for degradation in ERAD. Under these conditions, most molecules bearing M5 could have been rapidly degraded in the yeast cells, as occurs in mammalian cells (Frenkel, Gregory et al. 2003). In mutants that do not allow exposure of a non-reducing terminal α1,6-linked mannose, ERAD is blocked (Clerc, Hirsch et al. 2009), suggesting that this is an essential determinant, but perhaps also trimming of mannose residues a and d (Figure 1) is necessary in yeast. Consistent with a possible extensive trimming of most or all α1,2 mannose residues in *S.cerevisiae* as well as in mammalian cells, M5 binds with the highest affinity to the lectin Yos9, whereas M8B binding is negligible (Figure 2) (Quan, Kamiya et al. 2008).

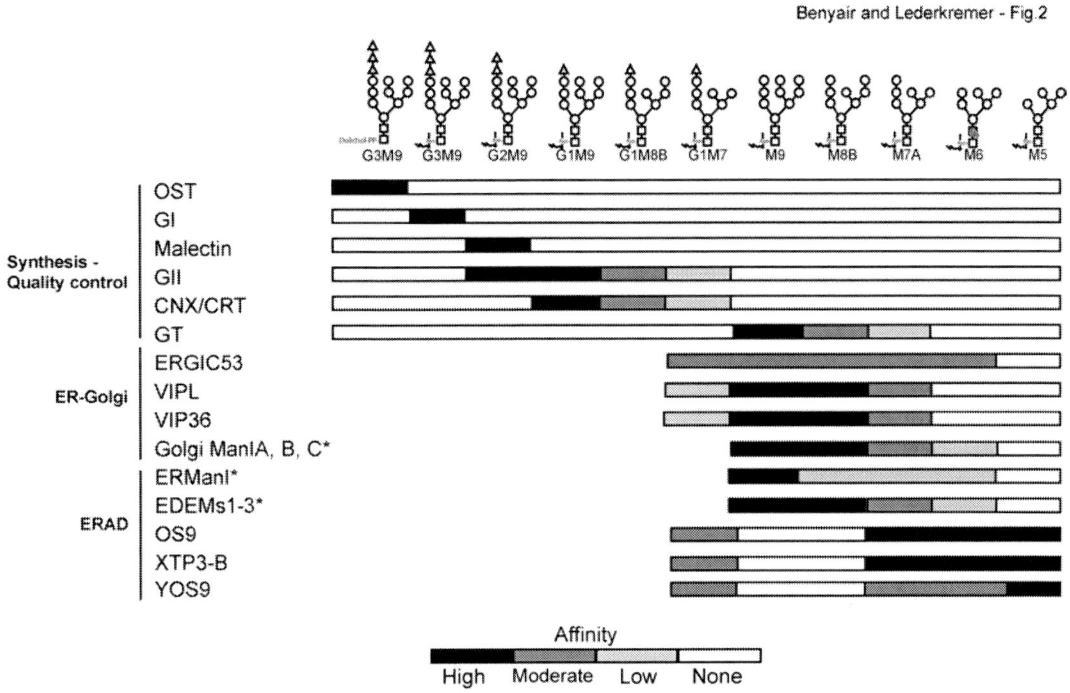

Figure 2. Binding affinities of lectins to the different N-linked oligosaccharide structures. Proteins which are involved in transfer of the oligosaccharide precursor to nascent glycoproteins, quality control, ER to Golgi transport and ERAD exhibit different affinities to specific N-linked glycan structures (Lederkremer 2009; Aebi, Bernasconi et al. 2010). It is interesting to note the complementary pattern of affinities between proteins involved in quality control, ER to Golgi transport and ERAD. This complementation allows specific identification of glycoproteins bound for different fates. The affinity of VIP36, VIPL and ERGIC53 is greater for M9-M7 and is non-existent for M5, allowing these lectins to bind properly folded glycoproteins in the ER for transport, while disregarding misfolded, M5-bearing proteins. This same specificity allows the release of properly folded glycoproteins in the Golgi, following trimming of their oligosaccharides to M5. *Affinities of proteins with asterisks are postulated.

Yeast Yos9 and its mammalian homologs OS9.1, OS9.2 and two forms of XTP3-B are ER lectins that interact with the membrane-bound ubiquitination machinery and are postulated ERAD substrate receptors which act before the retrotranslocation and ubiquitination of a glycoprotein targeted for ERAD, a role once proposed for EDEM1 (reviewed in (Kanehara, Kawaguchi et al. 2007)). Yos9 and its mammalian homologs are upregulated by the unfolded protein response (UPR) and are able to bind trimmed sugar chains as discussed above (Quan, Kamiya et al. 2008; Hosokawa, Kamiya et al. 2009; Yamaguchi, Hu et al. 2010) (Figure 2). Their high glycan specificity suggests their role in a decision-making checkpoint, where glycoproteins are dislocated to the cytosol only if most of their α1,2 mannose residues have been excised. Nevertheless, they can also bind unglycosylated proteins (Kanehara, Kawaguchi et al. 2007; Bernasconi, Pertel et al. 2008; Alcock and Swanton 2009). OS-9 and XTP3-B associate with the HRD1-SEL1L membrane-anchored ubiquitin ligase complex and with BiP as well as with GRP94, these two chaperones being capable of recruiting non-glycoprotein substrates (Christianson, Shaler et al. 2008; Hosokawa, Wada et al. 2008).

In another lower eukaryote, *Trichomonas*, the oligosaccharide transferred to glycoproteins is the M5-2 isomer mentioned above, lacking the B and C branches. Mannose residues are removed from this isomer to yield M3 (Banerjee, Vishwanath et al. 2007), similar to what was found in mutant mammalian cells (Ermonval, Kitzmuller et al. 2001), suggesting requirement for removal of mannose residues a and d (Figure 1).

Although there might be differences among species, altogether the recent studies suggest that the removal of all or most α1,2 mannose residues (a-c and perhaps also d) to yield M6 and M5 could be universal in ERAD. Removal of mannose-a excludes the glycoprotein irreversibly from the CNX cycle by eliminating the glucose acceptor, thus precluding CNX binding. Consistently, while M5 binds with high affinity to OS9 and the highest to yeast Yos9, it has no affinity for the lectins VIPL, VIP36 and ERGIC53, involved in ER-Golgi transport (Table 1 and Figure 2) thus providing the cell with an efficient manner in which to transport properly folded glycoproteins to the Golgi while targeting terminally misfolded glycoproteins for retrotranslocation and ERAD (Quan, Kamiya et al. 2008; Hosokawa, Kamiya et al. 2009; Yamaguchi, Hu et al. 2010).

The same M5 structure which is found in the ER is produced in the Golgi, but in the latter case, mannose trimming to M5 would release the glycoprotein from the lectins involved in ER-Golgi transport and allow further maturation through the secretory pathway (Figure 1). In contrast, for misfolded ER-retained glycoproteins, processing to M5 would prevent binding to the ER-Golgi transport lectins and would allow recognition by OS9 and XTP3-B and delivery to ERAD. Although mammalian OS9 can also bind M8C and M7A (both lacking the terminal mannose in branch C) *in vitro* (Figure 2) (Hosokawa, Kamiya et al. 2009), it is uncertain that these species would reach OS9 *in vivo*, as glycoproteins carrying them will be engaged in the CNX cycle, and mannose-a must probably be removed from them for efficient ERAD, similar to the case of M5-2 discussed above.

ER Mannosidase I and Mannose Trimming

An important question is which enzymes are in charge of the crucial decision-making process that targets a terminally misfolded glycoprotein to ERAD. Recent evidence points to an essential role of ER mannosidase I (ERManI) as the main α1,2 mannosidase involved in targeting glycoproteins to ERAD (Avezov, Frenkel et al. 2008). However, besides ERManI there is an apparent redundancy of α1,2 mannosidases and mannosidase-like proteins in mammalian cells, namely the EDEMs 1-3 (described later) and the Golgi α1,2 mannosidases, Golgi Man IA, IB and IC, all members of the glycosyl-hydrolase 47 family(Hebert and Molinari 2007; Molinari 2007). Nevertheless, ERManI may be able to act by itself in trimming M9 to M6-5. Although at low concentrations ERManI removes exclusively mannose-b, at high concentrations it can trim all α1,2 mannoses *in vitro* (Herscovics, Romero et al. 2002) or by overexpression *in vivo* (Hosokawa, Tremblay et al. 2003), proving that if concentrated, ERManI could act alone in ERAD targeting. In mammalian cells, ERManI has been shown to be concentrated in the ERQC (Figure 3 and (Avezov, Frenkel et al. 2008)), a compartment that will be described later, creating a high local concentration capable of removing all four α1,2 mannoses. Knockdown of ERManI blocks the production of M6-5 and inhibits ERAD, even under conditions where cleavage to M8B is not much affected

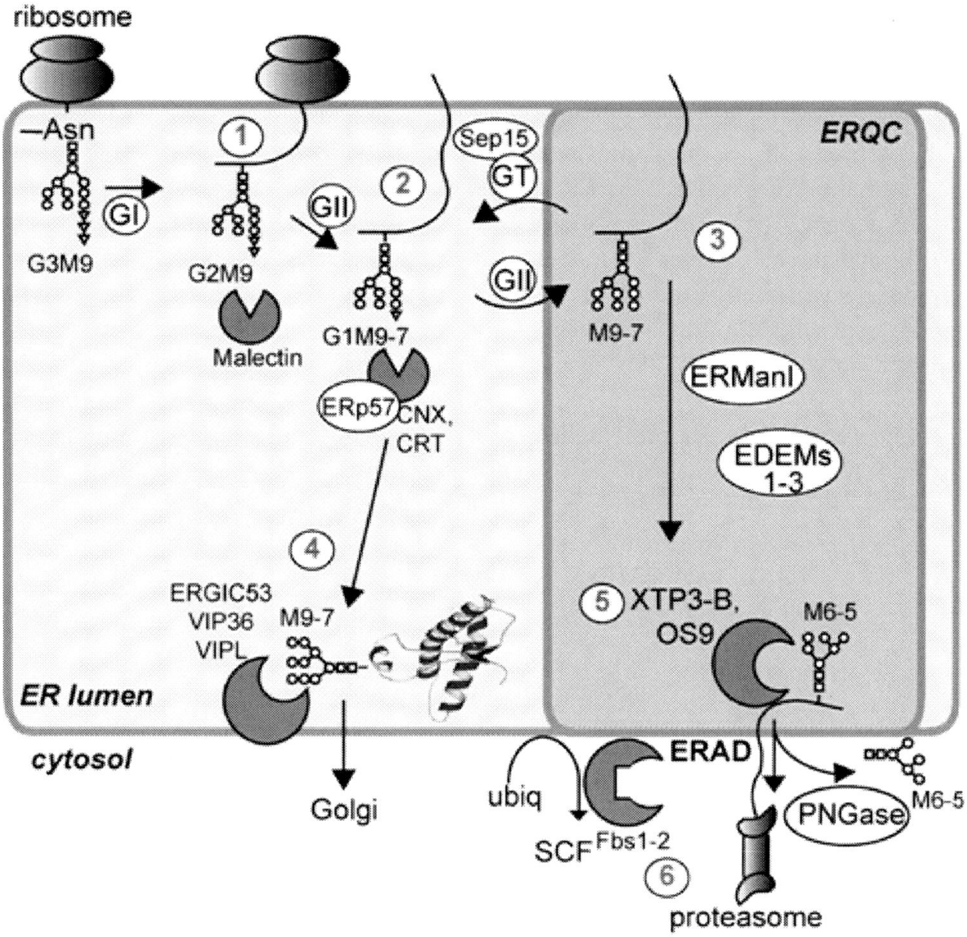

Figure 3. Glycoprotein quality control. 1) The N-linked oligosaccharide precursor Glc3Man9GlcNAc2 is transferred to a nascent polypeptide and is immediately trimmed by GI, yielding Glc2Man9GlcNAc2 which may bind to malectin. 2) Trimming of an additional glucose residue by GII yields a monoglucosylated oligosaccharide, allowing the lectin chaperones CNX or CRT to bind the glycoprotein. The bound glycoprotein cycles between the peripheral ER and the ERQC. GII can trim the remaining glucose residue, precluding the glycoprotein from CNX binding and allowing for a quality control examination by GT. GT reglucosylates incompletely folded glycoproteins, allowing them to once again bind CNX and re-attempt folding. CNX recruits the oxidoreductase ERp57, while GT recruits Sep15. The deglucosylation/ reglucosylation events, coupled with CNX binding are collectively known as the CNX cycle. 3) In the ERQC, glycoproteins are exposed to mannosidases and lectins. ERManI, with the possible assistance of EDEMs1-3, trims mannose residues from the incompletely folded glycoprotein. Mannose trimmed glycoproteins may be reglucosylated by GT only while the acceptor mannose is still present. 4) Properly folded glycoproteins are released from the CNX cycle and can be recognized by the lectins ERGIC53, VIP36 and VIPL which participate in the ER to Golgi transport. 5) Terminally misfolded glycoproteins undergo extensive mannose trimming to M5-6, preventing reglucosylation and allowing identification by the lectins OS9 and XTP3-B which transport the misfolded glycoprotein to a yet unidentified retrotranslocation complex. During or after retrotranslocation to the cytosol, the glycoprotein is deglycosylated by PNGAse, ubiquitinated by various ubiquitin ligases, among them SCF with the lectin components Fbs1 or 2, and degraded by the proteasome.

(Avezov, Frenkel et al. 2008). Overexpression of Golgi Man IA, IB or IC accelerated ERAD of mutant α1-antitrypsin, a glycoprotein that may exit to the Golgi and be retrieved (Hosokawa, You et al. 2007). However, the ER level of these enzymes may also be increased by their overexpression, leading to glycoprotein trimming and targeting to ERAD even if this is not the normal role of the Golgi mannosidases. Nevertheless, it is possible that these enzymes normally cycle between the ER and the Golgi or that, as in yeast (Haynes, Caldwell et al. 2002), certain ERAD substrates might cycle through the Golgi before delivery to ERAD also in mammalian cells. Knockdown of the Golgi mannosidases IB and IC had no effect on the degradation dynamics of the ERAD substrate asialoglycoprotein receptor (ASGPR) H2a. The knockdown of Golgi mannosidase IA however, had a hindering effect on ERAD, suggesting a possible role for this enzyme in pre-Golgi quality control (our unpublished results).

The ER mannosidase-like proteins EDEMs 1-3 also participate in the trimming process but it is not yet clear whether as mannosidases or as cofactors as discussed below.

Mannose Trimming in the ER-Derived Quality Control Compartment

The ERQC is a membrane-enclosed, microtubule-dependent subcompartment of the ER, which allows for a high local concentration of ERManI. This high local concentration allows ERManI to extensively trim mannose residues from terminally misfolded glycoproteins, targeting them to ERAD. The ERQC also acts to segregate misfolded proteins that could interfere with ER functions in a membrane-enclosed compartment, distinct from the bulk of the ER. Upon proteasomal inhibition or UPR induction, ERAD substrates, along with CNX, CRT and UPR components accumulate in the ERQC but not BiP, PDI, GT or ERp57 (Kamhi-Nesher, Shenkman et al. 2001; Frenkel, Shenkman et al. 2004; Kondratyev, Avezov et al. 2007). The ERAD components, needed for targeting and retrotranslocation of a misfolded glycoprotein concentrate at the ERQC, which would be a staging ground for ERAD (Kondratyev, Avezov et al. 2007) (Figure 3). A transmembrane protein, Bap31 associates with CNX and cycles between the peripheral ER and the ERQC (Wakana, Takai et al. 2008) aiding in the delivery of substrates to this compartment (Wang, Heath-Engel et al. 2008). Non-aggregated ERAD substrates are delivered to the ERQC in order to undergo degradation (Kamhi-Nesher, Shenkman et al. 2001), whereas aggregated glycoproteins like mutant α1-antitrypsin Z are segregated into CNX-depleted inclusion bodies (Granell, Baldini et al. 2008) and delivered to autophagy (Perlmutter 2009). Glycoproteins that fold properly and pass the quality control checkpoint are delivered to ER exit sites where they encounter ERGIC53 and other lectins involved in trafficking to the Golgi.

A Role for the Edems in Glycoprotein Quality Control

EDEM1 was initially thought to be the receptor lectin for glycoproteins which have undergone extensive mannose trimming, but recent data suggest that it participates in the trimming process itself and that it interacts with its substrate even before the trimming event has occurred, suggesting that EDEM1 has either direct, or indirect mannosidase activity

((Olivari and Molinari 2007; Hosokawa, Tremblay et al. 2010) and our unpublished data). The EDEMs (1-3 in mammalian cells and Htm1 in yeast) may act as cofactors, aiding the activity of ERManI, or may be true mannosidases as was suggested recently (Olivari and Molinari 2007; Quan, Kamiya et al. 2008; Clerc, Hirsch et al. 2009; Hosokawa, Tremblay et al. 2010). Evidence pointing to the activity of EDEM1 as a cofactor has accumulated and the mannosidase activity of the EDEM proteins has not been detected yet *in vitro*, except in the case of an EDEM-like protein of *T.cruzi* (Banerjee, Vishwanath et al. 2007). In addition, EDEM1 homologs in the yeasts *S.cerevisiae* and *S.pombe* do not trim mannose residues in mutants lacking the ERManI homolog (Mns1) (Movsichoff, Castro et al. 2005; Clerc, Hirsch et al. 2009), which could either mean that initial trimming by ERManI is required for EDEM1 to perform its activity, or that EDEM1 is a cofactor of ERManI and not a mannosidase *per se*. The debate regarding the true activity of EDEM1 is a complicated one, as new evidence to back both theories emerges. On one hand, the complete mannosidase-like domain of EDEM1 cannot even partially replace the whole protein for mannose trimming in yeast (Clerc, Hirsch et al. 2009) while on the other hand, overexpression of a point mutant that by homology with ERManI should affect the mannosidase activity of EDEM1, could not accelerate mannose trimming of ERAD substrates (Olivari and Molinari 2007; Clerc, Hirsch et al. 2009). However, it could be that the effect of the point mutation on EDEM1 slows dissociation of this mutant from its substrate, thus interfering with ERManI activity. Whether acting as an independent mannosidase or as a cofactor of ERManI, it was shown in yeast that EDEM1 is required for trimming of mannose-c (Figure 1) (Quan, Kamiya et al. 2008; Clerc, Hirsch et al. 2009). An ERAD substrate which does not contain mannose-c, carrying M5-2 or M7A is still degraded in yeast cells lacking a functional EDEM1 (Quan, Kamiya et al. 2008; Clerc, Hirsch et al. 2009). This suggests that although trimming of mannose residues a and d may still be required, as discussed above, this trimming can be achieved without EDEM1 intervention.

EDEM1 is usually kept at a low level by an autophagy-like degradation process (Cali, Galli et al. 2008). ERManI is also subjected to lysosomal degradation, perhaps through a similar mechanism. It was suggested that high levels of EDEM1, which is upregulated during UPR, may inhibit the degradation of ERManI, which is unaffected by the UPR (Avezov, Frenkel et al. 2008), by occupying the shared degradation machinery, thus indirectly accelerating ERAD (Termine, Moremen et al. 2009).

At its normal low levels, EDEM1 is located in LC-3 coated vesicles, distinct from the ER (Zuber, Cormier et al. 2007), but at high levels it is compartmentalized (Cali, Galli et al. 2008) in the ERQC (our unpublished results). This finding raises the possibility that substrate delivery by EDEM1 to ERManI would be accelerated upon UPR and EDEM1 overexpression.

Interestingly, EDEM1 forms a complex with ERdj5, a disulfide reductase required for ERAD (Ushioda, Hoseki et al. 2008), possibly replaced by PDI in yeast (Sakoh-Nakatogawa, Nishikawa et al. 2009). The disulfide reductase activity might participate in the unfolding process in the targeting of a glycoprotein to retrotranslocation and ERAD.

Retrotranslocation of Glycoproteins Targeted to ERAD

In order for glycoproteins to undergo ERAD, they must first be retrotranslocated from the ER to the cytosol before they can be processed by the ubiquitin-proteasome pathway. The mechanism of retrotranslocation of the misfolded glycoproteins to the cytosolic proteasomes is still unknown, despite the existence of several candidate polytopic membrane proteins, such as subunits of the Sec61 forward translocation machinery, the Derlins 1-3 and a growing number of transmembrane ubiquitin ligases, the best studied being HRD1 (reviewed in (Hirsch, Gauss et al. 2009)). Following retrotranslocation to the cytosol and prior to proteasomal degradation, the remnant oligosaccharide found on ERAD-bound glycoproteins is removed (Zhao, Zhou et al. 2007). To this end, the cytosolic AAA ATPase P97 (cdc48 in yeast) (Ballar and Fang 2008) associates with peptide:N-glycanase (PNGase) for glycoprotein deglycosylation (Figure 3). Deglycosylation takes place in some cases prior to complete dislocation (Baker and Tortorella 2007), but it is not absolutely required that it take place before glycoprotein degradation (Kario, Tirosh et al. 2008). The degradation machinery, including proteasomes and several other chaperones and cytosolic ubiquitin ligases are recruited to the ER membrane for ERAD (reviewed in (Brodsky 2007)). Surprisingly, one of these ligase families contains lectin components, Fbs1 and Fbs2, which recognize glycans on retrotranslocated glycoproteins (Yoshida 2007). Fbs2 is ubiquitous and can bind the glycan while HRD1 binds the polypeptide chain on the same glycoprotein substrate (Groisman, Avezov et al. 2006). The similar Fbs1 is neuron-specific and binds to the M3 oligosaccharide core (Mizushima, Yoshida et al. 2007). Fbs1 and 2 have a broad N-glycan specificity (Yoshida 2007) (Table 1), suggesting a role in disposal of any glycoprotein that has reached the cytoplasm, in contrast to the high specificity and sorting capability of OS9 and XTP3-B (Quan, Kamiya et al. 2008; Hosokawa, Kamiya et al. 2009; Yamaguchi, Hu et al. 2010), suggesting that the sorting of glycoproteins for ERAD takes place only in the ER, before retrotranslocation.

N-Linked Oligosaccharide Recognition and Processing in Correctly Folded Glycoproteins

Properly folded glycoproteins in the ER may carry a variety of oligosaccharide structures, depending on the speed with which they achieved proper folding. These different oligosaccharide structures vary in the amount of mannose residues they contain, a result of differential exposure to mannosidases in the ER. ERGIC53 is a lectin that participates in glycoprotein transport from the ER to the Golgi, although few of its specific substrates have so far been identified (Nyfeler, Reiterer et al. 2008). ERGIC53 is upregulated upon UPR (Renna, Caporaso et al. 2007), which suggests another mechanism to evacuate the ER of accumulated proteins, by accelerating their exit to the Golgi. ERGIC53 has a broad specificity, binding all large high-mannose chains including glucosylated ones, but not the extensively trimmed M5, which is found on misfolded glycoproteins in the ER (Figure 2) (Kamiya, Kamiya et al. 2008). Two other lectins, VIPL and VIP36 also appear to participate in glycoprotein transport to the Golgi, but their specificity is more restricted, binding with higher affinity M9, M8B, M8C and with much lower affinity glucosylated or more trimmed sugar chains and, similarly to ERGIC53, they do not binding M5 (Figure 2) (Kamiya, Kamiya et al. 2008). The specificity of these lectins, complementary to that of yos9, suggests that they avoid glycoproteins that have been extensively trimmed by the quality control machinery and only act to elicit transport of correctly folded glycoproteins. Recently, the exact role of VIP36 has come into question, as evidence has emerged linking the lectin to retrograde transport of misfolded glycoproteins that had somehow evaded ER quality control. These new findings raise a possible role for VIP36 in post-ER quality control, where the lectin could be responsible for retrograde transport to the ER from the Golgi (Reiterer, Nyfeler et al. 2010). Another novel intracellular lectin that might also be involved in ER-Golgi traffic is intracellular mannan-binding protein, which has been shown to bind high mannose glycoproteins (Nonaka, Ma et al. 2007).

Properly folded glycoproteins arrive at the cis-Golgi in COPII-coated vesicles, bearing oligosaccharides with varying amounts of mannose residues. Once in the cis-Golgi, all oligosaccharide species are cleaved to M5 by the Golgi α1,2 mannosidases. These M5 bearing glycoproteins will now arrive in the medial-Golgi, but it is yet unclear whether this happens by cisternal maturation or by vesicular transport. Either way, in the medial-Golgi, M5 bearing glycoproteins are recognized by GlcNAc transferase I, which catalyzes the addition of a GlcNAc residue from UDP-GlcNAc to the C branch of the oligosaccharide. Subsequently, Golgi mannosidase II recognizes the $GlcNAc_1Man_5GlcNAc_2$ oligosaccharide and catalyzes the removal of 2 α1,3 and α1,6 mannose residues, resulting in a glycoprotein bearing $GlcNAc_1Man_3GlcNAc_2$ (Figure 1). An additional GlcNAc residue can now be added by GlcNAc transferase II, yielding $GlcNAc_2Man_3GlcNAc_2$. Glycoproteins bearing this oligosaccharide move to the trans-Golgi where galactosyl transferase catalyzes the addition of a galactose on each of the GlcNAc residues previously added in the medial-Golgi. This reaction is followed by that of sialyl-transferases, which attach terminal sialic acid residues to the previously added galactoses. Glycoproteins that are bound for lysosomal localization undergo a different pathway in which they keep 5 mannose residues and receive phospho-GlcNAc residues on carbon 6 of mannose. These GlcNAc residues are then removed, exposing terminal phosphates on mannose residues. These forms of oligosaccharides are known as mannose-6-phosphate (M6P) bearing oligosaccharides and are recognized in the trans-Golgi by M6P receptors, which enable transport of M6P bearing glycoproteins to the lysosomes.

Partial activity, or lack of activity in enzymes along this pathway, such as mannosidase II, allow for a variety of end products that differ from the basic product mentioned above and microheterogeneity may exist between molecules of the same glycoprotein with different structures attached to its N-glycosylation sites. Another variation, which is common in vertebrates, is the addition of a fucose residue by a fucosyl transferase to the first GlcNAc that is N-linked to aspargine.

After passing through the Golgi, mature glycoproteins are secreted to their final destinations – either as soluble proteins which will be secreted from the cell, transmembrane proteins which will be transported to become part of the plasma membrane or lysosomal proteins, in the case of M6P tagged oligosaccharide bearing glycoproteins.

Conclusion

More clarity has been recently gained regarding the mechanism of glycoprotein targeting to productive folding and exit to the Golgi as opposed to ERAD by use of compartmentalization, differential N-glycan processing and selective lectin interactions

(Figure 2-3). Different lectins with different roles identify the oligosaccharide structures that are achieved by processing of the N-linked oligosaccharide precursor. At early stages, lectins bind to G2M9 (malectin) and G1M9-7 (CNX and CRT). After deglucosylation by GII, properly folded glycoproteins do not undergo reglucosylation by GT, and associate with the lectins VIPL, VIP36 or ERGIC53, which assist in glycoprotein delivery to the Golgi. There, mannose trimming to M5 by Golgi mannosidases IA, IB or IC would prevent reassociation of glycoproteins to these lectins allowing them to be released for further processing and delivery to their final destination. On the other hand, improper folding causes delivery of CNX-bound glycoproteins to the ERQC, association with EDEM1-3 and trimming to M6 and M5 by ERManI with the help of the EDEMs and possibly of recycled Golgi α1,2 mannosidases. M6 and M5 are recognized by OS9 and XTP3-B variants, which permit glycoprotein delivery to the retrotranslocation machinery. Cytosolic recognition of the sugar chains by Fbs1-2 helps in the ubiquitination process and delivery to the proteasomes for degradation. Basic features of the mechanism seem to be conserved also in *S.cerevisiae*, with fewer intervening proteins and an absence of the CNX cycle, but apparently with a similar extensive processing of ERAD substrate sugar chains. Although the outline of the overall process is being clarified, there are still several important unresolved issues: 1) The precise interactions between misfolded glycoproteins and the various mannosidases have yet to be completely clarified. It is not clear whether ERManI can, *in vivo* create the ERAD signal alone or whether other proteins are needed. The nature of the EDEMs must be clarified in order to understand their function, either as mannosidases in their own right or as cofactors for ERManI activity, and the intervention of Golgi mannosidases in ER and post ER quality control needs to be elucidated. 2) The roles and specific glyco-affinities of newly discovered lectins that are involved in glycoprotein quality control and proper maturation should be further studied. Understanding these roles and specificities could shed new light on the transport of glycoproteins to and from subcellular compartments. 3) The mechanism by which glycoproteins, bound for ERAD, are retrotranslocated from the ERQC to the cytosol still poses a major challenge. Although components involved in retrotranslocation have been identified and characterized, a retrotranslocation channel has yet to be discovered.

Acknowledgments

Research related to this work is supported by grants from the Israel Science Foundation (1229/07) and German-Israeli Project Cooperation (DIP).

References

Aebi, M., R. Bernasconi, et al. (2010). "N-glycan structures: recognition and processing in the ER." *Trends Biochem Sci.* 35(2): 74-82.

Alcock, F. and E. Swanton (2009). "Mammalian OS-9 is upregulated in response to endoplasmic reticulum stress and facilitates ubiquitination of misfolded glycoproteins." *J Mol Biol.* 385(4): 1032-1042.

Arnold, S. M., L. I. Fessler, et al. (2000). "Two homologues encoding human UDP-glucose:glycoprotein glucosyltransferase differ in mRNA expression and enzymatic activity." *Biochemistry,* 39(9): 2149-2163.

Avezov, E., Z. Frenkel, et al. (2008). "Endoplasmic reticulum (ER) mannosidase I is compartmentalized and required for N-glycan trimming to Man(5-6)GlcNAc(2) in glycoprotein ER-associated degradation." *Mol Biol Cell,* 19(1): 216-225.

Avezov, E., E. Ron, et al. (2010). "Pulse-chase analysis of N-linked sugar chains from glycoproteins in mammalian cells." *J Vis Exp.* (38): 1899.

Baker, B. M. and D. Tortorella (2007). "Dislocation of an endoplasmic reticulum membrane glycoprotein involves the formation of partially dislocated ubiquitinated polypeptides." *J Biol Chem.* 282(37): 26845-26856.

Ballar, P. and S. Fang (2008). "Regulation of ER-associated degradation via p97/VCP-interacting motif." *Biochem Soc Trans.* 36(Pt 5): 818-822.

Banerjee, S., P. Vishwanath, et al. (2007). "The evolution of N-glycan-dependent endoplasmic reticulum quality control factors for glycoprotein folding and degradation." *Proc Natl Acad Sci U S A,* 104(28): 11676-11681.

Bernasconi, R., T. Pertel, et al. (2008). "A dual task for the Xbp1-responsive OS-9 variants in the mammalian endoplasmic reticulum: inhibiting secretion of misfolded protein conformers and enhancing their disposal." *J Biol Chem.* 283(24): 16446-16454.

Bosis, E., E. Nachliel, et al. (2008). "Endoplasmic reticulum glucosidase II is inhibited by its end products." *Biochemistry,* 47(41): 10970-10980.

Brockmeier, A., U. Brockmeier, et al. (2009). "Distinct contributions of the lectin and arm domains of calnexin to its molecular chaperone function." *J Biol Chem.* 284(6): 3433-3444.

Brodsky, J. L. (2007). "The protective and destructive roles played by molecular chaperones during ERAD (endoplasmic-reticulum-associated degradation)." *Biochem J.* 404(3): 353-363.

Cali, T., C. Galli, et al. (2008). "Segregation and rapid turnover of EDEM1 by an autophagy-like mechanism modulates standard ERAD and folding activities." *Biochem Biophys Res Commun.* 371(3): 405-410.

Caramelo, J. J. and A. J. Parodi (2008). "Getting in and out from calnexin/calreticulin cycles." *J Biol Chem.* 283(16): 10221-10225.

Christianson, J. C., T. A. Shaler, et al. (2008). "OS-9 and GRP94 deliver mutant alpha1-antitrypsin to the Hrd1-SEL1L ubiquitin ligase complex for ERAD." *Nat Cell Biol.* 10(3): 272-282.

Clerc, S., C. Hirsch, et al. (2009). "Htm1 protein generates the N-glycan signal for glycoprotein degradation in the endoplasmic reticulum." *J Cell Biol.* 184(1): 159-172.

Ermonval, M., C. Kitzmuller, et al. (2001). "N-glycan structure of a short-lived variant of ribophorin I expressed in the MadIA214 glycosylation-defective cell line reveals the role of a mannosidase that is not ER mannosidase I in the process of glycoprotein degradation." *Glycobiology,* 11(7): 565-576.

Frenkel, Z., W. Gregory, et al. (2003). "Endoplasmic reticulum-associated degradation of mammalian glycoproteins involves sugar chain trimming to Man6-5GlcNAc2." *J Biol Chem.* 278(36): 34119-34124.

Frenkel, Z., M. Shenkman, et al. (2004). "Separate roles and different routing of calnexin and ERp57 in endoplasmic reticulum quality control revealed by interactions with asialoglycoprotein receptor chains." *Mol Biol Cell,* 15(5): 2133-2142. Epub 2004 Feb 2120.

Granell, S., G. Baldini, et al. (2008). "Sequestration of mutated alpha1-antitrypsin into inclusion bodies is a cell-protective mechanism to maintain endoplasmic reticulum function." *Mol Biol Cell,* 19(2): 572-586.

Groisman, B., E. Avezov, et al. (2006). "The E3 Ubiquitin Ligases HRD1 and SCF-Fbs2 Recognize the Protein Moiety and Sugar Chains, Respectively, of an ER-Associated Degradation Substrate." *Israel J Chem.* 46(2): 189-196.

Haynes, C. M., S. Caldwell, et al. (2002). "An HRD/DER-independent ER quality control mechanism involves Rsp5p-dependent ubiquitination and ER-Golgi transport." *J. Cell Biol.* 158(1): 91-102.

Hebert, D. N. and M. Molinari (2007). "In and Out of the ER: Protein Folding, Quality Control, Degradation, and Related Human Diseases." *Physiol. Rev.* 87(4): 1377-1408.

Herscovics, A., P. A. Romero, et al. (2002). "The specificity of the yeast and human class I ER alpha 1,2-mannosidases involved in ER quality control is not as strict previously reported." *Glycobiology,* 12(4): 14G-15G.

Hirsch, C., R. Gauss, et al. (2009). "The ubiquitylation machinery of the endoplasmic reticulum." *Nature,* 458(7237): 453-460.

Hosokawa, N., Y. Kamiya, et al. (2009). "Human OS-9, a lectin required for glycoprotein endoplasmic reticulum-associated degradation, recognizes mannose-trimmed N-glycans." *J Biol Chem.* 284(25): 17061-17068.

Hosokawa, N., L. O. Tremblay, et al. (2010). "EDEM1 accelerates the trimming of alpha1,2-linked mannose on the C branch of N-glycans." *Glycobiology,* 20(5): 567-575.

Hosokawa, N., L. O. Tremblay, et al. (2003). "Enhancement of endoplasmic reticulum (ER) degradation of misfolded Null Hong Kong alpha1-antitrypsin by human ER mannosidase I." *J Biol Chem.* 278(28): 26287-26294.

Hosokawa, N., I. Wada, et al. (2008). "Human XTP3-B forms an endoplasmic reticulum quality control scaffold with the HRD1-SEL1L ubiquitin ligase complex and BiP." *J Biol Chem.* 283(30): 20914-20924.

Hosokawa, N., Z. You, et al. (2007). "Stimulation of ERAD of misfolded null Hong Kong alpha1-antitrypsin by Golgi alpha1,2-mannosidases." *Biochem Biophys Res Commun.* 362(3): 626-632.

Ireland, B. S., U. Brockmeier, et al. (2008). "Lectin-deficient calreticulin retains full functionality as a chaperone for class I histocompatibility molecules." *Mol Biol Cell,* 19(6): 2413-2423.

Izquierdo, L., A. Atrih, et al. (2009). "Trypanosoma brucei UDP-glucose:glycoprotein glucosyltransferase has unusual substrate specificity and protects the parasite from stress." *Eukaryot Cell,* 8(2): 230-240.

Jessop, C. E., T. J. Tavender, et al. (2009). "Substrate specificity of the oxidoreductase ERp57 is determined primarily by its interaction with calnexin and calreticulin." *J Biol Chem.* 284(4): 2194-2202.

Kamhi-Nesher, S., M. Shenkman, et al. (2001). "A novel quality control compartment derived from the endoplasmic reticulum." *Mol Biol Cell,* 12(6): 1711-1723.

Kamiya, Y., D. Kamiya, et al. (2008). "Molecular basis of sugar recognition by the human L-type lectins ERGIC-53, VIPL, and VIP36." *The Journal of biological chemistry,* 283(4): 1857-1861.

Kanehara, K., S. Kawaguchi, et al. (2007). "The EDEM and Yos9p families of lectin-like ERAD factors." *Semin Cell Dev Biol.* 18(6): 743-750.

Kario, E., B. Tirosh, et al. (2008). "N-linked glycosylation does not impair proteasomal degradation but affects class I major histocompatibility complex presentation." *The Journal of biological chemistry,* 283(1): 244-254.

Kondratyev, M., E. Avezov, et al. (2007). "PERK-dependent compartmentalization of ERAD and unfolded protein response machineries during ER stress." *Exp Cell Res.* 313(16): 3395-3407.

Labunskyy, V. M., D. L. Hatfield, et al. (2007). "The Sep15 protein family: roles in disulfide bond formation and quality control in the endoplasmic reticulum." *IUBMB Life*, 59(1): 1-5.

Lederkremer, G. Z. (2009). "Glycoprotein folding, quality control and ER-associated degradation." *Curr Opin Struct Biol.* 19(5): 515-523.

Lederkremer, G. Z. and M. H. Glickman (2005). "A window of opportunity: timing protein degradation by trimming of sugars and ubiquitins." *Trends Biochem Sci.* 30(6): 297-303.

Michalak, M., J. Groenendyk, et al. (2009). "Calreticulin, a multi-process calcium-buffering chaperone of the endoplasmic reticulum." *Biochem J.* 417(3): 651-666.

Mizushima, T., Y. Yoshida, et al. (2007). "Structural basis for the selection of glycosylated substrates by SCFFbs1 ubiquitin ligase." *Proc Natl Acad Sci U S A*, 104(14): 5777-5781.

Molinari, M. (2007). "N-glycan structure dictates extension of protein folding or onset of disposal." *Nat Chem Biol.* 3(6): 313-320.

Movsichoff, F., O. A. Castro, et al. (2005). "Characterization of Schizosaccharomyces pombe ER alpha-mannosidase: a reevaluation of the role of the enzyme on ER-associated degradation." *Mol Biol Cell*, 16(10): 4714-4724.

Nonaka, M., B. Y. Ma, et al. (2007). "Subcellular localization and physiological significance of intracellular mannan-binding protein." *J Biol Chem.* 282(24): 17908-17920.

Nyfeler, B., V. Reiterer, et al. (2008). "Identification of ERGIC-53 as an intracellular transport receptor of alpha1-antitrypsin." *J Cell Biol.* 180(4): 705-712.

Olivari, S. and M. Molinari (2007). "Glycoprotein folding and the role of EDEM1, EDEM2 and EDEM3 in degradation of folding-defective glycoproteins." *FEBS Lett.* 581(19): 3658-3664.

Pearse, B. R., L. Gabriel, et al. (2008). "A cell-based reglucosylation assay demonstrates the role of GT1 in the quality control of a maturing glycoprotein." *J Cell Biol.* 181(2): 309-320.

Perlmutter, D. H. (2009). "Autophagic disposal of the aggregation-prone protein that causes liver inflammation and carcinogenesis in alpha-1-antitrypsin deficiency." *Cell Death and Differentiation*, 16(1): 39-45.

Quan, E. M., Y. Kamiya, et al. (2008). "Defining the Glycan Destruction Signal for Endoplasmic Reticulum-Associated Degradation." *Molecular Cell*, 32(6): 870-877.

Reiterer, V., B. Nyfeler, et al. (2010). "Role of the Lectin VIP36 in Post-ER Quality Control of Human alpha1-Antitrypsin." *Traffic*: (in press).

Renna, M., M. G. Caporaso, et al. (2007). "Regulation of ERGIC-53 gene transcription in response to endoplasmic reticulum stress." *J Biol Chem.* 282(31): 22499-22512.

Sakoh-Nakatogawa, M., S. Nishikawa, et al. (2009). "Roles of protein-disulfide isomerase-mediated disulfide bond formation of yeast Mnl1p in endoplasmic reticulum-associated degradation." *J Biol Chem.* 284(18): 11815-11825.

Sandhu, N., K. Duus, et al. (2007). "Peptide binding specificity of the chaperone calreticulin." *Biochim Biophys Acta*, 1774(6): 701-713.

Sanyal, S. and A. K. Menon (2009). "Specific transbilayer translocation of dolichol-linked oligosaccharides by an endoplasmic reticulum flippase." *Proceedings of the National Academy of Sciences of the United States of America*, 106(3): 767-772.

Schallus, T., C. Jaeckh, et al. (2008). "Malectin: A Novel Carbohydrate-binding Protein of the Endoplasmic Reticulum and a Candidate Player in the Early Steps of Protein N-Glycosylation." *Mol Biol Cell*, 19(8): 3404-3414.

Solda, T., C. Galli, et al. (2007). "Substrate-specific requirements for UGT1-dependent release from calnexin." *Mol Cell*, 27(2): 238-249.

Sousa, M. and A. J. Parodi (1995). "The molecular basis for the recognition of misfolded glycoproteins by the UDP-Glc:glycoprotein glucosyltransferase." *Embo J.* 14(17): 4196-4203.

Termine, D. J., K. W. Moremen, et al. (2009). "The mammalian UPR boosts glycoprotein ERAD by suppressing the proteolytic downregulation of ER mannosidase I." *J Cell Sci.* 122(Pt 7): 976-984.

Totani, K., Y. Ihara, et al. (2006). "Substrate specificity analysis of endoplasmic reticulum glucosidase II using synthetic high mannose-type glycans." *The Journal of biological chemistry*, 281(42): 31502-31508.

Totani, K., Y. Ihara, et al. (2006). "Substrate specificity analysis of endoplasmic reticulum glucosidase II using synthetic high mannose-type glycans." *J Biol Chem.* 281(42): 31502-31508.

Totani, K., Y. Ihara, et al. (2009). "The recognition motif of the glycoprotein-folding sensor enzyme UDP-Glc:glycoprotein glucosyltransferase." *Biochemistry*, 48(13): 2933-2940.

Ushioda, R., J. Hoseki, et al. (2008). "ERdj5 is required as a disulfide reductase for degradation of misfolded proteins in the ER." *Science*, 321(5888): 569-572.

Wakana, Y., S. Takai, et al. (2008). "Bap31 is an itinerant protein that moves between the peripheral endoplasmic reticulum (ER) and a juxtanuclear compartment related to ER-associated Degradation." *Mol Biol Cell*, 19(5): 1825-1836.

Wang, B., H. Heath-Engel, et al. (2008). "BAP31 interacts with Sec61 translocons and promotes retrotranslocation of CFTRDeltaF508 via the derlin-1 complex." *Cell*, 133(6): 1080-1092.

Yamaguchi, D., D. Hu, et al. (2010). "Human XTP3-B binds to {alpha}1-antitrypsin variant nullHong Kong via the C-terminal MRH domain in a glycan-dependent manner." *Glycobiology*, 20(3): 348-355.

Yoshida, Y. (2007). "F-box proteins that contain sugar-binding domains." *Biosci Biotechnol Biochem*. 71(11): 2623-2631.

Zhang, Y., G. Kozlov, et al. (2009). "ERp57 Does Not Require Interactions with Calnexin and Calreticulin to Promote Assembly of Class I Histocompatibility Molecules, and It Enhances Peptide Loading Independently of Its Redox Activity." *J Biol Chem*. 284(15): 10160-10173.

Zhao, G., X. Zhou, et al. (2007). "Studies on peptide:N-glycanase-p97 interaction suggest that p97 phosphorylation modulates endoplasmic reticulum-associated degradation." *Proc Natl Acad Sci U S A* 104(21): 8785-8790.

Zuber, C., J. H. Cormier, et al. (2007). "EDEM1 reveals a quality control vesicular transport pathway out of the endoplasmic reticulum not involving the COPII exit sites." *Proc Natl Acad Sci U S A* 104(11): 4407-4412.

Zuber, C., J. Y. Fan, et al. (2001). "Immunolocalization of UDP-glucose:glycoprotein glucosyltransferase indicates involvement of pre-Golgi intermediates in protein quality control." *Proceedings of the National Academy of Sciences of the United States of America* 98(19): 10710-10715.

In: Oligosaccharides: Sources, Properties and Applications ISBN 978-1-61122-193-0
Editor: Nicole S. Gordon

Chapter 6

OLIGOSACCHARIDES FROM SUCROSE VIA GLYCANSUCRASES

Gregory L. Côté*

Renewable Product Technology Research Unit
National Center for Agricultural Utilization Research
Agricultural Research Service
United States Department of Agriculture*
1815 N. University Street, Peoria, IL 61604, USA

*Mention of a trade name, proprietary product, or specific equipment does not constitute a guarantee or warranty by the United States Department of Agriculture and does not imply its approval to the exclusion of other products that may be suitable.

ABSTRACT

Glycansucrases are a class of microbial enzymes that polymerize either the fructosyl or the glucosyl moiety of sucrose to give β-D-fructans or α-D-glucans. They are also capable of transferring fructosyl or glucosyl units to acceptor molecules to yield oligosaccharides. Although the glycosyl donor specificity is limited to sucrose and related sugars, the acceptor specificity is very broad, and includes numerous carbohydrates as well as non-carbohydrate molecules. This chapter describes the enzymes and the variety of oligosaccharide structures that result from their acceptor reactions. Applications can range from modified drugs to food ingredients. Although very few of the products have thus far been commercialized, the potential is being actively studied and shows great promise.

* Email: greg.cote@ars.usda.gov
Telephone: 309-681-6319
Fax 309-681-6040

INTRODUCTION

Many bacteria, harmful and useful alike, make copious amounts of extracellular polysaccharides in the form of slime or capsules. Most of the polysaccharides are produced by similar pathways involving membrane-bound enzymes and nucleoside-activated sugars. However, a small but significant number produce extracellular polysaccharides directly from sucrose using a much simpler route. These bacteria typically grow on sucrose-rich substrates such as decaying plant matter and secrete extracellular enzymes known as glycansucrases. The enzyme name derives from the ability to directly produce glycans by splitting sucrose. They polymerize either the fructosyl or glucosyl portion, and release the other, thereby using the energy of the glucosyl-fructosyl bond instead of requiring that the donor sugar be activated by linkage to a nucleoside via diphosphate linkages:

Polymerization reaction of glucansucrases: Sucrose → glucan + fructose
Polymerization reaction of fructansucrases: Sucrose → fructan + glucose
Acceptor reaction of glucansucrases: Sucrose + R-OH → glucosyl-O-R + fructose + glucan
Acceptor reaction of fructansucrases: Sucrose + R-OH → fructosyl-O-R + glucose + fructan
(R-OH = hydroxyl-bearing acceptor molecule)

As one might expect, there are two broad classes of glycansucrases, based on whether they are fructosyltransferases or glucosyltransferases. Each class can be further subdivided according to the fructan or glucan structure formed. Levan is a β-(2→6)-linked D-fructan, and the fructansucrase responsible for its synthesis is known as levansucrase (EC 2.4.1.10) [Cote and Ahlgren, 1993]. Similarly, inulosucrase (EC 2.4.1.9) catalyzes the formation of the β-(2→1)-linked D-fructan known as inulin. On the glucan side, there are a greater number and variety of structures found in nature. The most commonly encountered is dextran, a predominantly α-(1→6)-linked D-glucan, produced by dextransucrase (EC 2.4.1.5) [Leathers, 2002]. Likewise, we find mutansucrase, producing the predominantly α-(1→3)-linked mutan [Guggenheim and Newbrun, 1969]; alternansucrase (EC 2.4.1.140), producing alternan with its alternating α-(1→3) and α-(1→6) linkages [Cote and Robyt, 1982a; Côté, 2002]; reuteransucrase, producing the α-(1→4) and α-(1→6) linkages of reuteran [Kralj, 2004], and amylosucrase, which produces an α-(1→4)-linked amylopectin-like or glycogen-like glucan [Okada and Hehre, 1974].

In addition to the reaction whereby polysaccharide is formed, the glycansucrases are also capable of transferring glycosyl units to a wide variety of acceptor molecules. These so-called acceptor reactions have been of increasing interest recently, due to their potential for producing a variety of oligosaccharides relatively cheaply and easily [Monsan et al., 2010]. We will address these reactions and the resulting oligosaccharides according to enzyme class.

1. FRUCTANSUCRASES

As early as 1942, it was known that the bacterial polysaccharide levan, a β-(2→6)-linked D-fructan, was produced directly from sucrose or raffinose by the enzyme known as levansucrase [Aschner et al., 1942]. By 1955, it had been shown that levansucrase was capable not only of transferring D-fructofuranosyl units to the growing polymer chain of levan, but also to exogenously added monosaccharides to form sucrose analogues [Hestrin et al., 1955]. For example:

Raffinose + Glucose ⇆ Sucrose + Melibiose

By substituting glucose with other aldoses, a variety of sucrose analogues were formed. A practical synthesis of β-D-fructofuranosyl-(2↔1)-α-D-xylopyranoside was subsequently reported, and the authors recognized the reaction as *"...a tool of oligosaccharide synthesis..."* at that time [Avigad et al., 1956]. In a series of papers, the same investigators described a variety of acceptor reaction products [Hestrin et al., 1956; Feingold et al., 1956] and some of their studies on the mechanism of the reactions [Hestrin and Avigad, 1958]. One of these products was lactosucrose (or lactsucrose), which is formed by the acceptor reaction with lactose, and has the structure β-D-galactopyranosyl-(1→4)-α-D-glucopyranosyl-(1↔2)-β-D-fructofuranoside [Avigad, 1957; Han et al., 2009]. This trisaccharide is of considerable interest today as a noncariogenic sweetener and a prebiotic [Choi et al., 2004; Kitahata and Fujita, 1993; Park et al., 2005; Sako et al., 1999].

Hestrin, Avigad and their coworkers used a levansucrase from *Aerobacter levanicum* (now known as *Erwinia herbicola*), but many other bacteria also produce levansucrase. One of the better known and more extensively studied has been *Bacillus subtilis*. Based on a very limited number of studies, it appears that the acceptor specificity is the same regardless of enzyme source [*e.g.*, Avigad and Feingold, 1957; Tanaka et al., 1981; Côté and Ahlgren, 1993; Kim et al., 2001; Park et al., 2003; Choi et al., 2004].

The acceptor reactions of *B. subtilis* levansucrase have been proposed as part of a larger scheme to produce modified sucrose analogues and subsequently use the sucrose analogues for the production of modified polysaccharides via glucansucrases [Homann and Seibel, 2009]. One such example of a modified sucrose analogue is gal-sucrose (α-D-galactopyranosyl-(1↔2)-β-D-fructofuranoside), formed via levansucrase acceptor reaction with galactose. It is similar in structure to sucrose, but the glucopyranosyl moiety of sucrose has been replaced by a galactopyranosyl moiety [Feingold et al., 1957; Baciu, 2004]. Xylsucrose, mentioned above, is another such example [Avigad et al., 1956; Kitahata and Fujita, 1993].

Although the acceptor reactions of levansucrase are of considerable interest, levan-oligosaccharides can be formed without the addition of exogenous acceptors. At sufficiently high concentrations, sucrose itself can act as both donor and acceptor, leading to a variety of trisaccharides and higher oligomers [Crittenden and Doelle, 1993; Euzenat et al., 1997; Tambara et al., 1999; Abdel-Fattah et al., 2005]. These fructose-containing oligosaccharides have not been commercialized, although they presumably have the same potential as those produced by fungal fructosyltransferases.

It is only much more recently that bacterial inulosucrases have become known. Similar to levansucrase, inulosucrases produce from sucrose the β-(2→1)-linked D-fructan known as inulin. An inulosucrase from *Lactobacillus reuteri* has been described which is capable of producing high molecular-weight inulin and inulo-oligosaccharides [van Hijum et al., 2002; Ozimek, 2005; Ozimek et al., 2006].

2. Glucansucrases

2.1. Dextransucrases (EC 2.4.1.5) that Produce Linear Products

Dextransucrase is a generic term that refers to any enzyme that polymerizes the D-glucopyranosyl moiety of sucrose and forms dextran [Leathers, 2002]. It is generic in the sense that the term "dextran" includes a wide range of structures whose common feature is the predominance of α-(1→6)-linked D-glucopyranosyl units, linked sequentially in the "backbone" of the polymer. In the early 1950s, a group of USDA researchers showed that dextrans varied widely in the type and proportion of branching [Jeanes et al., 1954]. In addition to the α-(1→6)-linked backbone, branches were found to be linked α-(1→2), α-(1→3), or α-(1→4) to the main chain, and the proportion of branching ranged from about 3% up to nearly 50%. The structures seemed to be specific to individual strains or species of organisms.

At that time, members of the same team noticed that unusual disaccharides were always formed in the reactions of dextransucrase with sucrose. They discovered that these arose from acceptor reactions with fructose that had been released from sucrose during the polymerization reaction. The main product was called leucrose (after *Leuconostoc*), and was found to be α-D-glucopyranosyl-(1→5)-D-fructopyranose [Stodola et al., 1952]. They later isolated and described a minor disaccharide from the same reaction mixtures, α-D-glucopyranosyl-(1→6)-D-fructofuranose, commonly known as isomaltulose or palatinose [Sharp et al., 1960]. Further studies on the acceptor reactions of dextransucrase from *Leuconostoc mesenteroides* NRRL strain B-512F [Koepsell et al., 1953] showed that oligosaccharide acceptor products were formed from glucose, maltose, isomaltose, fructose, leucrose, α-methyl glucoside, and melibiose. The concept of "better" or "more efficient" acceptors was also investigated and defined. A "good" acceptor is one which is very efficient at diverting glucosyl transfer away from polymer formation, resulting in larger amounts of oligosaccharide acceptor products. A "poor" acceptor is one which results in very little acceptor product. In their study, Koepsell et al., [1953] found that isomaltose and maltose were very good acceptors, resulting in high oligosaccharide yields and greatly diminished dextran yields. Furthermore, they discovered that most acceptor reactions resulted in a homologous series of oligosaccharides, differing one from the next by a single α-(1→6)-linked D-glucosyl unit. Some of the acceptors also affected the overall reaction rate of dextransucrase, by an unknown mechanism. The relative quantitative effects of various acceptors on the B-512F enzyme have been summarized [Robyt and Eklund, 1983].

Studies on reactions with many acceptors have been carried out. Thus, panose, α-D-glucopyranosyl-(1→6)-α-D-glucopyranosyl-(1→4)-D-glucose, has been prepared from maltose [Killey et al., 1955], and a series of α-methyl isomaltodextrins were prepared from α-methyl glucoside [Jones et al., 1956]. Higher maltodextrins have also been investigated as acceptors [Fu and Robyt, 1990]. Reactions with cellobiose and lactose were shown to yield the unusual trisaccharides shown in Figure 1 [Bailey et al., 1958]. Raffinose gave rise to an unusual tetrasaccharide originally identified as α-D-galactopyranosyl-(1→6)-[α-D-glucopyranosyl-(1→2)-]-α-D-glucopyranosyl-(1↔2)-β-D-fructofuranoside [Neely, 1959], but later correctly identified as α-D-glucopyranosyl-(1→4)-α-D-galactopyranosyl-(1→6)-α-D-glucopyranosyl-(1↔2)-β-D-fructofuranoside [Côté et al., 2009]. A survey of acceptor reactions of B-512F dextransucrase with all known glucose disaccharides confirmed the structure of the maltose product to be panose, and of the cellobiose product shown in Figure 1. It revealed that most of the products resulted from glucosylation of the free 6-OH group of the non-reducing end, the only exception being cellobiose [Yamauchi and Ohwada, 1969]. A survey of various methyl glycosides revealed that the 4-OH is particularly important in substrate binding and orientation [Fu et al., 1990]. Other carbohydrate acceptors have also been described, including sugar acids and alcohols [Demuth et al., 2002] and 1,5-anhydro-D-fructose [Richard et al., 2005].

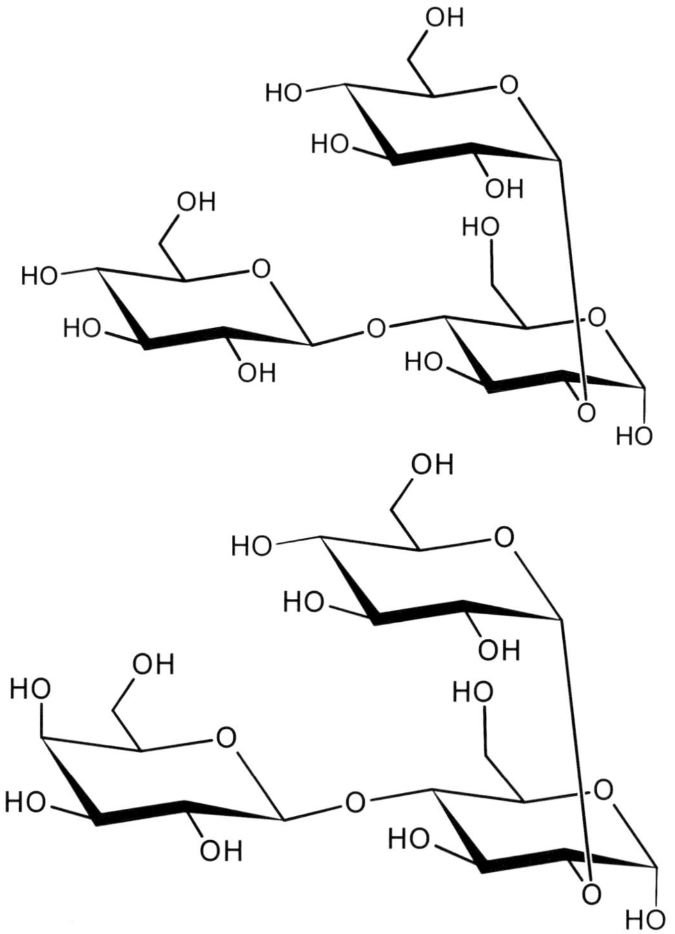

Figure 1. Dextransucrase trisaccharide acceptor products from cellobiose (top) and lactose (bottom).

As mentioned above, many acceptors give rise to products which themselves also serve as acceptors, leading to a homologous series of oligosaccharides. This phenomenon depends on the ability of each product compound to serve as an acceptor. As one might expect, it also depends on the amounts of each acceptor relative to the amount of donor (i.e., sucrose) initially present. A number of researchers have noted that a higher ratio of sucrose to acceptor results in the formation of higher DP (degree of polymerization) acceptor products, whereas a lower ratio of sucrose to acceptor results in mainly low DP products [Killey et al., 1955; Paul et al., 1986; Su and Robyt, 1993]. This phenomenon can be exploited to selectively synthesize mainly a single glucosylated product, such as panose from maltose, or a mixture of higher DP products [Pereira et al., 1998; Heincke et al., 1999; Rodrigues et al., 2006; Rabelo et al., 2006].

2.2. Dextransucrases that Produce Highly Branched Products

L. mesenteroides NRRL strain B-512F has been used for commercial production of dextran since the 1950s in the USA; strains used in other nations are similar if not identical. It was not until many years later that it was demonstrated that other strains made not only dissimilar dextrans, but dissimilar oligosaccharides via acceptor reactions, too. A recent survey of glucansucrase acceptor products from *L. mesenteroides* and related species showed that there are at least four distinct classes of glucansucrases based on their acceptor product profiles [Côté and Leathers, 2005]. These groupings are typified by strain NRRL B-512F, which makes linear α-(1→6)-linked products, strain NRRL B-1355, which makes alternating α-(1→3) and α-(1→6)-linked products, strain NRRL B-1299, which makes products bearing α-(1→6)-linkages with α-(1→2)-linked branch points, and NRRL B-742, which makes products bearing α-(1→6)-linkages with α-(1→3)-linked branch points.

It has been hypothesized that branches in dextrans are formed by acceptor reactions, and several reports support this [Robyt and Tanigichi, 1976; Côté and Robyt, 1983; Côté and Robyt, 1984; Brison et al., 2009]. As some dextrans are much more highly branched than others, it may be that some dextransucrases are capable of forming branches much closer to one another relative to other dextransucrases. If that is the case, then the difference between Class I and the other classes of Côté and Leathers, [2005] may be a matter of degree rather than a qualitative difference. Nonetheless, it still serves as a useful distinction over the size range of oligosaccharides studied, i.e., trisaccharides through octasaccharides.

A typical example of a dextransucrase that synthesizes highly branched dextrans and branched oligosaccharide acceptor products is that produced by *L. mesenteroides* NRRL strain B-1299. Most of the branching is through α-(1→2)-linked branch points [Slodki et al., 1986]. It has been demonstrated that a dextransucrase from this strain contains two catalytic domains, one responsible for the formation of α-(1→6) linear linkages, and one responsible for α-(1→2) branch linkages [Brison et al., 2009]. Dextransucrase preparations from strain B-1299 have been used to prepare a series of oligosaccharides by acceptor reactions with maltose [Remaud-Simeon et al., 1994]. In addition to the linear α-(1→6)-linked products common to dextransucrase reactions (see section 2.1), a second series of branched products were formed. The tetrasaccharide products in this series are α-D-glucopyranosyl-(1→6)-α-D-glucopyranosyl-(1→6)-α-D-glucopyranosyl-(1→4)-D-glucopyranose (as found with most glucansucrases) and α-D-glucopyranosyl-(1→2)-α-D-glucopyranosyl-(1→6)-α-D-glucopyranosyl-(1→4)-D-glucopyranose. There are three different pentasaccharides formed – one of the linear α-(1→6)-linked series, one bearing an α-(1→2)-linked D-glucosyl residue as the terminal unit, and one bearing a branched D-glucosyl residue in the penultimate position, bearing both an α-(1→6)-and an α-(1→2)-linked terminal D-glucosyl residue [Castillo et al., 1992; Quirasco et al., 1995; Dols et al., 1997; deSegura et al., 2006]. Presumably, the higher DP products follow a similar pattern.

In their survey of acceptor reactions, Côté and Leathers, [2005] found several other strains that produced patterns of products similar to strain B-1299, including NRRL strains B-1297, B-1298, B-1397, B-1399, B-1402, B-1422, B-1424, and ATCC strain 21436, to mention just a few. None of these have been studied in as much detail as B-1299. A mixture of oligosaccharides produced by B-1299 dextransucrase is marketed commercially (see section 3).

Côté and Leathers, [2005] also noted another class of acceptor reaction products, their Category III, typified by *L. citreum* strain NRRL B-742. These included B-1254, B-1374, B-1375, and B-1377. Only strain B-742 has been studied in detail. This strain produces a highly α-(1→3)-branched dextran, and it has been demonstrated that the branching can be produced by acceptor reactions [Côté and Robyt, 1983]. It has been shown that B-742 acceptor products contain not only α-(1→6)-linked linear structures, but above the level of pentasaccharides, structures with α-(1→3)-linked branches are also formed [Remaud et al., 1992]. The branched structures are resistant to hydrolysis by endodextranases, which may be used to enhance the production of lower-DP product mixtures [Remaud et al., 1992; Yoo et al., 2001]. These oligosaccharides are of interest as prebiotics [Chung, 2002].

2.3. Alternansucrase (EC 2.4.1.140)

L. mesenteroides (*L. citreum*) NRRL strains B-1355, B-1498, and B-1501 each produce at least two extracellular glucansucrases. One is a typical dextransucrase, whereas the other is responsible for the synthesis of an unusual glucan known as alternan [Côté and Robyt, 1982a]. Alternan is so named because of its unique structure consisting of alternating α-(1→3) and α-(1→6) linked D-glucose units. Alternansucrase is distinct from dextransucrase not only in the structure of the glucan it forms, but also in the structures of its acceptor products. Initial studies demonstrated that, although the enzyme could synthesize either α-(1→3) or α-(1→6) linkages, an α-(1→6) linkage was required before a subsequent α-(1→3) linkage was formed [Côté and Robyt, 1982b]. Thus, maltose gave rise to panose, but panose subsequently gave rise to two tetrasaccharides. Unlike the case with dextransucrase, in which maltose and isomaltose were approximately equal in their ability to act as acceptors, maltose and nigerose were much better acceptors for alternansucrase compared to isomaltose. Like dextransucrase, the ratio of sucrose (donor) to acceptor determines the size distribution of products [Pelenc et al., 1991].

Based on the structures identified in these initial studies, it was proposed that there would be three pentasaccharide products from maltose [Castillo et al., 1992]. Carrying out that logic, alternansucrase was hypothesized to produce five hexasaccharides, eight heptasaccharides, etc., the number of possible product structures increasing with degree of polymerization (DP) as a Fibonacci series. However, further studies showed that there was only one pentasaccharide formed, two hexasaccharides, one heptasaccharide, two octasaccharides, etc.

(Figure 2) [Côté and Sheng, 2006; Côté et al., 2008]. Based on the structures formed and on the observation that acceptors bearing α-(1→3) linkages at their non-reducing ends react more readily than acceptors with α-(1→6) linkages at their ends, it is currently believed that the sequence specificity of alternansucrase is kinetically and stereochemically controlled. The presence of a small proportion of sequential α-(1→6) linkages in alternan supports this hypothesis [Leathers et al., 2009]. The acceptor products from gentiobiose, also formed in good yields (nearly 90% in certain unoptimized reactions), follow a pattern similar to those formed from maltose [Côté, 2009].

Although maltose is the best naturally occurring acceptor known for alternansucrase, many other acceptors are also known [Côté et al., 2003]. Table 1 lists many of those examined and their relative acceptor strengths. Of particular note were the products arising from L-glucose and D-tagatose. In the presence of D-tagatose, alternansucrase produced the disaccharide α-D-glucopyranosyl-(1→1)-β-D-tagatopyranose. This disaccharide is analogous to trehalulose. We were unable to isolate a disaccharide product from L-glucose, but the trisaccharide α-D-glucopyranosyl-(1→6)-α-D-glucopyranosyl-(1→4)-L-glucose was isolated and identified. This is analogous to panose, in which the reducing-end D-glucose residue has been replaced by its L-enantiomer. The disaccharide, α-D-glucopyranosyl-(1→4)-L-glucose, was a better acceptor than maltose, previously the best known acceptor for alternansucrase. Thus, no disaccharide accumulated in the reaction mixture because it underwent subsequent acceptor reaction as quickly as it was formed [Côté et al., 2005].

Several methyl glycosides were examined and compared. The initial product arising from methyl β-D-glucopyranoside was methyl β-isomaltoside, which was subsequently glucosylated to yield methyl β-isomaltotrioside and methyl α-D-glucopyranosyl-(1→3)-α-D-glucopyranosyl-(1→6)-β-D-glucopyranoside. These products are analogous to those from methyl α-D-glucopyranoside. The major initial acceptor product from methyl α-D-mannopyranoside was methyl α-D-glucopyranosyl-(1→6)-α-D-mannopyranoside, but several minor products were also isolated and characterized, including a 3,6-di-*O*-substituted mannopyranoside. Methyl α-D-galactopyranoside yielded two initial products, methyl α-d-glucopyranosyl-(1→3)-α-D-galactopyranoside and methyl α-D-glucopyranosyl-(1→4)-α-D-galactopyranoside. Methyl D-allopyranosides were glucosylated primarily at position 6, yielding methyl α-D-glucopyranosyl-(1→6)-D-allopyranosides. The latter subsequently gave rise to methyl α-D-glucopyranosyl-(1→6)-α-D-glucopyranosyl-(1→6)-D-allopyranosides. In general, the methyl α-D-hexopyranosides were better acceptors than the corresponding β-glycosides [Côté and Dunlap, 2003].

The acceptor reaction of alternansucrase with raffinose provides another good illustration of the difference between dextransucrase and alternansucrase [Côté et al., 2009]. The main products were found to be the tetrasaccharides α-D-glucopyranosyl-(1→3)-α-D-galactopyranosyl-(1→6)-α-D-glucopyranosyl-(1↔2)-β-D-fructofuranoside and α-D-glucopyranosyl-(1→4)-α-D-galactopyranosyl-(1→6)-α-D-glucopyranosyl-(1↔2)-β-D-fructofuranoside in ratios ranging from 4:1 to 9:1, along with lesser amounts of α-D-glucopyranosyl-(1→6)-α-D-galactopyranosyl-(1→6)-α-D-glucopyranosyl-(1↔2)-β-D-fructofuranoside (Figure 3). Ten unusual pentasaccharide structures were isolated. Three of these arose from glucosylation of the major tetrasaccharide product, two each from the minor tetrasaccharides, and three were the result of glucosylations of the fructose acceptor products leucrose or isomaltulose.

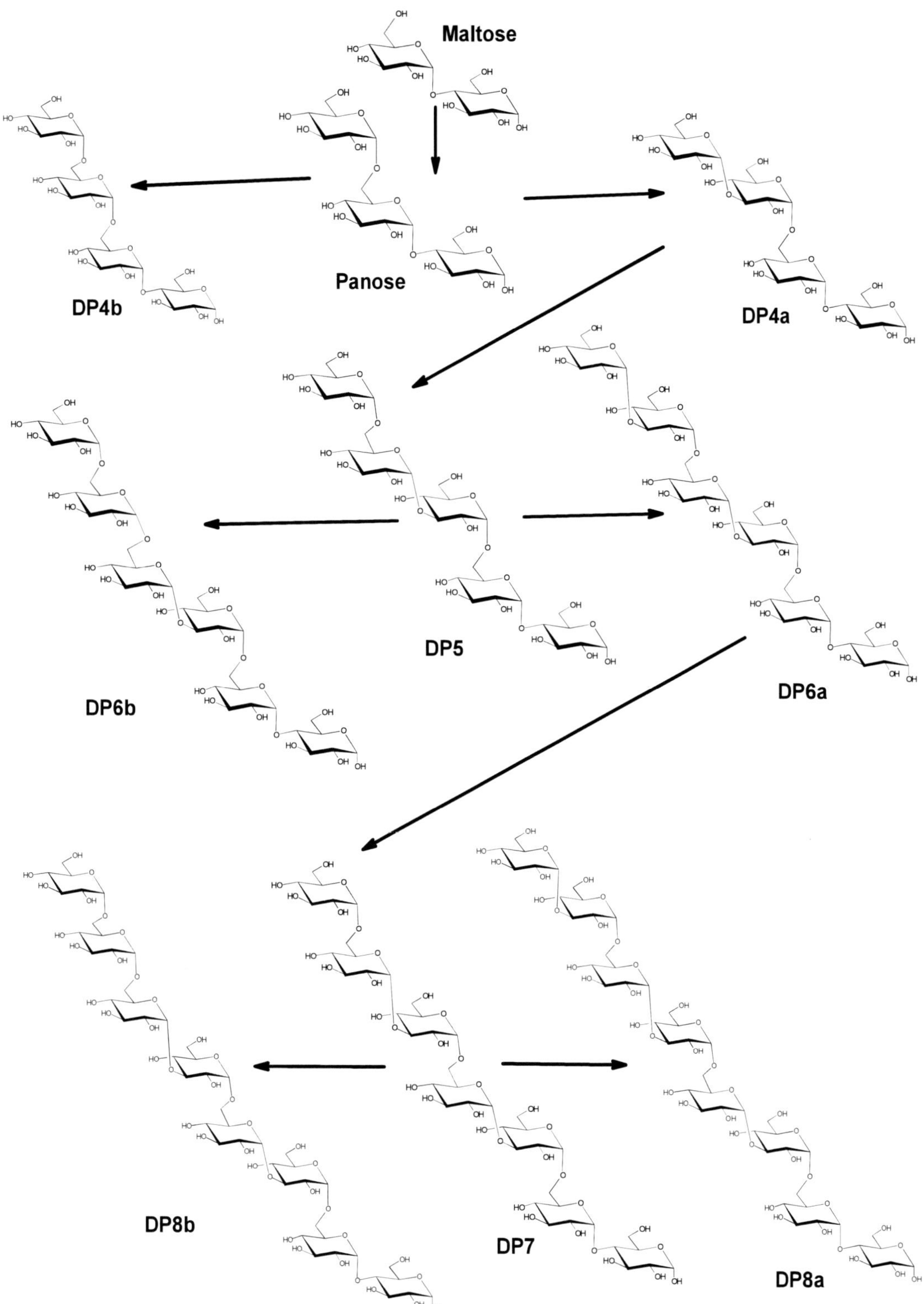

Figure 2. Alternansucrase acceptor products from maltose.

Table 1. Comparison of acceptor strengths with alternansucrase, as measured by relative amount of alternan formed

Monomers	Alternan (rel. %)	Di- & Oligomers	Alternan (rel. %)
dulcitol	101		
mannitol	100	α,α-trehalose	102
myo-inositol	99	lactulose	91
D-arabinose	99	galactosyl-arabinose	90
D-lyxose	99	melezitose	90
xylitol	97	cycloalternan tetrasaccharide	90
D-sorbose	97	stachyose	89
L-rhamnose	96	kojitriose	86
β-methyl-D-xyloside	96	β-dodecyl maltoside	83
β-methyl-D-galactoside	94	**leucrose**	81
L-fucose	94	sophorose	80
D-ribose	93	lactose	79
N-acetyl-D-mannosamine	93	planteose	76
L-lyxose	92	**cellobiose**	72
D-altrose	92	**palatinose**	69
D-mannose	90	**raffinose**	67
N-acetyl-D-glucosamine	90	gentianose	66
2-deoxy-D-galactose	89	maltotriose	65
β-methyl-D-mannoside	89	laminaribiose	58
N-acetyl-D-galactosamine	88	turanose	57
L-sorbose	88	6-O-α-D-Glc-trehalose	56
D-xylose	86	6,6'-di-O- α-D-Glc-trehalose	54
D-psicose	85	isomaltotriose	50
D-galactose	83	**melibiose**	45
sorbitol	82	**isomaltose**	44
2-deoxy-D-ribose	81	theanderose	42
D-fructose	78	6"-O-α-D-Glc panose	36
D-talose	77	3'-O-α-D-Glc isomaltose	31
L-arabinose	75	kojibiose	27
D-allose	75	**panose**	25
α-methyl-D-galactoside	73	**gentiobiose**	25
α-methyl-D-mannoside	70	**nigerose**	23
D-tagatose	65	maltitol	18
L-glucose	63	**maltose**	11
D-quinovose	61		
β-methyl-D-glucoside	60		
2-deoxy-D-glucose	59		
D-glucose	53		
β-octyl-D-glucoside	52		
α-methyl-D-glucoside	20		

Compounds in bold are those whose product structures have been determined.

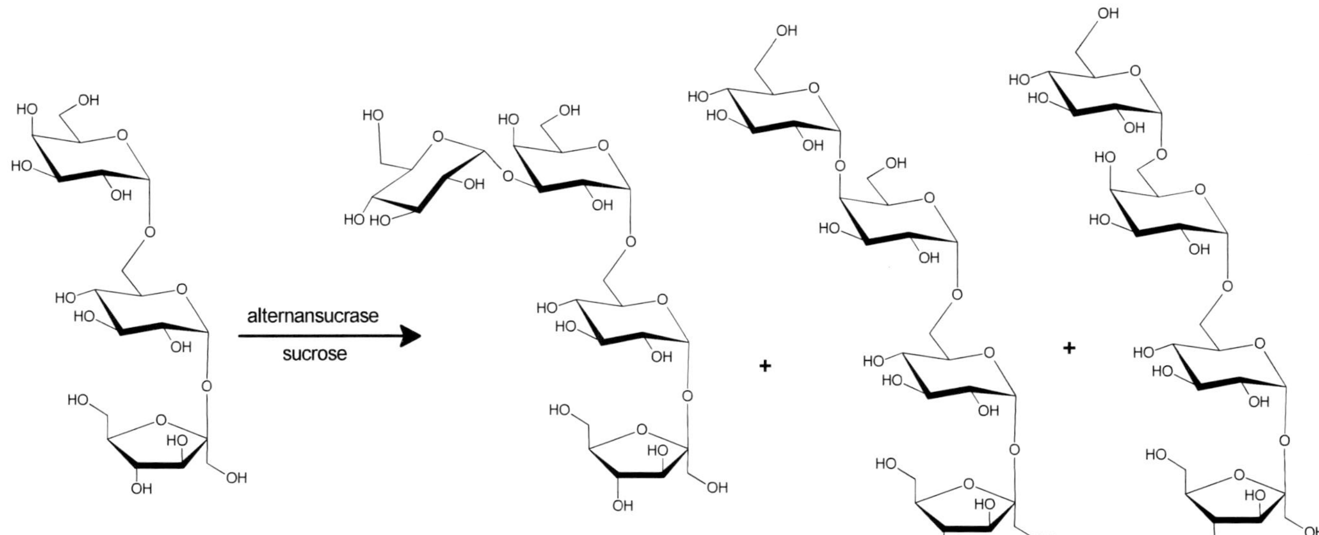

raffinose

1

2

3

Figure 3. Initial alternansucrase acceptor reaction with raffinose.

The major pentasaccharide product arose from glucosylation of the major tetrasaccharide at position 4 of the fructofuranosyl unit (Figure 4), to give a subunit structure analogous to that of maltulose. A number of hexasaccharides and higher oligosaccharides were also produced. Compared to dextransucrase, which gave only a single tetrasaccharide product, alternansucrase yielded a much greater number of products in greater quantities.

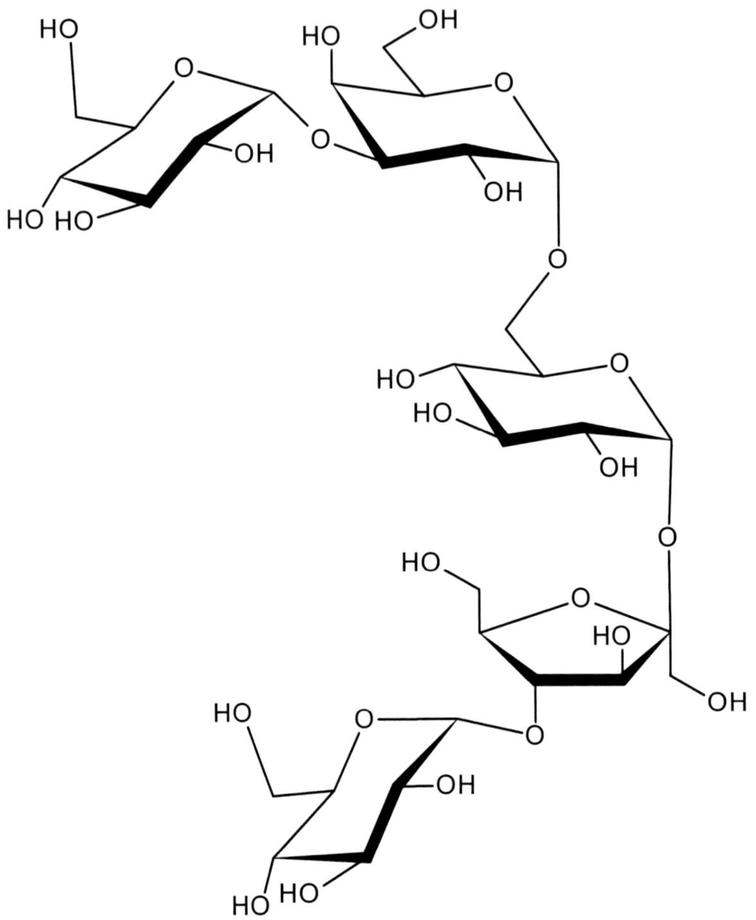

Figure 4. One of the major pentasaccharide products of alternansucrase acceptor reaction with raffinose.

As with dextransucrase, cellobiose behaved in a peculiar fashion when used as an acceptor for alternansucrase [Arguello-Morales et al., 2001]. Again, it was a more efficient acceptor for alternansucrase relative to dextransucrase. Although one of the trisaccharide products was the same as that observed for dextransucrase, namely α-D-glucopyranosyl-(1→2)-[β-D-glucopyranosyl-(1→4)]-D-glucopyranose (see Figure 1), a second product, α-D-glucopyranosyl-(1→6)-β-D-glucopyranosyl-(1→4)-D-glucopyranose, was also formed. This follows the pattern of multiple product formation via alternansucrase vs. single product formation via dextransucrase.

Fructose also reacted more readily with alternansucrase relative to dextransucrase. Although only leucrose and isomaltulose have been noted in dextransucrase reactions, a series of higher oligosaccharides are formed by alternansucrase [Côté et al., 2008]. These represent acceptor products arising from both disaccharides, and although yields are low, it still demonstrates the utility of alternansucrase over dextransucrase for the synthesis of oligosaccharides.

2.4. Reuteransucrase, Amylosucrase and Mutansucrase

Dextransucrases have been the most-studied glycansucrases as far as acceptor reactions are concerned, followed by levansucrase and alternansucrase. There are, however, a few published studies on the acceptor reactions of other glycansucrases. These include reuteransucrase, an enzyme from *Lactobacillus reuteri*, which makes a branched α-(1→6) glucan with α-(1→4) linkages in both linear and branching positions [Kralj, 2004]. Reuteransucrase can form both α-(1→6) and α-(1→4)-linked oligosaccharides [Kralj et al., 2004]. For example, reuteransucrase forms panose from maltose, just as dextransucrase and alternansucrase do (see Figure 2), but forms isopanose (α-D-glucopyranosyl-(1→4)-α-D-glucopyranosyl-(1→6)-D-glucopyranose) from isomaltose, in a manner more similar to alternansucrase than to dextransucrase [Kralj et al., 2005a]. Mutational modifications of reuteransucrase affect not only the structure of the glucan formed, but also the structures and/or proportions of the oligosaccharides formed by acceptor reactions [Kralj et al., 2005b]. Higher-DP oligosaccharides have been isolated from *Lactobacillus reuteri* glucansucrase reactions, but not structurally characterized [Tieking et al., 2005].

Amylosucrase is a glucansucrase found mainly intracellularly in bacteria not closely related to the lactic acid bacteria, which are the sources of many other glucansucrases. As the name implies, it produces an amylopectin-like or glycogen-like α-D-glucan from sucrose [Okada and Hehre, 1974]. The linkage type is predominantly α-(1→4), with α-(1→6) branching. Engineered variants of this enzyme have recently been used to synthesize D-glucosylated forms of L-rhamnose and *N*-acetyl-D-glucosamine which occur in bacterial antigenic determinants [Champion et al., 2009a, 2009b].

Streptococcus species are often involved in the process of tooth decay. This is in part mediated by their ability to synthesize exopolysaccharides from dietary sucrose, which then aid in the adhesion and aggregation of the cells to tooth surfaces. The exopolysaccharides are produced by glucansucrases, including dextransucrase and mutansucrase. These enzymes have often been referred to as GTF-S (glucosyltransferase, soluble-product) and GTF-I (glucosyltransferase, insoluble product), respectively. Despite the intense and widespread interest in streptococcal glucansucrases because of their role in dental caries, very few studies have focused on the acceptor reactions of these enzymes. Of those that have, most have described the reactions of their dextransucrases (GTF-S), which have not been remarkably different from the dextransucrases of *Leuconostoc* enzymes, mainly forming α-(1→6)-linked oligosaccharides [Walker, 1973; Mayer et al., 1981; Russell et al., 1990; Bhattacharjee and Mayer, 1991; Bhattacharjee et al., 1993; Lee et al., 1997].

Few studies have described the formation of α-(1→3)-linked oligosaccharides by *Streptococcus* glucansucrases. Fu and Robyt, [1991] studied the acceptor reactions of both GTF-S and GTF-I from *Streptococcus sobrinus* NIDR 6715 with various maltodextrins. Both enzymes formed α-(1→6)-linkages, but only GTF-I formed α-(1→3)-linkages. GTF-I was more active than GTF-S in catalyzing acceptor reactions. Interestingly, they observed glucosyl transfer not only to the non-reducing terminal glucose residues of acceptors, but also to the reducing-end glucose residues. The α-(1→3)-linked products are apparently poor acceptors themselves, as they did not give rise to higher-DP products. Lee et al., [1997] studied the acceptor products from maltose with both GTF-S and GTF-I from the same strain (*S. sobrinus* ATCC 27351 = NIDR 6715-7), and although they did not subject their products to structural analysis, they concluded from chromatographic mobilities that the GTF-S synthesized a series of α-(1→6)-linked oligosaccharides, and GTF-I synthesized a series of α-(1→3)-linked oligosaccharides.

A more rigorous structural analysis of *S. sobrinus* 6715 GTF-I acceptor reactions with nigerose (α-D-glucopyranosyl-(1→3)-D-glucose) showed the products to be a series of α-(1→3)-linked nigerooligosaccharides [Mukasa et al., 2000]. In addition, the oligosaccharides strongly activated the glucansucrase activity.

3. Applications

Initially, the acceptor reactions of glucansucrases were viewed mainly as scientific curiosities, with possible implications for the optimal production of dextran or as a source of information regarding the mechanisms of the enzymes. Since then, several applications have been either proposed or realized for oligosaccharides produced via glucansucrases.

3.1. Disaccharides: Leucrose and Sucrose Analogues

If we stretch the definition of oligosaccharides to include disaccharides, then we can include leucrose and xylsucrose as part of our discussion. As stated above, leucrose is formed whenever glucansucrases act on sucrose, due to the release of fructose and subsequent acceptor reaction with fructose. Leucrose has been produced on a pilot scale in Germany [Schwengers, 1991; Buchholz et al., 1998; Demuth et al., 1999]. The use of immobilized dextransucrase improves the economics of this process [Reh et al., 1996]. The susceptibility of leucrose to human digestive enzymes has been studied [Iizuka et al., 1990], and it has been proposed as a non-cariogenic sweetener [Ziesenitz et al., 1989; Peltroche-Llacsahuangal et al., 2001].

The sucrose analogue xylsucrose, or α-D-xylopyranosyl-(1↔2)-β-D-fructofuranoside, as we mentioned previously, can be produced by the acceptor reaction of levansucrase with xylose. It is a good inhibitor of the streptococcal glucansucrases involved in dental caries [Kitahata and Fujita, 1993], although it may be somewhat expensive.

3.2. Glucooligosaccharides (GOS or GlcOS)

The production of glucooligosaccharides by acceptor reactions of dextransucrase has been studied extensively over the past half-century. As maltose seems to be the best acceptor, many of these studies have focused on the glucooligosaccharides arising from maltose and its initial acceptor product, panose [Killey et al., 1955; Paul et al., 1986; Pereira et al., 1998; Heincke et al., 1999; Rodrigues et al., 2006; Rabelo et al., 2006]. Fewer studies have focused on the use of other acceptors, although the isomaltooligosaccharides arising from acceptor reactions of B-512F dextransucrase have been proposed as a useful way of producing these linear oligomers of dextran [Smiley et al, 1982]. A number of patents have described either actual or proposed uses for dextransucrase acceptor products from a variety of acceptors [Lamothe et al., 1996; Vercauteren and Nguyen, app 2006; Stahl et al., 2009], including their use as prebiotics in foods, feeds and cosmetics.

Although much of the academic interest has been focused on the dextransucrase from *L. mesenteroides* NRRL B-512F, there has been more commercial interest in the branched oligosaccharides produced by NRRL strains B-742 and B-1299.

Glucooligosaccharides produced by *L. mesenteroides* NRRL B-1299 are marketed in Europe by the Solabia Group (www.solabia.fr) under the trade name BioEcolians. A mixture primarily consisting of the branched pentasaccharide acceptor product from maltose, it is purported to act as a prebiotic in foods, feed, and cosmetics. Studies indicate that it does have the ability to enhance the growth of fecal *Bifidobacteria* spp. in dogs [Flickinger et al., 2000]. Studies with human fecal bacteria in rats suggest that it is capable of altering short-chain fatty acid profiles arising from fermentation by human intestinal bacteria [Valette et al., 1993]. It also served as a growth substrate for several strains of probiotic *Lactobacillus delbrueckii* from dairy cultures, indicating promise as a general prebiotic [Ignatova et al., 2009].

Leuconostoc citreum NRRL B-742 (*L. mesenteroides* ATCC 13146) produces two extracellular dextransucrases, one of which is capable of synthesizing a dextran with a very high percentage of $\alpha(1\rightarrow3)$ branch points [Côté and Robyt, 1983]. It can produce $\alpha(1\rightarrow3)$-branched oligosaccharides via acceptor reactions [Remaud et al., 1992]. Methods have been developed for the large scale production of branched oligosaccharides by this strain [Yoo et al., 2001; Chung, 2006] and it has been shown that they can act as prebiotics in chickens [Chung and Day, 2002; 2004]. Production has not yet been commercialized.

3.3. Alternan Oligosaccharides

The wide variety of alternansucrase-derived oligosaccharides discussed in section 2.3 were initially of interest as potential prebiotics [Côté et al., 2003]. Studies with pure cultures of prebiotic and pathogenic bacteria suggested that they might be useful in this regard [Côté et al., 2003; Holt et al., 2005, 2008]. Subsequent investigations using mixed cultures of human fecal bacteria lent further evidence for a potential prebiotic effect using the oligosaccharide mixtures arising from maltose and gentiobiose [Sanz et al., 2005, 2006a, 2006b]. The latter was especially interesting, as gentiooligosaccharides have been proposed as prebiotics [Rycroft et al., 2001], but face the difficulty that gentiobiose is bitter in taste. Glucosylation via acceptor reaction with alternansucrase eliminated the bitter taste of gentiobiose [Côté, 2009], and evidence suggested that the alternansucrase products may be more selective for *Bifidobacteria* spp. relative to the gentiooligosaccharides [Sanz et al., 2006b].

However, human feeding trials with the maltose-derived alternansucrase oligosaccharides showed that they do not act as prebiotics in humans, but instead are slowly converted to glucose, resulting in a slow increase in serum glucose levels [Grysman et al., 2008]. Studies in rats indicated a lack of toxicity [Eapen et al., 2007], and the mixture is now manufactured by Cargill [Carlson et al., 2009]. The acceptor reaction mixture is used as a low-glycemic sweetener in specialty foods for diabetics under the trade name Xtend™ sucromalt. Human feeding trials with other alternansucrase acceptor products have not been performed, but investigations on the metabolism of the maltose acceptor products in poultry are under way.

4. Outlook

Glycansucrase acceptor reactions represent a valuable and, until recently, underappreciated biocatalytic method for the production of large amounts of oligosaccharides from inexpensive carbohydrate sources. They are of particular value in the production of food, feed ingredients and cosmetics, especially in the rapidly growing field of nutraceuticals. Currently, several such products have been produced on an industrial scale. They include Xtend™ sucromalt and BioEcolians, both consisting of glucooligosaccharide mixtures for nutraceutical applications, and the disaccharides leucrose and xylsucrose, which have not seen as much commercial success.

Glycansucrases are currently limited to the production of oligosaccharides of glucose or fructose linked to various acceptors. An ambitious effort is underway in Europe to alter the donor specificity of glycansucrases to enable them to transfer glycosyl units from a variety of sucrose analogues [Homann and Seibel, 2009]. If successful, this may lead to methods for the custom synthesis of large quantities of a wider variety of oligosaccharides. Some success has already been reported in the use of glycansucrases with modified acceptor specificity, leading to custom-tailored products derived from cell-surface oligosaccharides of pathogenic bacteria [Champion et al., 2009a,b].

References

Abdel-Fattah, A.F.; Mahmoud, D.A.R.; Esawy, M.A.T. Production of levansucrase from *Bacillus subtilis* NRC 33a and enzymic synthesis of levan and fructo-oligosaccharides. *Curr. Microbiol.* 2005, *51,* 402–407.

Argüello Morales, M.A.; Remaud-Simeon, M.; Willemot, R.-M.; Vignon, M.R.; Monsan, P. Novel oligosaccharides synthesized from sucrose donor and cellobiose acceptor by alternansucrase. *Carbohydr. Res.* 2001, *331,* 403-411.

Aschner, M.; Avineri-Schapiro, S.; Hestrin, S. Enzymatic synthesis of levan. *Nature* 1942, *149*, 527.

Avigad, G. Enzymatic synthesis and characterization of a new trisaccharide, α-lactosyl-β-fructofuranoside. *J. Biol. Chem.* 1957, *229*, 121-129.

Avigad, G.; Feingold, D.S. Fructosides formed from sucrose by a Corynebacterium. *Arch. Biochem. Biophys.* 1957, *70*, 178-184.

Avigad, G.; Feingold, D.S.; Hestrin, S. An enzymic synthesis of a sucrose analog: α-D-xylopyranosyl-β-D-fructofuranoside. *Biochim. Biophys. Acta* 1956, *20,* 129-134.

Baciu, I.E. Extracted sugar-beet pulp and sucrose, two renewable materials as "hot" substrates for enzymatic synthesis of valuable saccharides. PhD Dissertation, Technischen Universität Carolo-Wilhelmina zu Braunschweig, Germany, 2004.

Bailey, R.W.; Barker, S.A.; Bourne, E.J.; Grant, P.M.; Stacey, M. Immunopolysaccharides. Part IX. The enzymic synthesis of trisaccharides containing the alpha-1,2-glucosidic linkage. *J. Chem. Soc.* 1958, 1895-1902.

Bhattacharjee, M.K.; Mayer, R.M. Monosaccharide acceptor specificity of dextransucrase. *Bioorg. Chem.* 1991, *19*, 445-455.

Bhattacharjee, M.K.; Mayer, R.M. Formation of α-(1→6), α-(1→3), and α-(1→2) glycosidic linkages by dextransucrase from *Streptococcus sanguis* in acceptor-dependent reactions. *Carbohydr. Res.* 1993, *242*, 191-201.

Brison, Y.; Fabre, E.; Moulis, C.; Portais, J.C.; Monsan, P.; Remaud-Siméon, M. Synthesis of dextrans with controlled amounts of *α*-1,2 linkages using the transglucosidase GBD–CD2. *Appl. Microbiol. Biotechnol.* 2010, *86,* 545–554.

Buchholz, K.; Noll-Borchers, M.; Schwengers, D. Production of leucrose by dextransucrase. *Starch/Stärke* 1998, *50,* 164–172.

Carlson, T.L.; Woo, A.; Zheng, G.-H. Methods of making syrups. *U.S. Pat. Appl.* 2009, *0123603.*

Castillo, E.; Iturbe, F.; Lopez-Munguia, A.; Pelenc, V.; Paul, F.; Monsan, P. Dextran and oligosaccharide production with glucosyltransferases from different strains of *Leuconostoc mesenteroides*. *Ann. NY Acad. Sci.* 1992, *672*, 425-430.

Champion, E.; Andre, I.; Mulard, L.; Monsan, P.; Remaud-Simeon, M.; Morel, S. Synthesis of L-rhamnose and *N*-acetyl-D-glucosamine derivatives entering in the composition of bacterial polysaccharides by use of glucansucrases. *J. Carbohydr. Chem.* 2009a, *28,* 142-160.

Champion, E.; Andre, I.; Moulis, C.; Boutet, J.; Descroix, K.; Morel, S.; Monsan, P.; Mulard, L.A.; Remaud-Simeon, M. Design of α-transglucosidases of controlled specificity for programmed chemoenzymatic synthesis of antigenic oligosaccharides. *J. Am. Chem. Soc.* 2009b, *131*, 7379–7389.

Choi, H.-J.; Kim, C.S.; Kim, P.; Jung, H.-C.; Oh, D.-K. Lactosucrose bioconversion from lactose and sucrose by whole cells of *Paenibacillus polymyxa* harboring levansucrase activity. *Biotechnol. Prog.* 2004, *20,* 1876-1879.

Chung, C.-H. Production of glucooligosaccharides and mannitol from *Leuconostoc mesenteroides* B-742 fermentation and its separation from byproducts. *J. Microbiol. Biotechnol.* 2006, *16,* 325–329.

Chung, C.-H. A potential nutraceutical from *Leuconostoc mesenteroides* B-742 (ATCC 13146); production and properties. PhD Dissertation, Louisiana State University, Baton Rouge, 2002.

Chung, C.-H.; Day, D.F Glucooligosaccharides from *Leuconostoc mesenteroides* B-742 (ATCC 13146): A potential prebiotic. *J. Indus. Microbiol. Biotechnol.* 2002, *29,* 196–199.

Chung, C.-H.; Day, D.F Efficacy of *Leuconostoc mesenteroides* (ATCC 13146) isomaltooligosaccharides as a poultry prebiotic. *Poultry Sci.* 2004, *83,* 1302–1306.

Côté, G.L. Alternan. In: *Biopolymers, Polysaccharides I: Polysaccharides from Prokaryotes, chapter 13.* Vandamme E.J.; DeBaets, S.; Steinbuchel, A., Eds.; Wiley, Weinheim, Germany, 2002, Vol 5. pp. 323-350.

Côté, G.L. Acceptor products of alternansucrase with gentiobiose. Production of novel oligosaccharides for food and feed and elimination of bitterness. *Carbohydr. Res.* 2009, *344,* 187-190.

Cote, G.L.; Ahlgren, J.A. Metabolism in Microrganisms. 1. Levan and Levansucrase, Chapter 5 In *Science and Technology of Fructans,* Suzuki, M.; Chatterton, N.J, Eds.; CRC Press, Inc.: Boca Raton, FL. 1993; pp.141-168.

Côté, G.L.; Dunlap, C.A. Alternansucrase acceptor reactions with methyl hexopyranosides. *Carbohydr. Res.* 2003, *338,* 1961-1967.

Coté, G.L.; Dunlap, C.A.; Appell, M.; Momany, F.A. Alternansucrase acceptor reactions with D-tagatose and L-glucose. *Carbohydr. Res.* 2005, *340,* 257-262.

Côté, G.L.; Dunlap, C.A.; Vermillion, K.E. Glucosylation of raffinose via glucansucrase acceptor reactions. *Carbohydr. Res.* 2009, *344*, 1951-1959.

Côté, G.L.; Holt, S.M.; Miller-Fosmore, C. Prebiotic oligosaccharides via alternansucrase acceptor reactions. *ACS Symp. Ser.* 2003, *849,* 75-89.

Côté, G.L.; Leathers, T.D. A method for surveying and classifying *Leuconostoc* spp. glucansucrases according to strain-dependent acceptor product patterns. *J. Indus. Microbiol. Biotechnol.* 2005, *32,* 53-60.

Cote, G.L.; Robyt, J.F. Isolation and partial characterization of an extracellular glucansucrase from *Leuconostoc mesenteroides* NRRL B-1355 that synthesizes an alternating (1→6),(1→3)-α-D-glucan. *Carbohydr. Res.* 1982a, *101,* 57-74.

Cote, G.L.; Robyt, J.F. Acceptor reactions of alternansucrase from *Leuconostoc mesenteroides* NRRL B-1355. *Carbohydr. Res.* 1982b, *111,* 127-142.

Cote, G.L.; Robyt, J.F. The formation of α-D-(1→3) branch linkages by an exocellular glucansucrase from *Leuconostoc mesenteroides* NRRL B-742. *Carbohydr. Res.* 1983, *119,* 141-156.

Cote, G.L.; Robyt, J.F. The formation of α-D-(1→3) branch linkages by a glucansucrase from *Streptococcus mutans* 6715 producing a soluble D-glucan. *Carbohydr. Res.* 1984, *127,* 95-107.

Côté, G.L.; Sheng, S. Penta-, hexa- and hepta-saccharide acceptor products of alternansucrase. *Carbohydr. Res.* 2006, *341,* 2066-2072.

Côté, G.L.; Sheng, S.; Dunlap, C.A. Alternansucrase acceptor products. *Biocatal. Biotrans.* 2008, *26,* 161-168.

Crittenden, R.G.; Doelle, H.W. Structural identification of oligosaccharides produced by *Zymomonas mobilis* levansucrase. *Biotechnol. Lett.* 1993, *15*, 1055-1060.

Demuth, B.; Jordening, H. J.; Buchholz, K. Modelling of oligosaccharide synthesis by dextransucrase. *Biotechnol. Bioeng.* 1999, *62,* 583-592.

Demuth, K.; Jördening, H.-J.; Buchholz, K. Oligosaccharide synthesis by dextransucrase: new unconventional acceptors. *Carbohydr. Res.* 2002, *337,* 1811–1820.

Dols, M.; Simeon, M.R.; Willemot, R.M.; Vignon, M.R.; Monsan, P.F. Structural characterization of the maltose acceptor-products synthesized by *Leuconostoc mesenteroides* NRRL B-1299 dextransucrase. *Carbohydr. Res.* 1997, *305,* 549-559.

Eapen, A.K.; Chengelis, C.P.; Jordan, N.P.; Baumgartner, R.E.; Zheng, G.-H.; Carlson, T. A 28-day oral (dietary) toxicity study of sucromalt in Sprague–Dawley rats. *Food Chem. Toxicol.* 2007, *45,* 2304–2311.

Euzenat, O.; Guibert, A.; Combes, D. Production of fructo-oligosaccharides by levansucrase from *Bacillus subtilis* C4. *Process Biochem.* 1997, *32,* 237-243.

Feingold, D.S.; Avigad, G.; Hestrin, S. The mechanism of polysaccharide production from sucrose. 4. Isolation and probable structures of oligosaccharides formed from sucrose by a levansucrase system. *Biochem. J.* 1956, *64*, 351-361.

Feingold, D.S.; Avigad, G.; Hestrin, S. Enzymatic synthesis and reactions of a sucrose isomer α-D- galactopyranosyl-β-D-fructofuranoside. *J. Biol. Chem.* 1957, *224*, 295-307.

Flickinger, E.A.; Wolf, B.W.; Garleb, K.A.; Chow, J.M; Leyer, G.J.; Johns, P.W.; Fahey, G.C., Jr. Glucose-based oligosaccharides exhibit different in vitro fermentation patterns and affect in vivo apparent nutrient digestibility and microbial populations in dogs. *J. Nutr.* 2000, *130,* 1267-1273.

Fu, D.; Robyt, J.F. Maltodextrin acceptor reactions of Streptococcus mutans 6715 glucosyltransferases. *Carbohydr. Res.* 1991, *217*, 201-211.

Fu, D.; Robyt, J.F. Acceptor reactions of maltodextrins with Leuconostoc mesenteroides B-512Fm dextransucrase. *Arch. Biochem. Biophys.* 1990, *283*, 379-387.

Fu, D.; Slodki, M.E.; Robyt, J.F. Specificity of acceptor binding to Leuconostoc mesenteroides B- 512F dextransucrase: Binding and acceptor product structure of alpha-methyl-D-glucopyranoside analogs modified at C-2, C-3, and C-4 by inversion of the hydroxyl and by replacement of the hydroxyl with hydrogen. *Arch. Biochem. Biophys.* 1990, *276*, 460-465.

Grysman, A.; Carlson, T.; Wolever, T.M.S. Effects of sucromalt on postprandial responses in human subjects. *Eur. J. Clin. Nutr.* 2008, *62,* 1364–1371.

Guggenheim, B.; Newbrun, E. Extracellular glucosyltransferase activity of an HS strain of *Streptococcus mutans*. *Helv. Odont. Acta* 1969, *13*, 84-97.

Han, W.-C.; Byun, S.-H.; Kim, M.-H.; Sohn, E.H.; Lim, J.D.; Um, B.H.; Kim, C.H.; Kang, S.A.; Jang, K.-H. Production of lactosucrose from sucrose and lactose by a levansucrase from *Zymomonas mobilis*. *J. Microbiol. Biotechnol.* 2009, *19,* 1153–1160.

Heincke, K.; Demuth, B.; Jordening, H.J.; Buchholz, K. Kinetics of the dextransucrase acceptor reaction with maltose - experimental results and modeling. *Enzyme Microb. Technol.* 1999, *24,* 523-534.

Hestrin, S.; Avigad, G. The mechanism of polysaccharide production from sucrose.5. Transfer of fructose to C-1 of aldose by levansucrase. *Biochem. J.* 1958, *69*, 388-398.

Hestrin, S.; Feingold, D.S.; Avigad, G. Synthesis of sucrose and other beta-D-fructofuranosyl aldosides by levansucrase. *J. Amer. Chem. Soc.* 1955, *77*, 6710.

Hestrin, S.; Feingold, D.S.; Avigad, G. The mechanism of polysaccharide production from sucrose. 3. Donor-acceptor specificity of levansucrase from Aerobacter levanicum. *Biochem. J.* 1956, *64*, 340-351.

van Hijum, S.A.F.T.; Geel-Schutten, G.H.; Rahaoui, H.; van der Maarel, M.J.E.C.; Dijkhuizen, L. Characterization of a novel fructosyltransferase from *Lactobacillus reuteri* that synthesizes high-molecular-weight inulin and inulin oligosaccharides. *Appl. Environ. Microbiol.* 2002, *68* (9), 4390-4398.

Holt, S.M.; Miller-Fosmore, C.M.; Côté, G.L. Growth of various intestinal bacteria on alternansucrase-derived oligosaccharides. *Lett. Appl. Microbiol.* 2005, *40,* 385-390.

Holt, S.M.; Teresi, J.M.; Côté, G.L. Influence of alternansucrase-derived oligosaccharides and other carbohydrates on enzyme activity in *Bifidobacterium adolescentis. Lett. Appl. Microbiol.* 2008, *6,* 73-79.

Homann, A.; Seibel, J. Towards tailor-made oligosaccharides—chemo-enzymatic approaches by enzyme and substrate engineering. *Appl. Microbiol. Biotechnol.* 2009, *83,* 209–216.

Ignatova, T.; Iliev, I.; Kirilov, N.; Vassileva, T.; Dalgalarrondo, M.; Haertlé, T.; Chobert, J.-M.; Ivanova, I. Effect of oligosaccharides on the growth of *Lactobacillus delbrueckii* subsp. *bulgaricus* strains isolated from dairy products. *J. Agric. Food Chem.* 2009, *57,* 9496–9502.

Iizuka, M.; Hiyama, M.; Itaya, K.; Furuichi, K.; Ann, Y.-G.; Minamiura, N.; Yamamoto, T. Susceptibility of leucrose to carbohydrases. *J. Ferment. Bioeng.* 1990, *70*, 277-279.

Jeanes, A.; Haynes, W.C.; Wilham, C.A.; Rankin, J.C.; Melvin, E.H.; Austin, M.J.; Cluskey, J.E.; Fisher, B.E.; Tsuchiya, H.M.; Rist, C.E. Characterization and classification of dextrans from 96 strains of bacteria. *J. Amer. Chem. Soc.* 1954, *76*, 5041-5052.

Jones, R.W.; Jeanes, A.; Stringer, C.S.; Tsuchiya, H.M. Crystalline methyl alpha-isomaltoside and its homologs obtained by synthetic action of dextransucrase. *J. Amer. Chem. Soc.* 1956, *78*, 2499-2502.

Killey, M.; Dimler, R.J.; Cluskey, J.E. Preparation of panose by the action of NRRL B-512 dextransucrase on a sucrose-maltose mixture. *J. Amer. Chem. Soc.* 1955, *77*, 3315-3318.

Kim, Y.M.; Park, J.P.; Sinha, J.; Lim, K.H.; Yun, J.W. Acceptor reactions of a novel transfructosylating enzyme from *Bacillus* sp. *Biotechnol. Lett.* 2001, *23,* 13–16.

Kitahata, S.; Fujita, K. Xylsucrose, isomaltosucrose and lactosucrose. Chapter 10. In: *Oligosaccharides. Production, Properties, and Application. Japanese Technology Reviews, Section E: Biotechnology.* Nakakuki, T., Ed.; Gordon and Breach Science Publishers, Yverdon, Switzerland, 1993, Vol. 3, pp.158-174.

Koepsell, H.J.; Tsuchiya, H.M.; Hellman, N.N.; Kazenko, A.; Hoffman, C.A.; Sharpe, E.S.; Jackson, R.W. Enzymatic synthesis of dextran. Acceptor specificity and chain initiation. *J. Biol. Chem.* 1953, *200*, 793-801.

Kralj, S. Glucansucrases of lactobacilli: Characterization of genes, enzymes and products synthesized. PhD Dissertation, University of Groningen, The Netherlands, 2004.

Kralj, S.; van Geel-Schutten, I.G.H.; Faber, E.J.; van der Maarel, M.J.E.C.; Dijkhuizen, L. Rational transformation of *Lactobacillus reuteri* 121 reuteransucrase into a dextransucrase. *Biochemistry* 2005b, *44,* 9206-9216.

Kralj, S.; van Geel-Schutten, G.H.; van der Maarel, M.J.E.C.; Dijkhuizen, L. Biochemical and molecular characterization of *Lactobacillus reuteri* 121 reuteransucrase. *Microbiology* 2004, *150,* 2099–2112.

Kralj, S.; Stripling, E.; Sanders, P.; van Geel-Schutten, G.H.; Dijkhuizen, L. Highly hydrolytic reuteransucrase from probiotic *Lactobacillus reuteri* strain ATCC 55730. *Appl. Environ. Microbiol.* 2005a, *71,* 3942–3950.

Kubik, C.; Sikora, B.; Bielecki, S. Immobilization of dextransucrase and its use with soluble dextranase for glucooligosaccharides synthesis. *Enzyme Microbial Technol.* 2004, *34,* 555–560.

Lamothe, J.-P.H.G.; Marchenay, Y.G.; Monsan, P.F.; Paul, F.M.B.; Pelenc, V. Cosmetic compositions containing oligosaccharides. *US Pat. 5,518,733,* 1996.

Leathers, T.D. Dextran. In: *Biopolymers, Polysaccharides I: Polysaccharides from Prokaryotes, chapter 12.* Vandamme E.J.; DeBaets, S.; Steinbuchel, A., Eds.; Wiley, Weinheim, Germany, 2002, Vol 5. pp 299–321.

Leathers, T.D.; Nunnally, M.S.; Côté, G.L. Modification of alternan by dextranase. *Biotechnol. Lett.* 2009, *31,* 289-293.

Lee, C.Y.; Yun, C.Y.; Yun, J.W.; Oh, T.K.; Kim, C.J. Production of glucooligosaccharides by an acceptor reaction using two types of glucansucrase from *Streptococcus sobrinus. Biotechnol. Lett.* 1997, *19,* 1227–1230.

Mayer, R.M.; Matthews, M.M.; Futerman, C.L.; Parnaik, V.K.; Jung, S.M. Dextransucrase: Acceptor substrate reactions. *Arch. Biochem. Biophys.* 1981, *208*, 278-287.

Monsan, P.; Remaud-Siméon, M.; André, I. Transglucosidases as efficient tools for oligosaccharide and glucoconjugate synthesis. *Curr. Opin. Microbiol.* 2010, *13,* 1–8.

Mukasa, H.; Shimamura, A.; Tsumori, H. Nigerooligosaccharide acceptor reaction of *Streptococcus sobrinus* glucosyltransferase GTF-I. *Carbohydr. Res.* 2000, *326,* 98–103.

Neely, W.B. Studies on the enzyme dextransucrase. II. The role of raffinose as an acceptor. *Arch. Biochem. Biophys.* 1959, *79*, 154-161.

Okada, G.; Hehre, E.J. New studies on amylosucrase, a bacterial α-D-glucosylase that directly converts sucrose to a glycogen-like α-glucan. *J. Biol. Chem.* 1974, *249*, 126-135.

Ozimek, L.K. Structure – function relationships in fructosyltransferase enzymes from *Lactobacillus reuteri* 121. PhD Dissertation, University of Groningen, The Netherlands, 2005.

Ozimek, L.K.; Kralj, S.; van der Maarel, M.J.E.C.; Dijkhuizen, L. The levansucrase and inulosucrase enzymes of *Lactobacillus reuteri* 121 catalyse processive and non-processive transglycosylation reactions. *Microbiology* 2006, *152,* 1187–1196.

Park, N.-H.; Choi, H.-J.; Oh, D.-K. Lactosucrose production by various microorganisms harboring levansucrase activity. *Biotechnol. Lett.* 2005, *27,* 495-497.

Park, H.-E.; Park, N.H.; Kim, M.J.; Lee, T.-H.; Lee, H.G.; Yang, J.-Y.; Cha, J. Enzymatic synthesis of fructosyl oligosaccharides by levansucrase from *Microbacterium laevaniformans* ATCC 15953. *Enz. Microbial Technol.* 2003, *32,* 820–827.

Paul, F.; Oriol, E.; Auriol, D.; Monsan, P. Acceptor reactions of a highly purified dextransucrase with maltose and oligosaccharides: Application to the synthesis of controlled-molecular-weight dextrans. *Carbohydr. Res.* 1986, *149*, 433-441.

Pelenc, V.; Lopez-Munguia, A.; Remaud, M.; Biton, J.; Michel, J. M.; Paul, F.; Monsan, P. Enzymatic synthesis of oligoalternans. *Sciences des Aliments* 1991, *11*, 465-476.

Peltroche-Llacsahuangal, H.; Hauk, C.J.; Kock, R.; Lampert, F.; Lütticken, R.; Haase, G. Assessment of acid production by various human oral micro-organisms when palatinose or leucrose is utlized. *J. Dent. Res.* 2001, *80,* 378-384.

Pereira, A.M.; Costa, F.A.A.; Rodrigues, M.I.; Maugeri, F. In vitro synthesis of oligosaccharides by acceptor reaction of dextransucrase from *Leuconostoc mesenteroides*. *Biotechnol. Lett.* 1998, *20,* 397-401.

Quirasco, M.; Lopez-Munguia, A.; Pelenc, V.; Remaud, M.; Paul, F.; Monsan, P. Enzymatic production of glucooligosaccharides containing α-(1→2) osidic bonds: Potential application in nutrition. *Ann. NY Acad. Sci.* 1995, *750*, 317-320.

Rabelo,M.C.; Honorato, T.L.; Gonçalves, L.R.B.; Pinto, G.A.S.; Rodrigues, S. Enzymatic synthesis of prebiotic oligosaccharides. *Appl. Biochem. Biotechnol.* 2006, *133,* 31-40.

Reh, K.D.; NollBorchers, M.; Buchholz, K. Productivity of immobilized dextransucrase for leucrose formation. *Enzyme Microb. Technol.* 1996, *19,* 518-524.

Remaud-Simeon, M.; Lopez-Munguia, A.; Pelenc, V.; Paul, F.; Monsan, P. Production and use of glucosyltransferases from *Leuconostoc mesenteroides* NRRL B-1299 for the synthesis of oligosaccharides containing α-(1→2) linkages. *Appl. Biochem. Biotechnol.* 1994, *44*, 101-117.

Remaud, M.; Paul, F.; Monsan, P.; Lopezmunguia, A.; Vignon, M. Characterization of α-(1→3) branched oligosaccharides synthesized by acceptor reaction with the extracellular glucosyltransferases from *L. mesenteroides* NRRL B-742. *J. Carbohyd. Chem.* 1992, *11*, 359-378.

Richard, G.; Yu, S.; Monsan, P.; Remaud-Simeon, M.; Morel, S. A novel family of glucosyl 1,5-anhydro-D-fructose derivatives synthesised by transglucosylation with dextransucrase from *Leuconostoc mesenteroides* NRRL B-512F. *Carbohydr. Res.* 2005, *340,* 395-401.

Robyt, J.F.; Eklund, S.H. Relative, quantitative effects of acceptors in the reaction of Leuconostoc mesenteroides B-512F dextransucrase. *Carbohydr. Res.* 1983, *121*, 279-286.

Robyt, J.F.; Taniguchi, H. The mechanism of dextransucrase action. Biosynthesis of branch linkages by acceptor reactions with dextran. *Arch. Biochem. Biophys.* 1976, *174*, 129-135.

Rodrigues, S.; Lona, L.M.F.; Franco, T.T. Optimizing panose production by modeling and simulation using factorial design and surface response analysis. *J. Food Eng.* 2006, *75,* 433–440.

Russell, R.R.B.; Gilpin, M.L.; Hanada, N.; Yamashita, Y.; Shibata, Y.; Takehara, T. Characterization of the product of the gtfS gene of *Streptococcus downei*, a primer-independent enzyme synthesizing oligo-isomaltosaccharides. *J. Gen. Microbiol.* 1990, *136,* 1631-1637.

Sako, T.; Matsumoto, K.; Tanaka, R. Recent progress on research and applications of non-digestible galacto-oligosaccharides. *Int. Dairy J.* 1999, *9,* 69-80.

Sanz, M.-L.; Côté, G.L.; Gibson, G.R.; Rastall, R.A. Prebiotic properties of alternansucrase maltose-acceptor oligosaccharides. *J. Agric. Food Chem.* 2005, *53,* 5911-5916.

Sanz, M.-L.; Côté, G.L.; Gibson, G.R.; Rastall, R.A. Influence of glycosidic linkages and molecular weight in fermentation process of maltose-based oligosaccharides. *J. Agric. Food Chem.* 2006a, *54,* 9779-9784.

Sanz, M.-L.; Côté, G.L.; Gibson, G.R.; Rastall, R.A. Selective fermentation of gentiobiose-derived oligosaccharides by human gut bacteria and influence of molecular weight. *FEMS Microbiol. Ecol.* 2006b, *56,* 383-388.

Schwengers, D. Leucrose, a ketodisaccharide of industrial design. *In Carbohydrates as Organic Raw Materials*, Lichtenthaler, F.W., Ed.; VCH: Weinheim, Germany, 1991, 183-195.

de Segura, A.G.; Alcalde, M.; Bernabé M.; Ballesteros, A.; Plou, F.J. Synthesis of methyl α-D-glucooligosaccharides by entrapped dextransucrase from *Leuconostoc mesenteroides* B-1299. *J. Biotechnol.* 2006, *124,* 439–445.

Sharpe, E.S.; Stodola, F.H.; Koepsell, H.J. Formation of isomaltulose in enzymatic dextran synthesis. *J. Org. Chem.* 1960, *25*, 1062-1063.

Slodki, M.E.; England, R.E.; Plattner, R.D.; Dick, W.E. Methylation analyses of NRRL dextrans by capillary GLC. *Carbohydr. Res.* 1986, *156*, 199-206.

Smiley, K.L.; Slodki, M.E.; Boundy, J. .; Plattner, R.D. A simplified method for preparing linear isomalto-oligosaccharides. *Carbohydr. Res.* 1982, *108*, 279-283.

Stahl, B.; Kliem, M.; Farwer, S.; Sawatzki, G.; Boehm, G. Food for diabetics. *US Pat. 7,498,318,* 2009.

Stodola, F.H.; Koepsell, H.J.; Sharpe, E.S. A new disaccharide produced by *Leuconostoc mesenteroides. J. Am. Chem. Soc.* 1952, *74*, 3202-3203.

Su, D.; Robyt, J.F. Control of the synthesis of dextran and acceptor-products by *Leuconostoc mesenteroides* B-512FM dextransucrase. *Carbohydr. Res.* 1993, *248,* 339-348.

Tambara, Y.; Hormaza, J.V.; Perez, C.; Leon, A.; Arrieta, J.; Hernandez, L. Structural analysis and optimised production of fructo- oligosaccharides by levansucrase from *Acetobacter diazotrophicus* SRT4. *Biotechnol Lett.* 1999, *21,* 117-121.

Tanaka, T.; Yamamoto, S.; Oi, S.; Yamamoto, T. Structures of heterooligosaccharides synthesized by levansucrase. *J. Biochem.* 1981, *90*, 521-526.

Tieking, M.; Kaditzky, S.; Valcheva, R.; Korakli, M.; Vogel, R.F.; Gänzle, M.G. Extracellular homopolysaccharides and oligosaccharides from intestinal lactobacilli. *J. Appl. Microbiol.* 2005, *99,* 692–702.

Valette, P.; Pelenc, V.; Djouzi, Z.; Andrieux, C.; Paul, F.; Monsan, P.; Szylit, O. Bioavailability of new synthesised glucooligosaccharides in the intestinal tract of gnotobiotic rats. *J. Sci. Food Agric.* 1993, *62*, 121-127.

Vercauteren, R.L.M.; Nguyen, V.S. Process for preparing isomalto-oligosaccharides with elongated chain and low-glycemic index. *US Pat. 7,670,811*, 2010.

Walker, G.J. Preparation of isomaltose oligosaccharides labelled with 14C in the non-reducing terminal unit, and their use in studies of dextranase activity. *Carbohydr. Res.* 1973, *30*, 1-10.

Yamauchi, F.; Ohwada, Y. Synthesis of oligosaccharides by growing culture of *Leuconostoc mesenteroides.* Part IV. Oligosaccharide formation in the presence of various types of glucobioses as acceptors. *Agric. Biol. Chem.* 1969, *33*, 1295-1300.

Yoo, S.; Kim, D.; Day, D.F. Highly branched glucooligosaccharide and mannitol production by mixed culture fermentation of *Leuconostoc mesenteroides* and *Lipomyces starkeyi*. *J. Microbiol. Biotechnol.* 2001, *11,* 700-703.

Ziesenitz, S.C.; Siebert, G.; Schwengers, D.; Lemmes, R. Nutritional assessment in humans and rats of leucrose [D-glucopyranosyl-alpha(1→5)-D-fructopyranose] as a sugar substitute. *J. Nutr.* 1989, *119,* 971-978.

In: Oligosaccharides: Sources, Properties and Applications ISBN 978-1-61122-193-0
Editor: Nicole S. Gordon

Chapter 7

ELECTROSPRAY MASS SPECTROMETRY OF OLIGOSACCHARIDES OF PLANT ORIGIN

Maria do Rosário M. Domingues[1], Fernando M. Nunes[2] and Manuel A. Coimbra[1]*

[1]Departamento de Química, Universidade de Aveiro, 3810-193 Aveiro, Portugal
[2] Departamento de Química, Universidade de Trás-os-Montes e Alto Douro, 5000-911 Vila Real, Portugal

ABSTRACT

Carbohydrates are widely distributed in nature, being responsible for different functions in almost all living organisms. They are present in the living systems as polysaccharides, glycoproteins, and glycolipids. Their properties and applications are dependent on their detailed structure, namely, their composition in monosaccharides, type of linkages, branching, and anomeric configurations. Mass spectrometry (MS) is a useful tool for the study of carbohydrates, especially when soft ionization methods are used. For the analysis of oligosaccharides, electrospray mass spectrometry (ESI-MS) and tandem mass spectrometry (ESI-MS^n) has been elected in the last recent years as the method of choice, having the advantage of using underivatised oligomers, even when present in mixtures and with low abundance.

The analysis of polysaccharides in their native form is difficult due to their high molecular weight. In order to overcome this, polysaccharides are usually cleaved into oligosaccharides.

In the present work the structural features of glucuronoxylans, galactomannans, and type-II arabinogalactans achieved by the analysis by ESI-MS of their oligosaccharides are discussed. In addition, it is shown how can the anomeric configurations of reducing disaccharides of glucose using quadrupole time-of-flight (Q-TOF2), linear ion trap (LIT), and triple quadrupole (QqQ) mass spectrometers be obtained.

1. Introduction.

Carbohydrates are the most abundant biomolecules in nature. The biological importance and the chemical behaviour of carbohydrates, namely, of polysaccharides, glycoproteins, and glycolipids, are related to their structure. Their properties and applications are dependent on their detailed structure, namely, their composition in monosaccharides, type of linkages, degree and type of branching, and anomeric configurations. Mass spectrometry (MS) is a useful tool for the study of carbohydrates, and has been used during the last decade to the structural characterization of oligosaccharides [1, 2]. For the analysis of oligosaccharides, electrospray mass spectrometry (ESI-MS) and tandem mass spectrometry (ESI-MS^n) has been elected as the method of choice, having the advantage of using underivatised oligomers, even when present in mixtures and with low abundance [1, 3]. Using MS, it is possible to obtain information about monosaccharide sequence, branching pattern, and presence of modifying chemical groups, allowing identifying the primary structure of the oligosaccharides (OS). The advantage of this method is it high sensibility, allowing to analyse OS in amounts of picomole level, and in mixtures.

Prior to MS analysis and in order to obtain lower molecular weight oligosaccharides, polysaccharides are usually subjected to partial hydrolysis, using acidic conditions or enzymatic procedures. The hydrolysate obtained is then fractionated by size exclusion chromatography and selected fractions are analysed by ESI-MS. Under ESI-MS conditions, oligosaccharides ionize as sodium adducts, $[M+Na]^+$, even without the addition of any salt to the analyte solution. No fragmentation is observed under ESI conditions. Thus, the analysis of the ESI-MS spectrum allows to obtain a direct information about the molecular weight of the oligosaccharides present in the fraction under analysis.

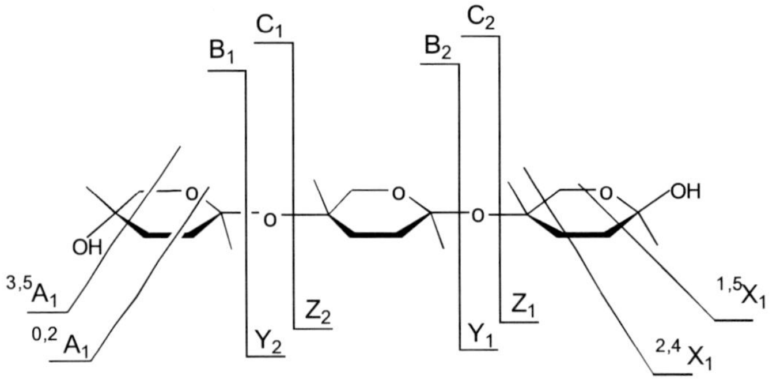

Figure 1. Nomenclature for fragment-ions from carbohydrates, according to Domon & Costello [4], as described by Reis *et al.* [5].

To confirm the structures proposed based on the information of the different $[M+Na]^+$ ions identified in the ESI mass spectrum, MS/MS need to be performed on each of these ions. The fragmentation pattern of oligosaccharides under MS/MS conditions results mainly of glycosidic cleavages between two sugar residues, and of cross-ring cleavages with the cleavage of two bonds within the sugar ring. The fragment ions formed have been named according to the nomenclature proposed by Domon and Costello [4] for the fragmentation of oligosaccharides (Figure 1). Oligosaccharide fragment ions that retain the charge at the reducing end are designated X (cross-ring cleavage), Y, and Z (glycosidic cleavage) whereas the complementary ions are numbered from the non-reducing end for A (cross-ring cleavage), B, and C (glycosidic cleavage). Following the letter that shows the fragment type is a subscript number that identifies the sugar residue. Cross-ring fragments also exhibit a superscript number that represents the bonds cleaved.

2. GLUCURONOXYLANS

Glucuronoxylans (GX) are one of the major components of the hemicellulosic fraction of plants cell wall material [6]. These polysaccharides are composed of β-(1→4)-linked xylopyranose residues forming linear chains. These chains are frequently substituted at *O*-2 by acidic residues of glucuronic (GlcA) or 4-*O*-methyl-glucuronic acid (MeGlcA), and at *O*-2 and *O*-3 by neutral arabinofuranose residues (Ara*f*), and by acetyl groups (Ac) (Figure 2). The presence of these substituent residues is dependent on the origin of the polysaccharides.

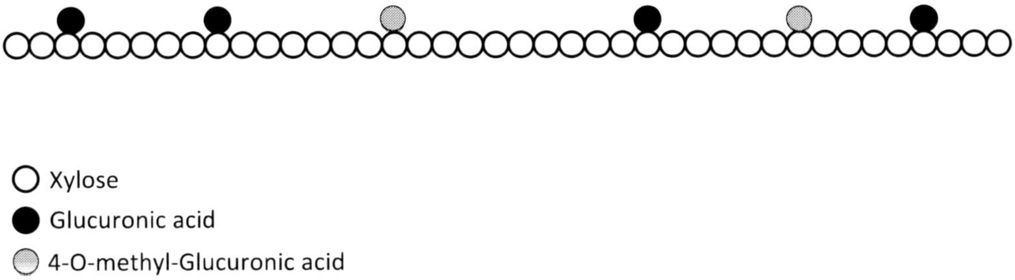

Figure 2. Schematic structure of a glucuronoxylan.

The physicochemical properties of cell walls are strongly affected by GX structural features (nature and frequency of the branching side chains, degree of acetylation, molecular weight, etc.), which knowledge is crucial regarding the digestibility and processability behaviour of plant/fruit/vegetable tissues [7, 8]. Commonly, GX structures, as other hemicelluloses, are elucidated by a couple of chemical analysis and NMR techniques [9, 10]. However, these methods have several limitations, such as not enough sensitivity towards important structural features of minor abundance (lower than 2 mol%) and the poor information revealed on the primary structure of glucuronoxylans.

Detailed structural features of glucuronoxylans can also be achieved by ESI-MS and MS/MS analysis, namely, for determination of the main xylose backbone and substitution with glucuronic acid. For that, the polysaccharides need to be converted into oligosaccharides. Usually, this can be performed by mild acid hydrolysis with 50 mM TFA during 45 min at 100ºC [5, 11]. The mixture of oligosaccharides obtained need to be fractionated, *e.g.*, by size exclusion chromatography in order to enriched fractions in small groups of molecules.

Figure 3 shows an ESI-MS spectrum, in positive mode, representative of oligosaccharides released from a glucuronoxylan [5]. This ESI-MS spectrum shows the $[M+Na]^+$ ions of xylo-oligosaccharides in a mass range below *m/z* 1000. The xylo-oligosaccharide structures identified corresponds to the series of neutral oligosaccharides of xylose at *m/z* 437, 569, 701, 833, 965 (Xyl_n, n=3-7), of acidic oligosaccharides substituted by one glucuronic acid (Xyl_nGlcA, n=3-5) at *m/z* 481, 613, 745 and two glucuronic acid residues (Xyl_nGlcA_2, n=2 and 3) at *m/z* 789, 921, and also of acidic oligosaccharides substituted with one 4-*O*-methyl-glucuronic acid residue ($Xyl_nMeGlcA$, n= 2-4) at *m/z* 495, 627 and 759. The presence of these structures needs to be confirmed by MS/MS spectra, by collisional induced dissociation of the observed ions.

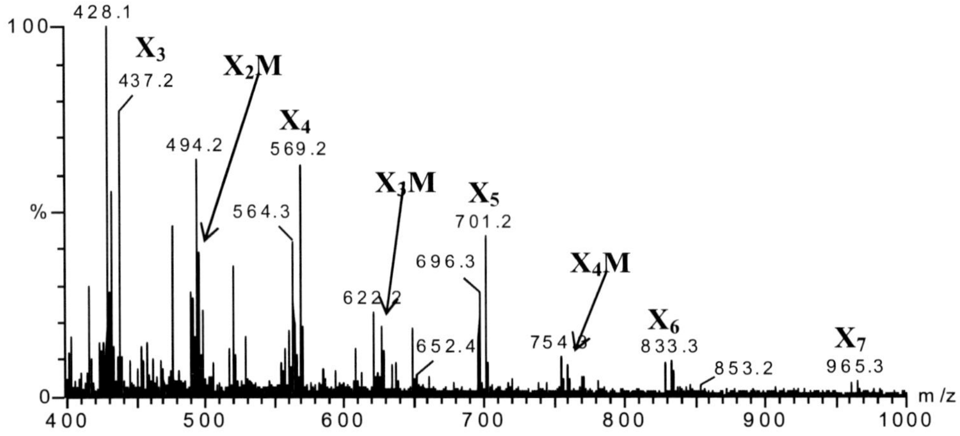

Figure 3. Electrospray mass spectrum of olive pulp xylo-oligosaccharides [5], where X is Xyl and M is MeGlcA.

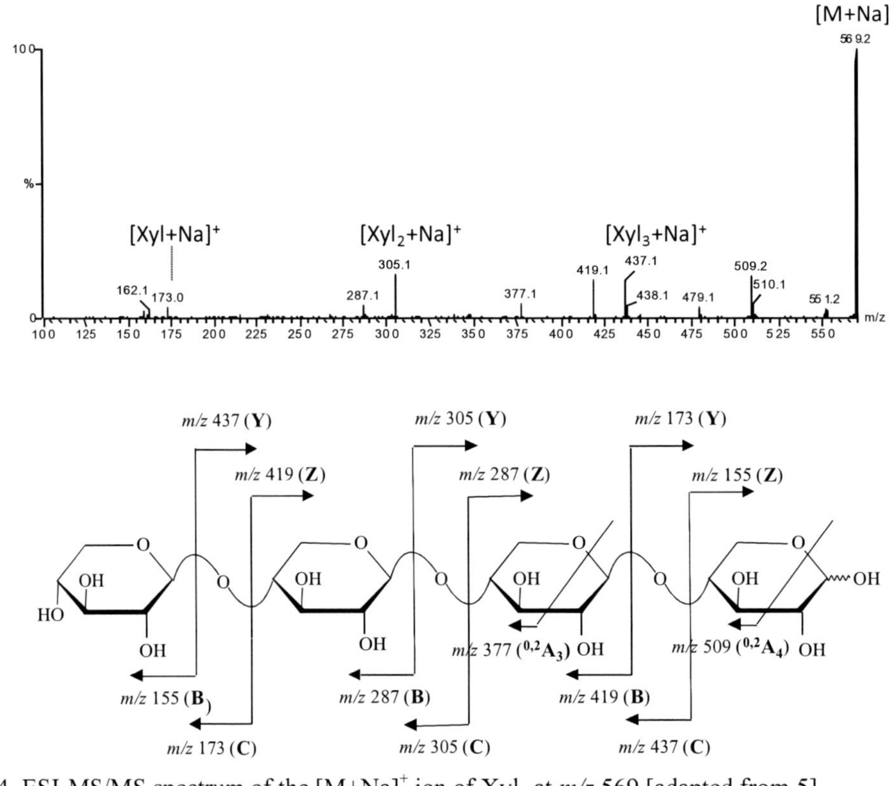

Figure 4. ESI-MS/MS spectrum of the $[M+Na]^+$ ion of Xyl_4 at *m/z* 569 [adapted from 5].

2.1. ESI-MS/MS of Linear Xylo-Oligosaccharides

The fragmentation pathways observed in the ESI-MS/MS spectra of the $[M+Na]^+$ ions of Xyl_n allows to confirm the composition of the oligosaccharides and to identify the number of xylose units, as well as confirmation of the type of linkage [5, 12]. As can be seen in Figure 4, the ESI-MS/MS of the $[M+Na]^+$ ion of Xyl_4 at *m/z* 569 shows the loss of one xylose residue (-Xyl_{res}, -132 Da), with formation of the ion $[Xyl_3+Na]^+$, at *m/z* 437, loss of two Xyl_{res} (-2x132 Da), with formation of the ion $[Xyl_2+Na]^+$, at *m/z* 305, and combined loss of three Xyl_{res} with formation of the ion $[Xyl+Na]^+$ at *m/z* 173, confirming that the oligosaccharide observed at *m/z* 569 is composed by four xylose units. Also loss f $C_2H_4O_2$ (-60Da) with formation of the ion at *m/z* 509, occurring by cross ring cleavage ($^{0,2}A_4$), confirms that xylose units are linked by a (1→4)-linkage. The similar information can be obtained for the other Xyl_n oligosaccharides using the same methodologies based on the ESI-MS/MS interpretation.

2.2. ESI-MS/MS of Linear Methyl-Glucuronic Acid Substituted Xylo-Oligosaccharides

The structural features of the oligosaccharides bearing a methyl-glucuronic acid (MeGlcA) can be inferred by the analysis of the correspondent ESI-MS/MS spectra [5, 11]. In figure 5 it is shown, as an example, the ESI-MS/MS spectrum of $[M+Na]^+$ ions of $Xyl_nMeGlcA$ (n=2) that show a major loss of the MeGlcA residue (-190 Da) with formation of the ion $[Xyl_2+Na]^+$ at *m/z* 305, confirming that this oligosaccharide has a methyl-glucuronic acid substituting linked to a disaccharide of xylose. The similar information can be obtained for the other $Xyl_nMeGlcA$ oligosaccharides using the same methodology based on the ESI-MS/MS interpretation.

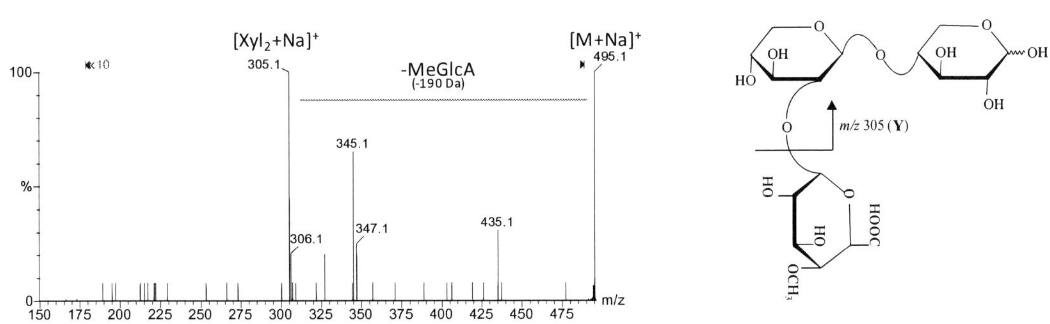

Figure 5. ESI-MS/MS spectrum of the $[M+Na]^+$ ion of $Xyl_2MeGlcA$ at *m/z* 495 [adapted from 5].

2.3. ESI-MS/MS of Linear Glucuronic Acid Substituted Xylo-Oligosaccharides and Distinction of Isomeric Structures

The structural features of the oligosaccharides bearing a glucuronic acid (GlcA) can be inferred by the analysis of the correspondent ESI-MS/MS spectra of the $[M+Na]^+$ adduct of the oligosaccharides from the Xyl_nGlcA [5, 12]. The ESI-MS/MS spectrum of $[M+Na]^+$ ion of the Xyl_3GlcA is shown in figure 6, as an case in point. In this spectrum it is possible to see a major ion formed due to the loss of the GlcA residue (-176 Da) with formation of the ion $[Xyl_3+Na]^+$ at *m/z* 437, confirming that this oligosaccharide has a glucuronic acid substitution linked to a trisaccharide of xylose. Similar information can be obtained for the other Xyl_nGlcA oligosaccharides using the same methodology based on the ESI-MS/MS interpretation.

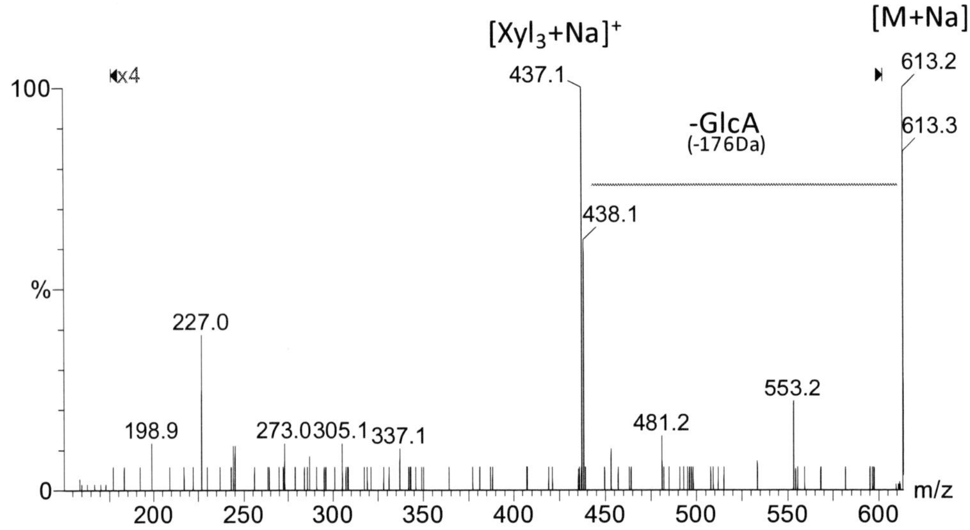

Figure 6. ESI-MS/MS spectrum of the $[M+Na]^+$ ion of Xyl_3GlcA at *m/z* 613 [adapted from 5].

The tetramer Xyl_2GlcA_2 was also identified by ESI-MS/MS in the oligosaccharides released from a xylan from olive, by tandem mass spectrometry confirming the presence of vicinal GlcA in these polymers. The MS/MS spectrum of the $[M+Na]^+$ ion of Xyl_2GlcA_2 (Figure 7A) show the ions at *m/z* 481 identified as $[Xyl_2GlcA +Na]^+$ and ion at *m/z* 305 identified as $[Xyl_2+Na]+$, formed by the loss of one and two GlcA, respectively and confirming the presence of two GlcA in the molecule, as shown in Figure 8 [12]. The glycosidic cleavage between the two Xyl leads to the ions $[GlcA\text{-}Xyl+Na]^+$ at *m/z* 349 , and $[GlcA\text{-}Xyl\text{-}H_2O+Na]^+$ at *m/z* 331. The loss of GlcA can occur combined with loss of CO_2 leading to the ion at *m/z* 419. Ions at *m/z* 613 and 569 are due to the loss of one and two CO_2 molecules.

This tetrasaccharide Xyl_2GlcA_2 can be distinguished from another tetramer, Xyl2MeGlcAHex, with the same molecular weight. The ESI- MS/MS spectrum of $Xyl_2MeGlcAHex$ (Figure 7B), shows the ion at *m/z* 495, due to the loss of 162 Da, indicative of the presence of a Hex, the ions at *m/z* 597 (loss of $C_2H_4O_2$), 567 (loss of $C_3H_6O_3$) and 537 (loss of $C_4H_8O_4$), due to cross-ring cleavages of the Hex. Loss of Xyl (- 150 Da) and loss of Xyl_{res} (-132 Da) with formation of the ions at *m/z* 507 and 525 respectively are also observed. The absence of a fragment ion formed by loss of MeGlcA (190 Da) from the precursor ion supports the presence of MeGlcA between Xyl and Hex, as shown in Figure 8. The ion at *m/z* 375, $[Hex\text{-}MeGlcA\text{-}H_2O+Na]^+$, also confirms the proposed structure.

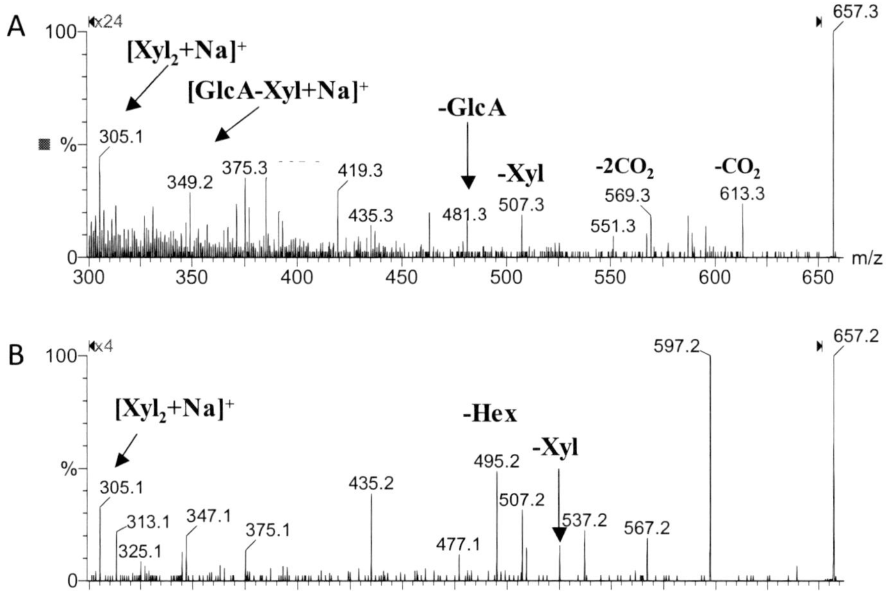

Figure 7. ESI-MS/MS spectra of $[M+Na]^+$ ions at *m/z* 657 of A) Xyl_2GlcA_2 and B) $Xyl_2MeGlcAHex$ [12].

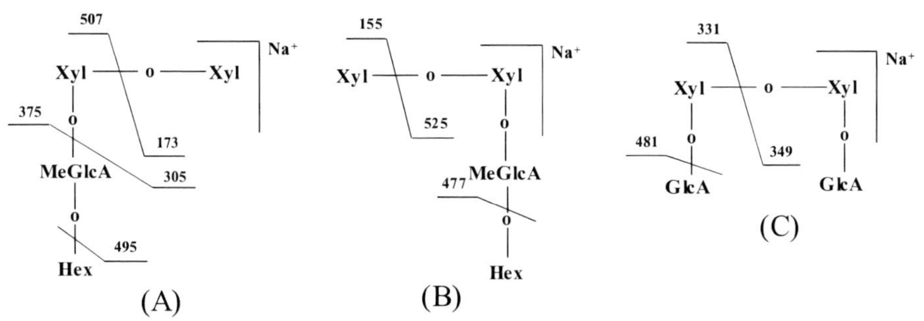

Figure 8. Proposed fragmentation pathways for the $[M+Na]^+$ of $Xyl_2MeGlcAHex$ (A and B) and Xyl_2GlcA_2 (C) [12].

Overall, ESI-MS and MS/MS analysis allows an accurate identification of uronosyl substitutions in complex mixtures of XOS, not possible by the usual methodologies in carbohydrate structural analysis. By using ESI-MS and MS/MS it were identified neutral $Xyl_{2\text{-}13}$ and acidic xylo-oligosaccharides, namely $Xyl_{2\text{-}11}MeGlcA$, and $Xyl_{2\text{-}3}GlcA$ and Xyl_2GlcA_2. The occurrence of higher molecular weight oligosaccharides with a low substitution pattern, such as Xyl_{13} and $Xyl_{11}MeGlcA$, allows to infer a scatter and random distribution of MeGlcA along the xylan backbone. Since these structures occur in small amounts in complex acidic XOS mixtures, and are very difficult, if possible, to isolate, tandem mass spectrometry revealed to be a powerful tool for the characterisation of substitution patterns present in glucuronoxylans.

3. Structural Features of Coffee Polysaccharides and Effect of the Roasting Process

Polysaccharides are important constituents of green and roasted coffee beans and coffee infusions. Roasted coffee beans comprise 30–43% of polysaccharides [13] and are composed mainly by galactomannans (48%), type II arabinogalactans (34%), and cellulose (18%). Coffee infusion polysaccharides are mainly composed by galactomannans (68%) and type II arabinogalactans [14, 15, 16]. A significant amount of coffee infusion galactomannans are acetylated [17]. During the roasting process, 35% of the mannose (Man) residues are degraded [18, 19] and their fates are largely unknown. Several modifications in coffee polysaccharides structure occur during the roasting process such as depolymerization, debranching, Maillard reaction, caramelization, isomerisation, oxidation, and decarboxylation of their reducing end [20]. Various new structural details of water soluble coffee polysaccharides have been determined, many of them with the help of structural studies by ESI-MS/MS. Also, some reaction pathways of coffee galactomannans have been elucidated by the detection and structural determination of the new derivatives by ESI-MS/MS. The following sections show the use of ESI-MS/MS in the structural elucidation of the coffee polysaccharides features, namely galactomannans and type II arabinogalactans, and also the roasting induced structural changes in coffee galactomannans. ESI-MS analysis offers the advantage of allowing the study of intact oligosaccharides, even when present in mixtures and with low abundance, without any manipulation/derivatization being required.

3.1. Coffee Galactomannans

The galactomannans of green and roasted coffee infusions are composed by a backbone of β-(1→4)-linked D-mannopyranosyl residues substituted at O-6 by single α-D-galactopyranosyl residues and arabinosyl residues. The mannan backbone also contains acetyl groups at O-2 and/or O-3 position of the mannosyl residues. Single acetylated mannosyl residues, di-acetylated mannosyl residues and consecutively acetylated mannosyl residues are present in coffee galactomannans (Figure 9).

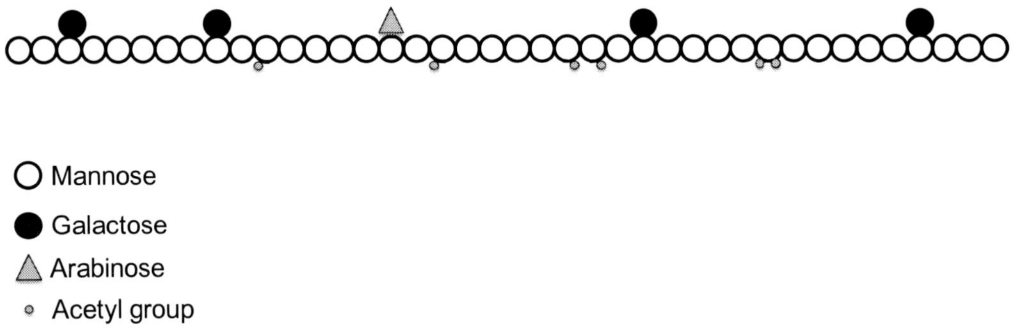

Figure 9. One of the many possible arrangements of coffee infusion galactomannans.

These structural features can be elucidated by ESI-MS/MS analysis of the oligosaccharides obtained by specific enzymatic hydrolysis of the mannan backbone with the

Aspergillus niger endo-β-mannanase [17]. The enzymatic hydrolysis of coffee galactomannans yields a range of oligosaccharides, the most abundant having two (DP2) or three (DP3) sugar residues. These oligosaccharides are the most informative ones when analyzed by ESI-MS/MS. As an example, the ESI-MS spectra of a DP3 fraction obtained from green coffee galactomannans is present in Figure 10. Predominant oligosaccharide $[M+Na]^+$ ions are observed at *m/z* 365 and 527, for di- and trihexoses, respectively. The corresponding potassium adduct ions ($[M+K]^+$) can also be observed at *m/z* 381 and 543. The ion at *m/z* 497, with a mass increase of 132 Da relative to *m/z* 365, is the $[M+Na]^+$ ion of the pentose–dihexose trisaccharide. It can also be observed intense peaks at *m/z* 407 and 569, with a mass 42 Da higher relative to the di- and trihexoses, corresponding, to $[M+Na]^+$ ions of the acetylated di- and trisaccharides and the ion at *m/z* 449 corresponding to the $[M+Na]^+$ of diacetylated-dihexoses. The nature and further structural details of the above mentioned ions are obtained by MS/MS analysis.

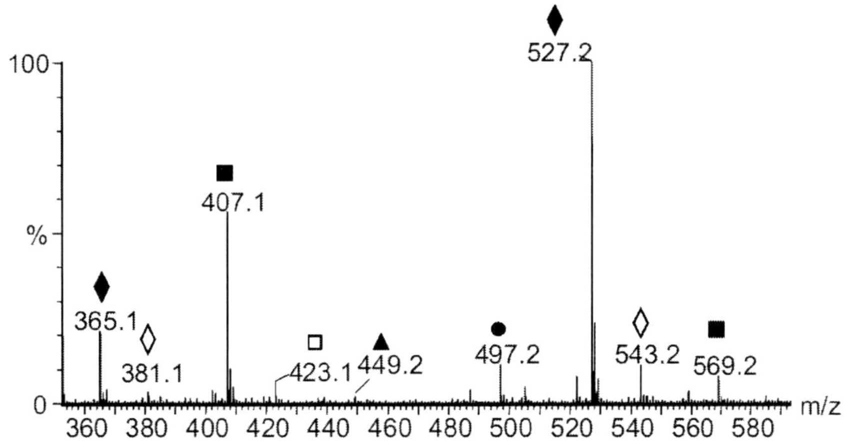

Figure 10. Positive ESI-MS spectrum of a DP3 fraction obtained after enzymatic hydrolysis of water soluble green coffee galactomannan. •: $[M+Na]^+$ adduct of $PentHex_2$; ■: $[M+Na]^+$ adduct of $AcHex_{2-3}$; □: $[M+K]^+$ adduct of $AcHex_{2-3}$; ♦:$[M+Na]^+$ adduct of Hex_{2-3}; ◊ $[M+K]^+$ adduct of Hex_{2-3}; ▲: $[M+Na]^+$ adduct of Ac_2Hex_2 [17].

The ESI-MS/MS spectrum of $[Hex_3+Na]^+$ (*m/z* 527) is shown in Figure 11. The predominant ions observed are attributed to C/Y-type glycosidic bond cleavage yielding the ion at *m/z* 365 and the ion at *m/z* 203 corresponding to the loss of one and two Hex_{res} (162 Da), respectively. Ions at *m/z* 467 and *m/z* 305 are attributed to a $^{0,2}A_3$ and $^{0,2}A_2$ (60 Da) cross-ring fragments. $^{2,4}X_1$ cross-ring cleavage can also explain the product ion at *m/z* 305. However, since the low energy CID ion spectrum showed predominant fragmentation from the reducing end [1] it allows to attribute this fragmentation to $^{0,2}A$ cross-ring fragmentation. The loss of a neutral fragment with 60 Da and the absence of fragmentations due to losses of 90 Da and 120 Da, is a characteristic of the MS/MS of (1→4)-linked hexoses (this concept will be detailed later).

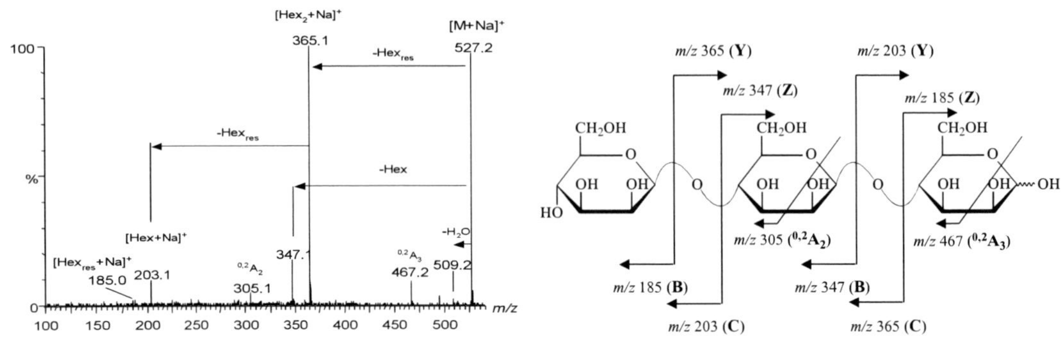

Figure 11. ESI-MS/MS spectrum of $[M+Na]^+$ adducts and schematic fragmentation pathways of (1→4)-linked Hex_3 [17].

The presence of an acetyl group and its location in the oligosaccharides can be easily identified by ESI-MS/MS analysis of the acetylated oligosaccharides observed in the ESI-MS spectra. The ESI-MS/MS spectrum of $[AcHex_2+Na]^+$ (*m/z* 407), shows the presence of ions at *m/z* 245, corresponding to $[AcHex+Na]^+$, due to loss of 162 Da (Hex_{res}) by a C/Y-type cleavage (Figure 12). Also, the abundant ion at *m/z* 305, corresponding to a loss of 102 Da, is attributed to an $^{0,2}A_2$ cross-ring fragment of a hexose residue containing one acetyl group at O-2. Another feature of the MS/MS spectra of $[AcHex_2+Na]^+$ is the occurrence of fragment ion at *m/z* 364, due to a loss of 43 Da, attributed to a homolytic cleavage of the acetate group. The ion at *m/z* 347 (loss of a neutral of 60 Da), can be due to cross-ring cleavage of hexose residue and also to loss of acetic acid, from fragmentation of an acetyl substituent.

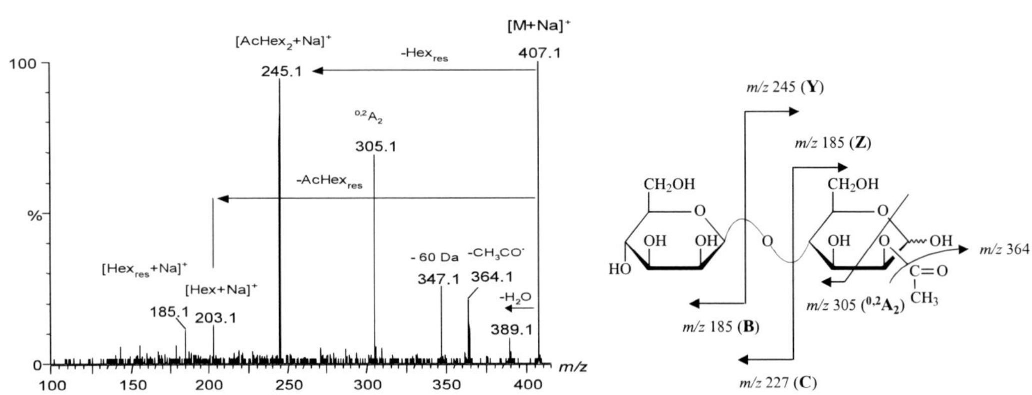

Figure 12. ESI-MS/MS spectrum of $[M+Na]^+$ adducts and schematic fragmentation pathways of (1→4)-linked $AcHex_2$ [17].

The presence of consecutively acetylated mannose residues and di-acetylated mannose residues can be also differentiated by the ESI-MS/MS analysis of ions *m/z* 449 and *m/z* 611. Figure 13 shows the ESI-MS/MS spectrum of $[Ac_2Hex_2+Na]^+$ (*m/z* 449). Fragment ion at *m/z* 287 corresponds to $[Ac_2Hex+Na]^+$ ion, resulting from loss of an Hex_{res} by a C/Y-type cleavage from $[Ac_2Hex_2+Na]^+$. Fragment ion at *m/z* 245 corresponds to $[AcHex+Na]^+$, resulting from loss of an $AcHex_{res}$ by a C/Y-type cleavage. The presence of the fragment ions at *m/z* 287 and *m/z* 245 in the MS/MS spectrum of this trisaccharide shows that the ion with *m/z* 449 is composed by two isomers, one hexose disaccharide bearing two acetyl groups that can occur in a single diacetylated residue or in two consecutive acetylated residues. However, the ion at *m/z* 245, formed by loss of $AcHex_{res}$ through a C/Y-type cleavage from a diacetylated disaccharide (*m/z* 449), obtained by endo-β-(1→4)-mannanase hydrolysis (Figure 13), is not compatible with the known mechanism of the hydrolysis by this enzyme [21], due to the fact that the presence of acetyl groups hinders the enzyme hydrolysis of those linkages. The detection of acetyl groups in the non-reducing end of these oligosaccharides can be attributed to acetyl migration during the freeze drying process as was observed in acetylated xylo-oligosaccharides [22]. Fragment ions at *m/z* 347 and 406 are identified, respectively, as $^{0,2}A$ type fragment ion and due to the homolytic cleavage of the acetyl ester group (loss of 43 Da). The occurrence of fragmentation pathway by loss of Hex, AcHex and Ac_2Hex residues in the MS/MS spectra and the absence of fragments due to loss of 90 and 120 Da (characteristic of (1→6)-linked Hex residues), shows that the majority of the acetyl groups, if not all, when present are substituent of the (1→4)-linked Hex residues. The fragment ions resulting from loss of 102 Da, observed in the MS/MS spectrum of the acetylated and diacetylated oligosaccharides, show that 2-O-acetylhexose residues are present. Nevertheless, it does not allow to exclude the presence of other isomeric structures with substitution at O-3 and O-6 position of the (1→4)-linked Hex residues.

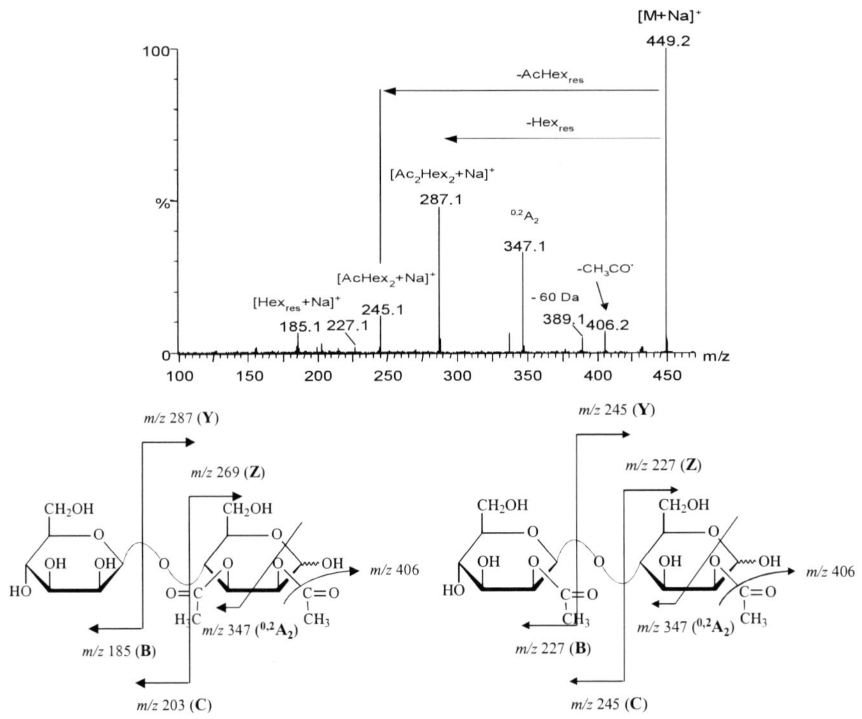

Figure 13. ESI-MS/MS spectrum of $[M+Na]^+$ adducts and schematic fragmentation pathways of (1→4)-linked Ac_2Hex_2 [17].

The MS/MS spectrum of $[Pent.Hex_2+Na]^+$ (m/z 497) is shown in Figure 14. The major fragment ion at *m/z* 335, attributed to $[Pent.Hex+Na]^+$, results from the loss of a hexose residue (Hex_{res}). Also observed with a high intensity is the ion at *m/z* 365, attributed to $[Hex_2+Na]^+$, resulting from loss of 132 Da that is due to the loss of a pentose residue ($Pent_{res}$).

The ions at *m/z* 173 $[Pent+Na]^+$ and 155 $[Pent_{res}+Na]^+$, resultant from Bα- and Cα-type cleavages, respectively, confirm the presence of the pentose residues in the structure of this oligosaccharide. The loss of 60, 90 and 120 Da from $[Pent.Hex+Na]^+$, at *m/z* 335, leading to the ions, respectively, at *m/z* 275, 245 and 215, confirms the (1→6)-linkage of the pentose residue to the Hex.

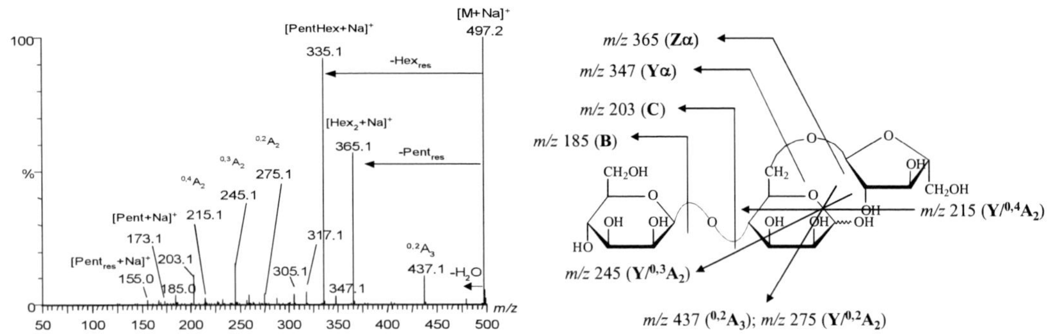

Figure 14. ESI-MS/MS spectrum of $[M+Na]^+$ adducts and schematic fragmentation pathways of (1→4)-linked $PentHex_2$ [17].

3.2. Roasted Induced Structural Changes of Galactomannans

The roasting process induces structural changes in coffee galactomannans, many of them at their reducing end. The reaction pathways of galactomannans during the coffee roasting process were inferred from the detection of specific chemical markers by ESI-MS/MS. Maillard reaction, caramelization, oxidation, and decarboxylation pathways were identified by detection of Amadori compounds, 1,6-β-anhydromannose, mannonic acid and arabinonic acid, among other modifications, in the reducing end of the oligosaccharides obtained by specific enzymatic hydrolysis of the mannan backbone [20]. The chemical structures of these structural changes at the reducing end of galactomannans can be determined as described in the previous section. As this structural changes occur at the low abundant reducing end, the ions observed, after enzymatic hydrolysis, of these modifications are also of low abundance. In Figure 15, it can be observed the spectra of fractions containing ions that correspond to roasting induced modification, absent in the ESI-MS spectra obtained for the green coffee infusion galactomannans.

One of the most abundant modifications is shown by the occurrence of 1,6-β-anhydromannose residues, a typical caramelization product. A series of ions are present at *m/z* 347, 509, 671, and 833 ($[Hex_{1\text{-}4}AnHex+Na]^+$). Figure 16 shows the ESI-MS/MS spectra of $[Hex_2AnHex+Na]^+$ ions. Predominant Y type ions are observed, formed by glycosidic bond cleavage, with successive loss of one and two Hex_{res} (-162 Da and -324 Da). Loss of anhydrosugar residue ($AnHex_{res}$, 144 Da) is also observed in the spectra, occurring from either precursor ion or combined with loss of Hex_{res}. The ion at *m/z* 185 corresponds to the sodium adduct of AnHex.

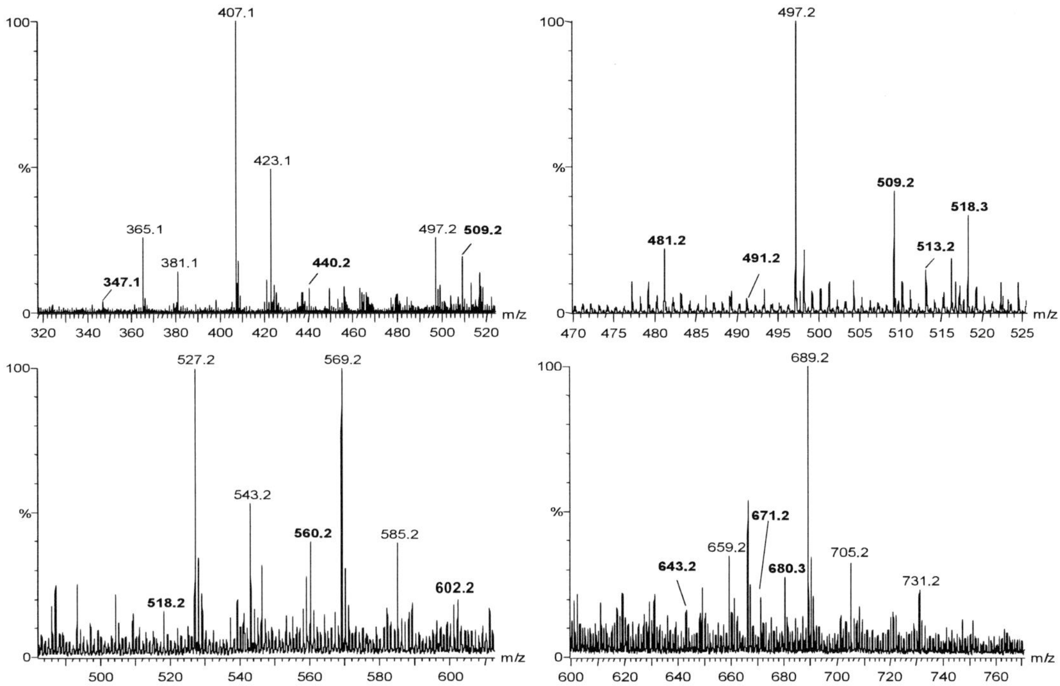

Figure 15. Enlargement of the positive ESI-MS spectra of DP3 fractions obtained after enzymatic hydrolysis of roasted coffee infusions galactomannans.

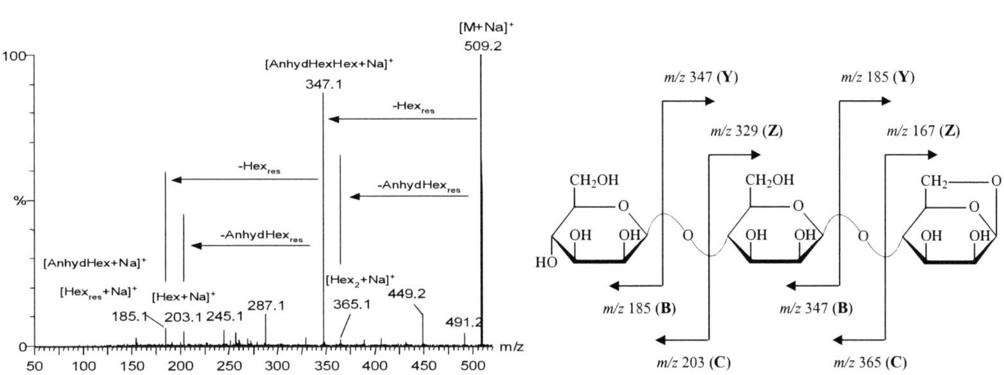

Figure 16. ESI-MS/MS spectrum of $[M+Na]^+$ adducts and schematic fragmentation pathways of (1→4)-linked Hex_2AnHex.

Other two series of relatively abundant ions are identified as modified oligosaccharides by oxidation/decarboxylation: (a) hexonic acid residues (HxoA), present at *m/z* 381 and 543 ($[Hex_{2\text{-}3}HxoA+Na]^+$); and (b) pentonic acid residues (PentA), present at *m/z* 513 ($[Hex_2PentA+Na]^+$). The detection of the oligosaccharides with HxoA in the ESI-MS spectra do not allow easy assignment because its sodium adducts has the same *m/z* value as the potassium adducts of di- and trihexose oligosaccharides ($[Hex_{2\text{-}3}+ K]^+$). To confirm the presence of $[Hex_nHxoA+Na]^+$, the oligosaccharides can be reduced with sodium borohydride. The alditol derivatives obtained show *m/z* values 2 Da higher than the original corresponding oligosaccharides, but the *m/z* value of the oligosaccharides containing HxoA residues do not change. The reduction of the DP3 fractions resulted in the appearance of ions at *m/z* 367 and *m/z* 529 instead of those at *m/z* 365 and *m/z* 527. Also present are the ions at *m/z* 381 and *m/z* 543, confirming the presence of the sodium adducts of hexonic acid containing oligosaccharides, $[HexHxoA+Na]^+$and $[Hex_2HoxA+Na]$+, respectively.

The ESI-MS/MS of the oligosaccharides with pentonic (PentA) and hexonic acids (HxoA) are shown in Figure 17 and Figure 18, respectively. The ESI-MS/MS spectrum of $[Hex_2PentA+Na]^+$ ion at *m/z* 513 (Figure 17) shows predominant ions formed by Y type glycosidic cleavage, with loss of one and two Hex_{res} leading to $[HexPentA+Na]^+$ and $[PentA+Na]^+$ ions at *m/z* 351 and 189, respectively. The fragment at *m/z* 469, due to loss of CO_2 (-44 Da), confirms the presence of a carboxylic acid [12]. Furthermore, the ion due to the loss of -166 Da, with formation of the ion $[Hex+Na]^+$, confirms the presence of PentA at the reducing end.

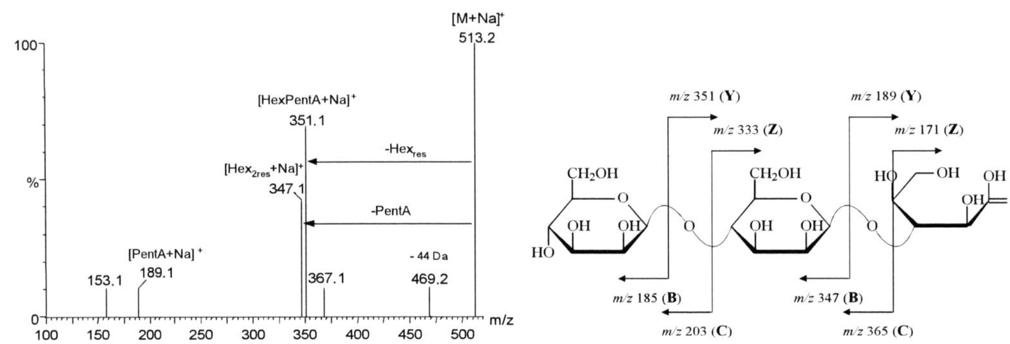

Figure 17. ESI-MS/MS spectrum of $[M+Na]^+$ adducts and schematic fragmentation pathways of (1→4)-linked Hex_2PentA [20].

The oligosaccharides with HxoA have a fragmentation pattern similar to the oligosaccharides with PentA (Figure 18). Their ESI-MS/MS spectra show a main loss of one and two Hex with formation of $[HxoA+Na]^+$ at *m/z* 219 and also fragment ions at *m/z* 347 due to the loss of HxoA (-196 Da), confirming the occurrence of HxoA at the reducing end of the galactomannan-derived oligosaccharides. For this type of oligosaccharide, it is possible to observe an ion due to loss of HCOOH (46 Da) confirming the presence of a carboxylic acid. The loss of HCOOH in HxoA residues occurs probably because the acidity of the C-3 hydrogen, eliminated with the acid group, is higher than in the case of PentA.

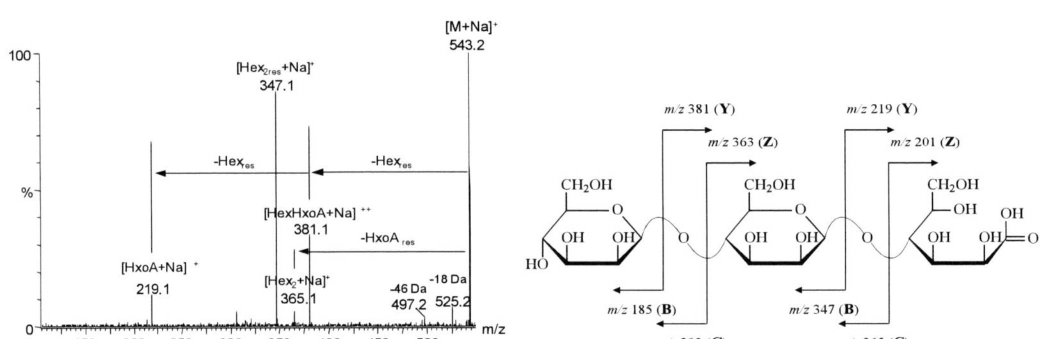

Figure 18. ESI-MS/MS spectrum of $[M+Na]^+$ adducts and schematic fragmentation pathways of (1→4)-linked Hex_2HxoA [20].

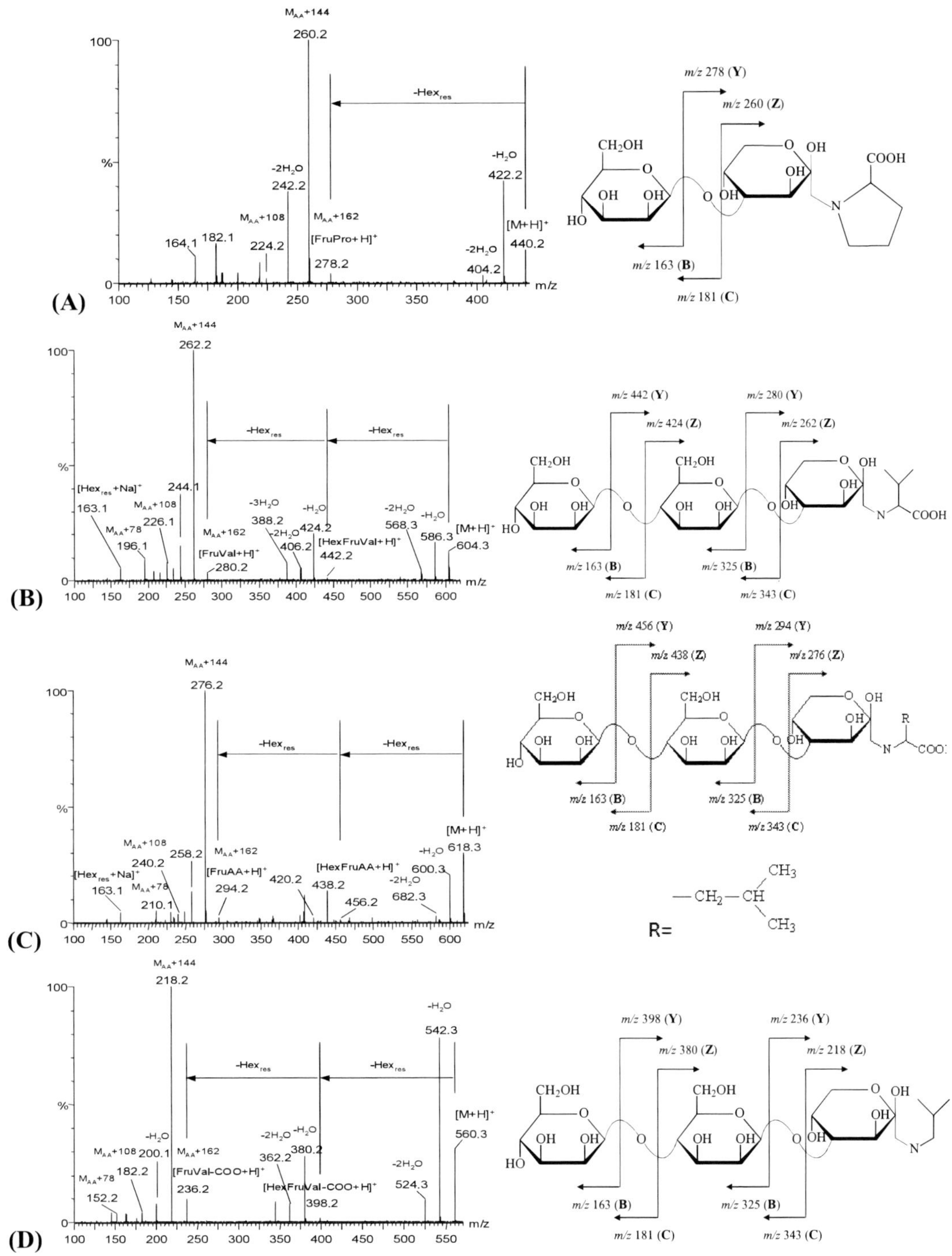

Figure 19. ESI-MS/MS spectra of $[M+H]^+$ adducts and schematic fragmentation pathways of 1→4 linked a) HexFruPro, b) Hex_2FruVal, c) Hex_2FruAA (AA= Leu/Ile or Hyp) and d) Hex_2FruVal-CO_2 [20].

Other series of ions can be identified in the ESI-MS spectra of DP3 fraction, showing even *m/z* values, thus indicating the presence of an odd number of nitrogen atoms in the molecules (MacLafferty-nitrogen rule [23]. The ESI-MS/MS spectra obtained for the ions at *m/z* 440, 604, 618, and 560, identified as the $[M+H]^+$ ions of oligosaccharide-containing Amadori compounds of proline, valine, leucine/isoleucine, or 4-hydroxyproline and decarboxylated valine, are shown in Figure 19. As can be seen in each spectrum, the oligomeric structure is evidenced by the loss of one or two Hex_{res} (-162 Da), leading to the fragments at *m/z* 278 (Figure 19a), *m/z* 280 (Figure 19b), *m/z* 294 (Figure 19c), and *m/z* 236 (Figure 19d) that are the Amadori products fructosylproline, fructosylvaline, fructosylleucine/ isoleucine, or 4-hydroxyproline and fructosyl-decarboxylated valine, respectively. All of the MS/MS spectra of these fructosylamino acid derivatives showed abundant loss of one and two water molecules. Also observed in each ESI-MS/MS spectrum of these derivatives are the diagnostic fragment ions named as M_{AA}+ 144, M_{AA}+ 108, and M_{AA}+ 78 (Figure 20), which are diagnostic ions for the presence of Amadori compounds [24].

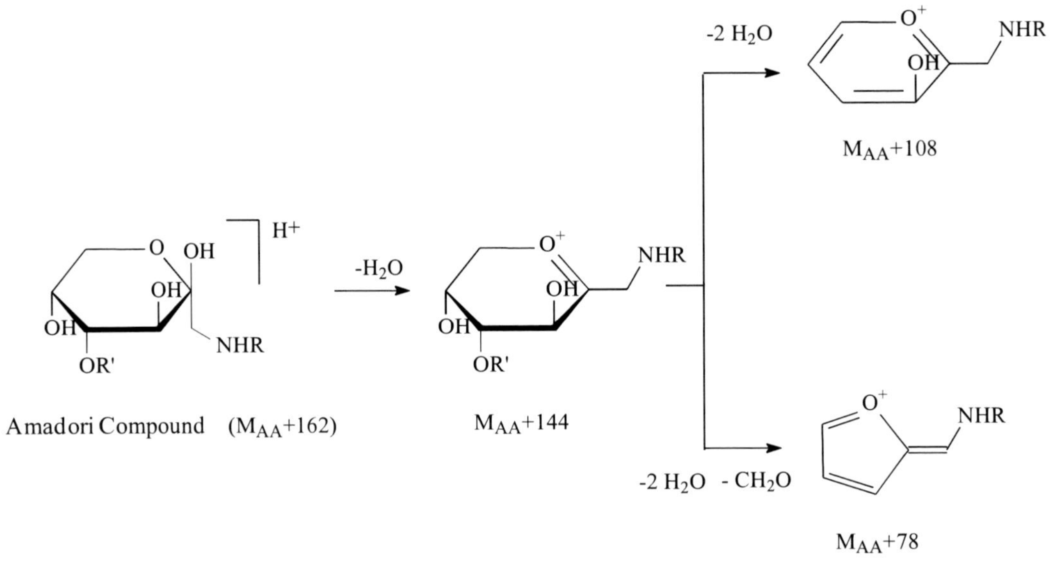

Figure 20. Nomenclature of fragment ions from Amadori compounds according Jeric et al. [24]. M_{AA}, molecular weight of amino acid involved in the formation of the Amadori compound.

4. Type II Arabinogalactans

The second most abundant polysaccharides in green and roasted coffee are type II arabinogalactans. A variety of green coffee type II arabinogalactans have been isolated and structurally characterized [13, 25]. These polysaccharides are composed by a main backbone of β-(1→3)-linked D-galactosyl residues, some of them substituted at the O-6 position with short chains of β-(1→6)-linked D-galactosyl residues (Figure 21). The galactosyl residues of these β-(1→6)-linked D-galactosyl side chains can be substituted at the O-3 position with single α-arabinosyl residues and (1→5)-linked arabinosyl disaccharides. Terminally-linked to these β-(1→6)-linked D-galactosyl side chains can also be glucuronic acid residues [13], and in some hot water soluble type II arabinogalactans, rhamnoarabinosyl and rhamnoarabinoarabinosyl side chains are also linked to the O-3 position of the β-(1→6)-linked D-galactosyl side chains (Figure 21).

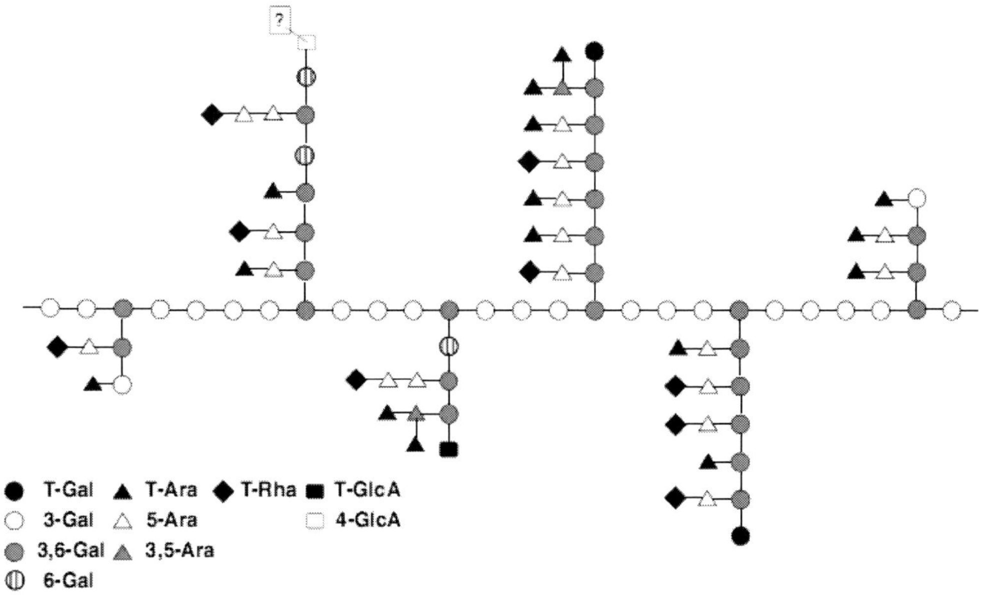

Figure 21. One of the many possible structural arrangements for the carbohydrate moiety of a water soluble type II arabinogalactan-protein isolated by hot water extraction of the green coffee. (adapted from Nunes et al. [25]).

The structural characterization of type II arabinogalactans can be accomplished by ESI-MS/MS after partial acid hydrolysis [25], in order to render this high molecular weight polysaccharides into oligosaccharides, amenable to be analyzed by ESI-MS. Taking advantage of the known relative acid stabilities [26] of the glycosidic-linkages known to occur in type II arabinogalactans, as furanosyl linkages are more acid labile than pyranosyl ones, and α-linkages are more acid labile than β-linkages, a sequential partial acid hydrolysis can be performed using 50 mM TFA followed by 250 mM TFA.

For the 50 mM TFA partial hydrolysate, the ions observed at *m/z* 173, 305, 187, 319, and 451 are the sodium adducts of pentose (Pent) and deoxyhexose (dHex) monomers, dimers, and trimers, namely $[\text{Pent+Na}]^+$, $[\text{Pent}_2\text{+Na}]^+$, $[\text{dHex+Na}]^+$, $[\text{dHexPent+Na}]^+$, and $[\text{dHexPent}_2\text{+Na}]^+$, respectively (Figure 22).

Structural information on the arabinosyl side chains can be obtained by analysis of the ESI-MS/MS spectra of $[\text{Pent}_2\text{+Na}]^+$ (*m/z* 305) (Figure 23). The predominant ions observed at *m/z* 173 and 155, attributed to the ions $[\text{Pent+Na}]^+$ and $[\text{Pent}_{res}\text{+Na}]^+$ are formed by a C/Y-type and B/Z-type glycosidic bond cleavage, by the loss of a pentose residue and a pentose, respectively. Ions at m/z 245 and m/z 215 were attributed to and $^{0,3}A_2$ cross-ring fragments. The loss of 60 and 90 Da from the precursor ion shows the occurrence of a terminally linked pentose residue at the O-5 position of the reducing end pentose.

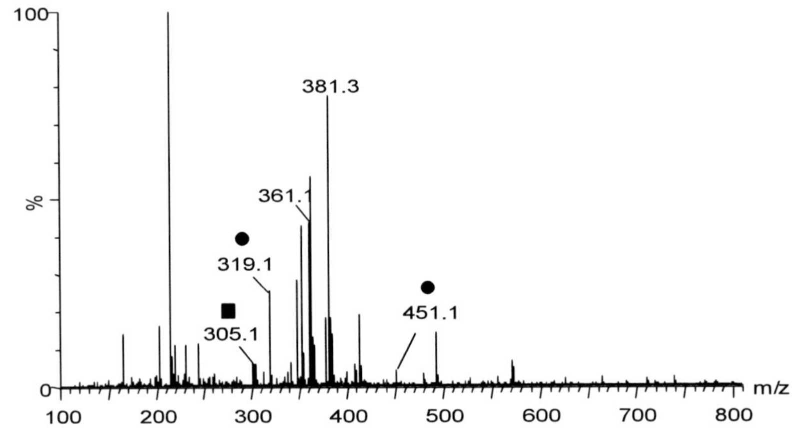

Figure 22. Positive ESI-MS of the sugars released by a 50 mM partial acid hydrolysis of a type II arabinogalactan extracted with hot water from green coffee beans. ■ - $[M+Na]^+$ adducts of $Pent_2$; ● - $[M+Na]^+$ adducts of $dHexPent_{1\text{-}2}$.

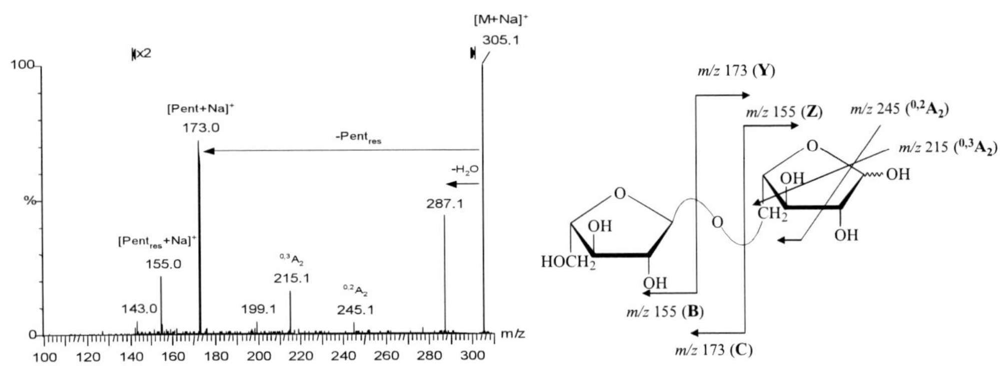

Figure 23. ESI-MS/MS spectrum of $[M+Na]^+$ adducts and schematic fragmentation pathways of (1→5)-linked $Pent_2$ [25].

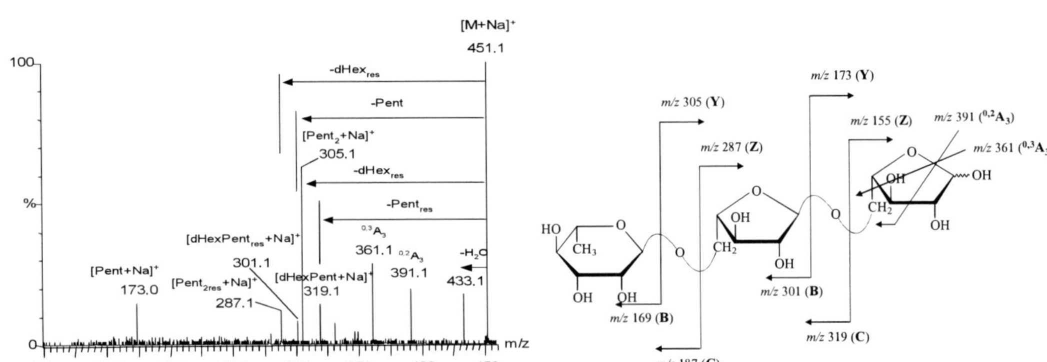

Figure 24. ESI-MS/MS spectrum of $[M+Na]^+$ adducts and schematic fragmentation pathways of (1→5)-linked $RhaPent_2$ [25].

The ESI-MS/MS spectrum of ion at *m/z* 451 (Figure 24), corresponding to $[dHexPent_2+Na]^+$, shows the presence of ions at *m/z* 319 and 305, identified as

$[dHexPent+Na]^+$ and $[Pent_2+Na]^+$, resultant of a loss of 132 Da ($Pent_{res}$) and 146 Da ($dHex_{res}$), respectively. Also present are the ions at *m/z* 301 and *m/z* 287, corresponding to losses of 150 and 164 Da, and identified as $[dHexPent_{res}+Na]^+$ and $[Pent_{2res}+Na]^+$, respectively. The ion present at *m/z* 173 corresponds to $[Pent+Na]^+$. The ions at *m/z* 391 and 361 are attributed to $^{0,2}A_3$ and $^{0,3}A_3$ cross-ring fragments. These results show that the rhamnosyl residues are at the non-reducing end and that the arabinosyl residues at the reducing end are substituted on the O-5 position by another arabinosyl residue.

Further structurally informative oligosaccharides can be obtained by a second partial acid hydrolysis with 250 mM TFA. For the 250 mM TFA partial hydrolysate, the ions detected in the mass spectrum (Figure 25) are attributed to the sodium adducts of Pent, dHex, and hexose (Hex) monomers, dimers, trimers, and tetramers of Hex_2, trimers, tetramers, and pentamers of PentHex, a trimer of $Pent_2Hex$, and trimers, tetramers, and pentamers containing uronic acid (HexA), HexAHex.

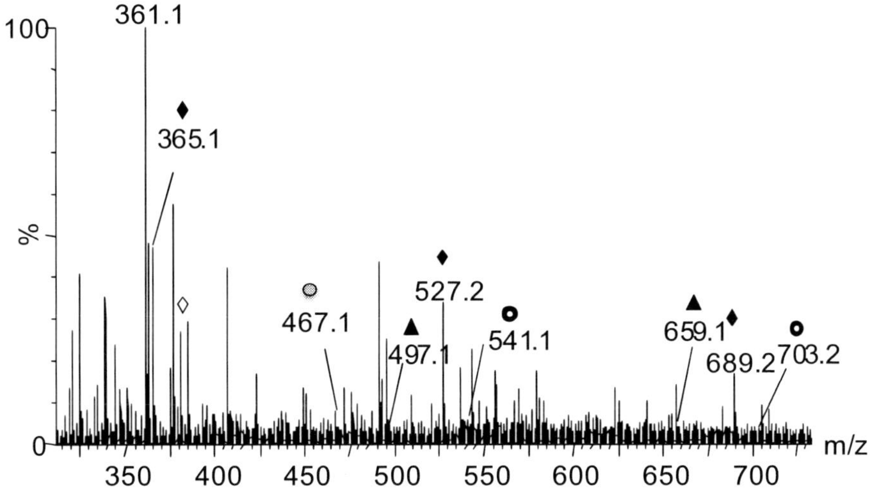

Figure 25. Positive ESI-MS spectrum of the sugars released by a 250 mM partial acid hydrolysis of a type II arabinogalactan extracted with hot water from green coffee beans: ♦ – $[M+Na]^+$ adducts of Hex_{2-4}; ◊ – $[M+K]^+$ adducts of Hex_2; ▲ - $[M+Na]^+$ adducts of $PentHex_{2-3}$; ● - $[M+Na]^+$ adducts of $Pent_2Hex$; ◘ - $[M+Na]^+$adducts of $HexAHex_{2-3}$.

The MS/MS spectrum of ion *m/z* 467, corresponding to $[Pent_2Hex+Na]^+$, is shown in Figure 26. The major fragment ion at *m/z* 335, attributed to $[PentHex+Na]^+$, results from the loss of a pentose residue ($Pent_{res}$) from the precursor ion. Also observed with a high intensity is the ion at *m/z* 305, attributed to $[Pent_2+Na]^+$, resultant from a loss of 162 Da corresponding to the loss of an hexose residue (Hex_{res}). The loss of 60, 90, and 120 Da from $[Pent_2Hex+Na]^+$, resulting in ions at *m/z* 407, 377, and 347, and from $[PentHex+Na]^+$, resulting in ions at *m/z* 275, 245, and 215, suggests that the reducing end hexose residue is substituted at the O-6 position. Nevertheless, this result does not exclude the possible occurrence of arabinosyl disaccharides linked to the O-3 position of the reducing hexose residue. This is another possibility of linkage of the arabinose to the galactose residues, as their presence would result in a fragmentation pattern with a cross-ring cleavage with the loss of only 90 Da that would overlap with the fragmentation pattern observed.

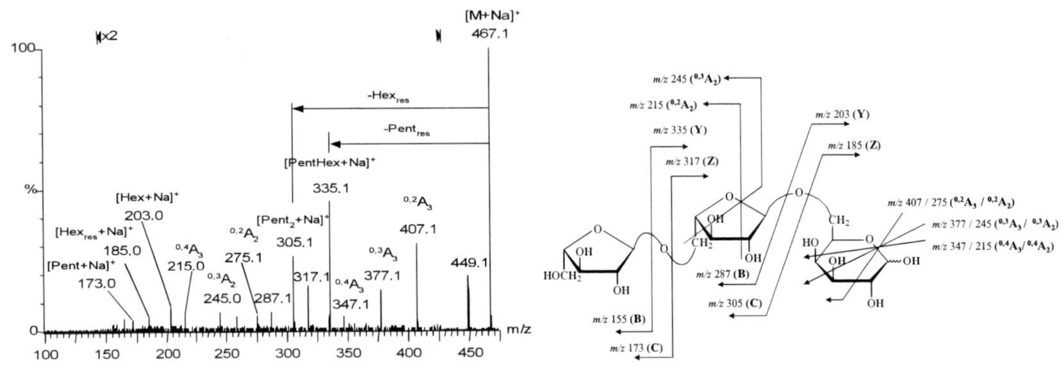

Figure 26. ESI-MS/MS spectrum of $[M+Na]^+$ adducts and schematic fragmentation pathways of $Pent_2Hex$ [25].

The MS/MS spectrum of ion at *m/z* 497, attributed to $[PentHex_2+Na]^+$, is shown in Figure 27. The major fragment ion at *m/z* 365, attributed to $[Hex_2+Na]^+$, is resultant from the loss of the pentose residue, indicating that the two hexoses are linked to each other. Also observed with a high intensity is the ion at *m/z* 335, attributed to $[PentHex+Na]^+$, resultant from a loss of 162 Da, that can be attributed to the loss of an hexose residue (Hex_{res}). The loss of 60, 90, and 120 Da from $[PentHex_2+Na]^+$, resulting in ions at *m/z* 437, 407, and 377, indicates that the reducing end hexose residue is substituted at the O-6 position [27].

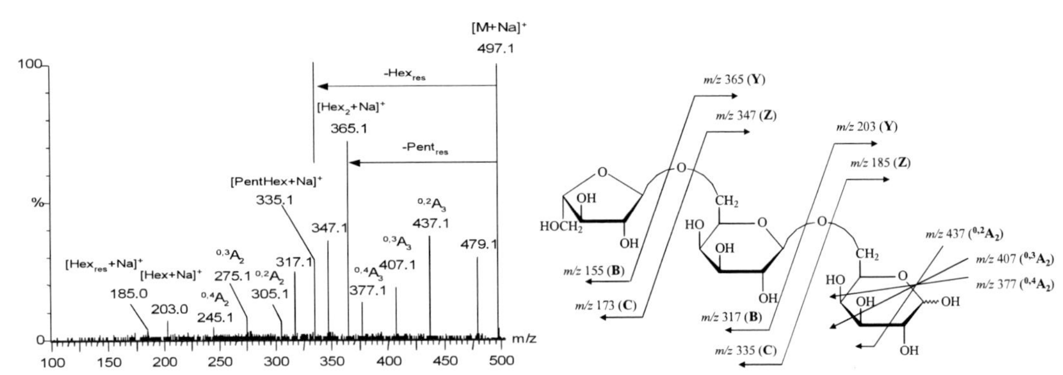

Figure 27. ESI-MS/MS spectrum of $[M+Na]^+$ adducts and schematic fragmentation pathways of $PentHex_2$ [25].

The MS/MS spectrum of ion at *m/z* 541, attributed to $[HexAHex_2+Na]^+$, is shown in Figure 28. The major fragment ion at *m/z* 365, attributed to $[Hex_2+Na]^+$, results from the loss of an uronic acid residue ($HexA_{res}$). Also observed is the ion at *m/z* 379, attributed to $[HexAHex+Na]^+$, resultant from a loss of 162 Da, that can be attributed to the loss of an hexose residue (Hex_{res}). The ions at *m/z* 305, 275, and 245 are attributed to $^{0,2}A_2$, $^{0,3}A_2$, and $^{0,4}A_2$ cross ring fragments from ion *m/z* 365 $[Hex_2+Na]^+$. This fragmentation pattern shows that ion $[Hex_2+Na]^+$ contains a reducing end hexose residue substituted at the O-6 position.

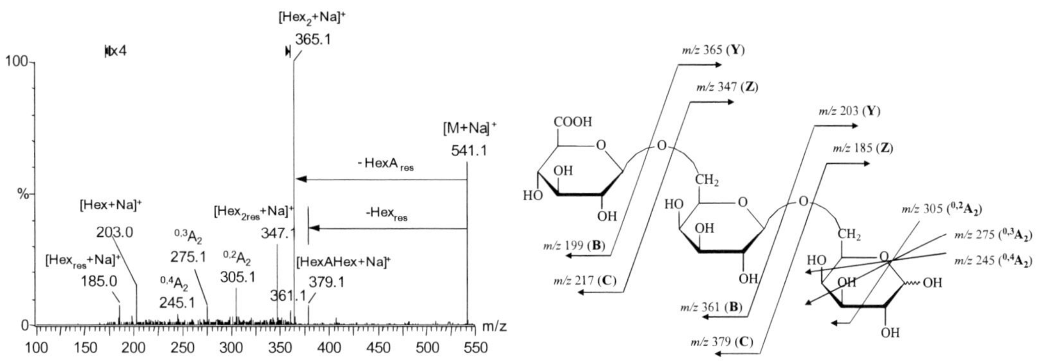

Figure 28. ESI-MS/MS spectrum of $[M+Na]^+$ adducts and schematic fragmentation pathways of $HexAHex_2$ [25].

5. Differentiation of Anomeric Configurations of Reducing Disaccharides of Glucose by ESI-MS/MS

The evaluation of the anomeric configuration of the oligosaccharides is a relevant issue since oligosaccharides with the same sugar composition but different anomeric configuration show distinct properties. The classical example is amylose and cellulose, two polysaccharides with very dissimilar behaviour and biological functions that are composed, respectively, by maltose repeating units, α–(1→4)-glucosyl-glucose, and cellobiose repeating units, β–(1→4)-glucosyl-glucose. Furthermore, these two disaccharides show also distinct physical and chemical features. For example, maltose is sweet, easily digested by humans, and fermented by yeast, while cellobiose has virtually no taste, is indigestible by humans, and is not usually fermented by yeast. Many other examples can be found in nature reflecting this issue. Usually the identification of the anomeric configuration is achieved by NMR spectroscopy that has the inconvenient of requiring large amount of pure samples, often a very hard task considering the biological origin, diversity and heterogeneity of most of the carbohydrates.

The possibility of discrimination of the anomeric configuration (α or β) of underivatized reducing glucopyranosyl-glucose disaccharides with all possible types of linkages (Figure 29) was shown to be possible by mass spectrometry. This discrimination was demonstrated to be independent of the type of mass spectrometer used: 1. hybrid mass spectrometer Q-TOF, 2. a linear ion trap, and 3. a triple quadrupole, with an Electrospray source (ESI) [28].

ESI-MS/MS spectra of $[M+Na]^+$ and $[M+Li]^+$ ions show fragment ions due to glycosidic (-162 Da and -180 Da) and cross ring cleavages using a Q-TOF [28]. Table 1 summarizes the product ions observed in the product ion mass spectra of $[M+Li]^+$, obtained using different collision energies (25, 30, and 33 eV). Loss of water is also observed in the product ion spectra of $[M+Na]^+$ and $[M+Li]^+$ ions of β-(1→6)-, β-(1→4)- and β-(1→3)-linked disaccharides, and with low relative abundance in the product ion spectra of [M+Li]+ of α-(1→2)- and α-(1→6)-linked disaccharides [28].

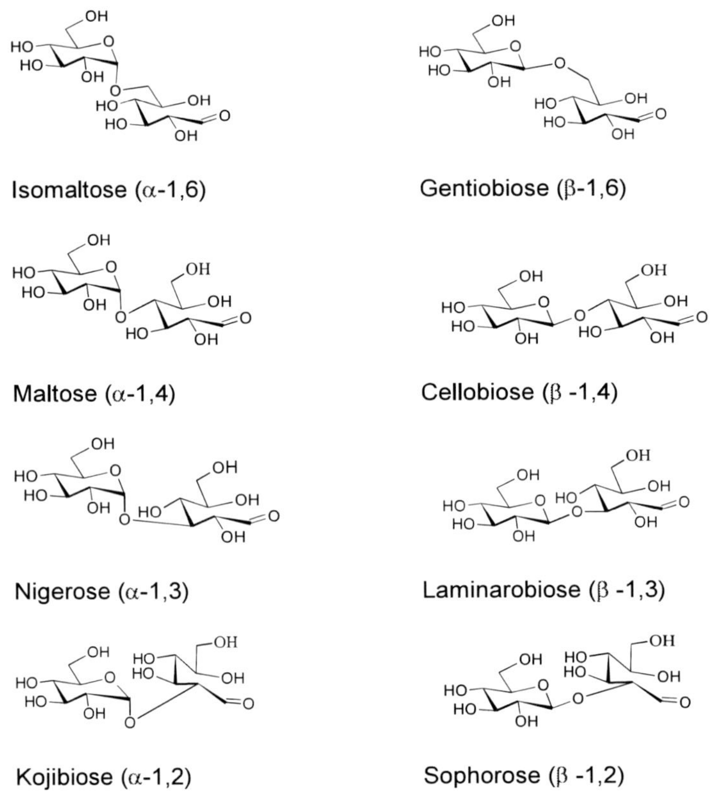

Figure 29. Reducing glucopyranosyl-glucose disaccharides [28].

The cross ring fragmentations observed allow to infer the type of linkage of the disaccharide [29, 30, 31]. The (1→6)-linked disaccharides fragment by loss of $C_2H_4O_2$ ($^{0,2}A_2$ or $^{2,4}A_2$, -60 Da), $C_3H_6O_3$ ($^{0,3}A_2$, -90 Da), $C_4H_8O_8$ ($^{0,4}A_2$, -120Da), (1→3)-linked disaccharides fragment by loss of $C_3H_6O_3$ ($^{0,3}A_2$, -90 Da) and (1→2)-linked disaccharides fragment by loss of $C_4H_8O_8$ ($^{0,3}A_2$, -120Da). Furthermore, the β–(1→4)-linked disaccharides, in addition to the loss of $C_2H_4O_2$ ($^{0,2}A_2$ or $^{2,4}A_2$, -60 Da), showed the product ions formed by elimination of $C_4H_8O_4$ (-2 x 60 Da). The $[M+Na]^+$ ions of β–(1→3)-linked disaccharides also fragment by loss of $C_4H_8O_4$, but with very low relative abundance.

Depending on the anomeric configuration (α or β), the product ion spectra of $[M+Na]^+$ and $[M+Li]^+$ show significant differences (p=0.05) in the relative abundances of some product ions. From the ESI-MS/MS spectra of the $[M+Na]^+$ adducts, it is possible to observe that the most abundant product ion (base peak) for the α-anomers, is the ion at *m/z* 203 ($[Hex+Na]^+$), formed by glycosidic cleavage with loss of glucose residue (-162 Da). For the β-anomers, the base peak of the spectra is due to cross ring cleavages: loss of $C_2H_4O_2$ for β-(1→6), and β–(1→4), or loss of $C_4H_8O_4$ for β–(1→2), or due to loss of H_2O for β–(1→3)-linked disaccharides. The experiments done by changing the collision energy (25, 30, 33, and 35 eV, Table 1), show that the differences described for the fragmentation pathways of the different disaccharides still persist. The interday reproducibility was also observed. These differences can be used to differentiate the α and β configuration of the anomeric carbon in the reducing glucosyl disaccharides [28].

Table 1. Product ions observed in the ESI-MS/MS spectra of the $[M+Li]^+$ adducts of glucopyranosyl-glucose disaccharides obtained in Q-TOF 2, using different collision energies, showing their *m/z* values and the relative abundances [28]

Disaccharide	Colision Energy	Product ions *m/z* 331	*m/z* 289	*m/z* 259	*m/z* 229	*m/z* 205	*m/z* 187	*m/z* 169
Isomaltose	25	8 ± 4	23 ± 5	12 ± 4		32 ± 13	100 ± 0	17 ± 5
α-1,6	30	5 ± 1	20 ± 2	13 ± 2	7 ± 2	31 ± 17	100 ± 0	35 ± 7
	33	3 ± 2	18 ± 4	9 ± 1	8 ± 2	24 ± 14	100 ± 0	43 ± 10
Maltose	25	41 ± 4	17 ± 3			22 ± 9	100 ± 0	47 ± 7
α-1,4	30	18 ± 1	11 ± 1			19 ± 7	100 ± 0	50 ± 4
	33	11 ± 3	9 ± 2			25 ± 6	100 ± 0	55 ± 2
Nigerose	25	99 ± 1		8 ± 1		13 ± 1	100 ± 1	52 ± 1
α-1,3	30	22 ± 1		5 ± 0		12 ± 0	100 ± 0	61 ± 1
	33	11 ± 1		3 ± 0		13 ± 1	100 ± 0	65 ± 2
Kojibiose	25	3 ± 3			26 ± 5	26 ± 6	100 ± 0	20 ± 6
α-1,2	30				21 ± 16	24 ± 8	100 ± 0	32 ± 1
	33				12 ± 3	23 ± 3	100 ± 0	38 ± 8
Gentiobiose	25	25 ± 3	100 ± 0	49 ± 3	20 ± 8	17 ± 8	57 ± 7	43 ± 9
β-1,6	30	17 ± 3	100 ± 0	61 ± 6	61 ± 6	23 ± 5	89 ± 5	90 ± 3
	33	9 ± 1	72 ± 9	46 ± 8	56 ± 10	20 ± 7	84 ± 10	100 ± 0
Cellobiose	25	100 ± 0	27 ± 9		5 ± 1	8 ± 4	45 ± 10	53 ± 11
β-1,4	30	78 ± 12	31 ± 1		11 ± 2	11 ± 6	63 ± 2	100 ± 0
	33	42 ± 9	22 ± 4		13 ± 5	6 ± 0	51 ± 4	100 ± 0
Laminaribiose	25	100 ± 0		9 ± 2		6 ± 1	41 ± 1	57 ± 8
β-1,3	30	49 ± 2		11 ± 1		8 ± 2	68 ± 3	100 ± 0
	33	28 ± 5		8 ± 2		8 ± 2	68 ± 3	100 ± 0
Sophorose	25	3 ± 3			100 ± 0	10 ± 2	53 ± 5	23 ± 6
β-1,2	30	6 ± 1			100 ± 0	16 ± 3	69 ± 11	*63* ± 4
	33	4 ± 4			100 ± 0	13 ± 5	85 ± 7	78 ± 19

The ESI-MS/MS spectra of the $[M+Li]^+$ adducts of α and β anomers show variation, but in this case the main difference in the relative abundances occurs between loss of glucose from the reducing end (-180 Da) and loss of a glucosyl residue (-162 Da) from non-reducing end (Table 1). For the α-anomers, the most abundant product ion is at *m/z* 187 (-162 Da), while for β-anomers, the most abundant product ion is at *m/z* 169 (-180 Da). One exception is observed for the β–(1→2)-linked disaccharide (sophorose).

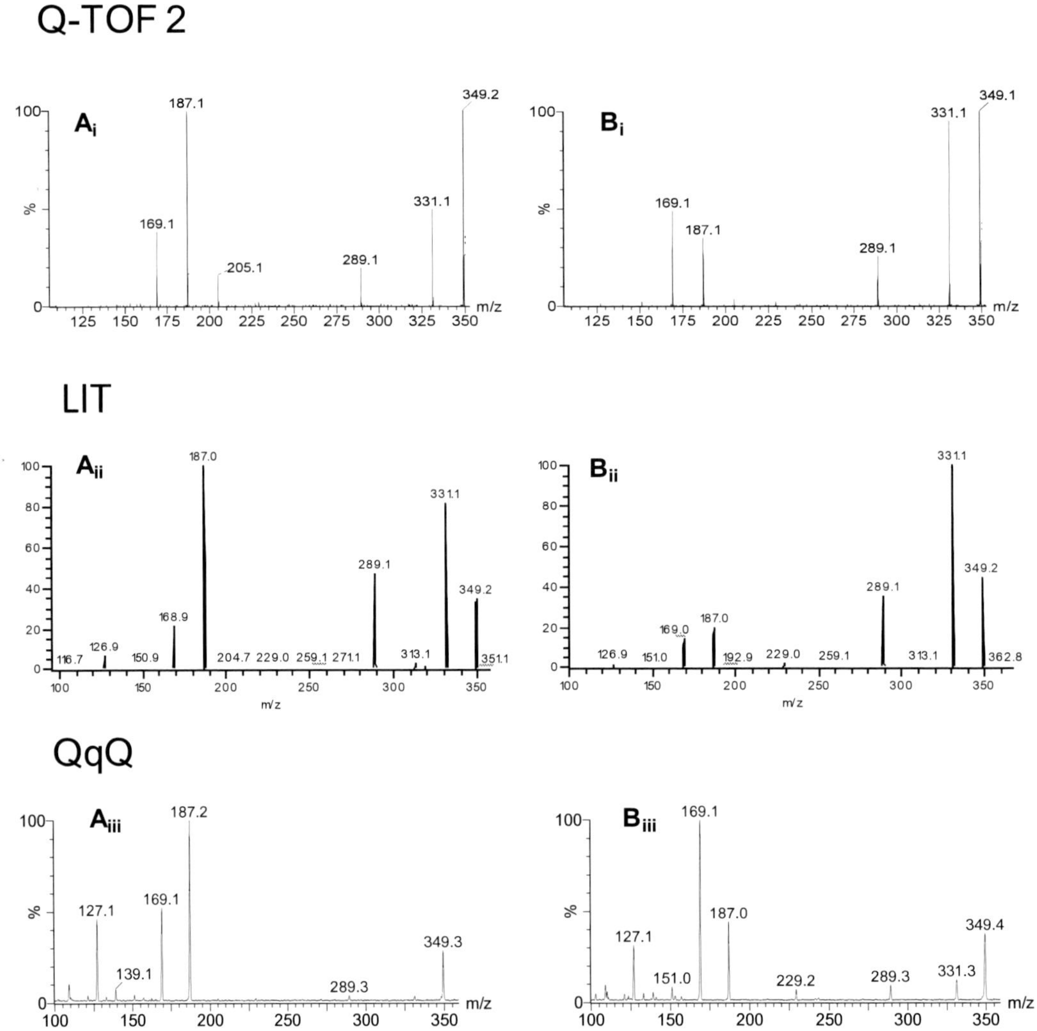

Figure 30. ESI-MS/MS spectra of A) α-(1→4) and B) β-(1→4)-linked $[M+Li]^+$ ions of glucopyranosyl-glucose disaccharides acquired using (i) Q-TOF 2 (CE 25 eV), (ii) LIT (CE 22), and (iii) QqQ (CE 33 eV) [28].

In this case, the most abundant ion is the one formed by loss of $C_4H_8O_4$ (-120 Da), at *m/z* 229, when collision energy was set between as 25 and 30 eV. For higher collision energies (33-35 eV), the differences in the relative abundances of the product ions due to loss of 120, 162, and 180 Da, are not so evident, and the α and β configuration of the anomeric carbon were more difficult to be assigned . Differentiation is also possible to achieve by the product ion spectra of the same type of ions obtained in a triple quadrupole or an ion trap mass spectrometers, as exemplified in Figure 30, that show the ESI-MS/MS spectra of A) α-(1→4) and B) β-(1→4)-linked $[M+Li]^+$ ions of glucopyranosyl-glucose disaccharides acquired using (i) Q-TOF 2 (CE 25 eV), (ii) LIT (CE 22), and (iii) QqQ (CE 33 eV).

Electropray MS/MS spectra of $[M+Li]^+$ and $[M+Na]^+$ ions of reducing glucosyl disaccharides allows to identify their anomeric configuration. These diagnostic differences, based on the differences in the relative abundances of some product ions are even more relevant considering that no derivatization is needed for obtaining this structural information.

REFERENCES

[1] Zaia, J. (2004). Mass Spectrometry of oligosaccharides. *Mass Spectrometry Reviews*, 23, 161–227.

[2] Harvey, D. J. (2006). Analysis of carbohydrates and glycoconjugates by matrix-assisted laser desorption/ionization mass spectrometry: an update covering the period 1999-2000. *Mass Spectrometry Reviews*, 25, 595-662.

[3] Fernández, L. E. M. (2007). Introduction to ion trap mass spectrometry: Applications to the structural characterization of plant oligosaccharides. *Carbohydrate Polymers*, 68, 797-807.

[4] Domon. B. and Costello, C. E. (1998). A systematic nomenclature for carbohydrate fragmentations in FAB-MS/MS spectra of glycoconjugates, *Glycoconjugate Journal*, 5, 397- 409.

[5] Reis, A., Coimbra, M.A., Domingues, P., Ferrer-Correia, A. J. and Domingues, M.R.M. (2002). Structural characterisation of underivatised olive pulp xylo-oligosaccharides by Mass Spectrometry using Matrix Assisted Laser Desorption/Ionisation and Electrospray Ionisation. *Rapid Communications in Mass Spectrometry*, 16, 2124-2132.

[6] Selvendran, R.R. (1985). Developments in the chemistry and biochemistry of pectic and hemicellulosic polymers. *Journal of Cell Science Supplement*, 2, 51-88.

[7] Shimizu, K. *In* Wood and Cellulosic Chemistry, Hon D. N.-S., Shiraishi, N., Eds., Marcel Dekker: New York, 1991, pg. 177-214.

[8] Ralph, J. and Helm, R. F. (1993) Forage Cell Wall Structure and Digestibility; Jung, G. H., Buxton, D. R., Hatfield, R. D., Ralph, J., Eds; ASA-CSSA-SSSA: Madison, WI, pp. 201-246.

[9] Shalatov, A. A., Evtuguin, D. V., Neto, C. P. (1999). (2-O-α-D-Galactopyranosyl-4-O-methyl-α-D-glucurono)–D-xylan from *Eucalyptus globulus* Labill. *Carbohydrate Research*, 320, 93-99.

[10] Teleman, A., Lundqvist, J., Tjerneld, F.; Stalbrand, H. and Dahlman, O. (2000). Characterization of acetylated 4-*O*-methylglucuronoxylan from aspen employing ^1H and ^{13}C NMR spectroscopy. *Carbohydrate Research*, 329, 807-815.

[11] Reis, A., Domingues, M.R.M., Domingues, P., Ferrer-Correia, A.J. and Coimbra, M.A. (2003) Positive and negative Electrospray Ionisation tandem Mass Spectrometry as a tool for structural characterisation of acid released oligosaccharides from olive pulp glucuronoxylans. Carbohydrate Research, 338, 1497-1505.

[12] Reis, A., Pinto, P., Coimbra, M. A., Evtuguin, D. V., Neto, C. P., Ferrer Correia, A. J. and Domingues, M. R. M. (2004). Structural differentiation of uronosyl substitution patterns in acidic heteroxylans by electrospray tandem mass spectrometry. *Journal of the American Society for Mass Spectrometry*, 15,43-47.

[13] Redgwell, R., Curti, D., Fischer, M., Nicola, P. and Fay, L. (2002). Coffee arabinogalactans: acidic polysaccharides covalently linked to proteins. *Carbohydrate Research*, 337, 239–253.

[14] Nunes, F. M. and Coimbra, M. A. (2001). Chemical Characterization of the High Molecular Weight Material Extracted with Hot Water from Green and Roasted Arabica Coffee. *Journal of Agricultural and Food Chemistry*, 49, 1773–1782.

[15] Nunes, F. M. and Coimbra, M. A. (2002). Chemical characterization of the high molecular weight material extracted with hot water from green and roasted robusta coffee as affected by the degree of roast. *Journal of Agricultural and Food Chemistry*, 50, 7046–7052.

[16] Nunes, F. M. and Coimbra, M. A. (2002). Chemical characterization of galactomannans and arabinogalactans from two arabica coffee infusions as affected by the degree of roast. *Journal of Agricultural and Food Chemistry*, 50, 1429–1434.

[17] Nunes, F. M., Domingues, M. R. and Coimbra, M. A. (2005). Arabinosyl and Glucosyl Residues as Structural Features of Acetylated Galactomannans from Green and Roasted Coffee Infusions. *Carbohydrate Research*, 340, 1689–1698.

[18] Redgwell, R. J., Trovato, V., Curti, D. and Fischer, M. (2002). Effect of roasting on degradation and structural features of polysaccharides in Arabica coffee beans. *Carbohydrate Research*, 337, 421–431

[19] Oosterveld, A., Voragen, A. G. J. and Schols, H. A. (2003). Effect of roasting on the carbohydrate composition of *Coffea arabica* beans. *Carbohydrate Polymers*, 54, 183-192.

[20] Nunes, F. M., Reis, A., Domingues, M. R. and Coimbra, M. A. (2006). Characterization of Galactomannan Derivatives in Roasted Coffee Beverages. *Journal of Agricultural and Food Chemistry*, 54, 3428–3439.

[21] Daas, P. J. H., Schols, H. A. and de Jongh, H. H. J. (2000). On the galactosyl distribution of commercial galactomannans. *Carbohydrate Research*, 329, 609-619.

[22] Kabel, M. A., de Waard, P., Schols, H. A. and Voragen, A. G. J. (2003). Location of *O*-acetyl substituents in xylo-oligosaccharides obtained from hydrothermally treated *Eucalyptus* wood. *Carbohydrate Research*, 338, 69–77.

[23] McLafferty, F. W., Turecek, F., *Interpretation of Mass Spectra*; University Science Books: Mill Valley, CA, 1993; pp 37-38.

[24] Jeric, I., Versluis, C., Horvat, S. and Heck, A. J. R. (2002). Tracing glycoprotein structures: Electron ionization tandem mass spectrometric analysis of sugar-peptide adducts. *Journal of Mass Spectrometry*, 37, 803-811.

[25] Nunes, F. M., Reis, A., Silva, A. M. S., Domingues, M. R. M. and Coimbra, M. A. (2008). Rhamnoarabinosyl and rhamnoarabinoarabinosyl side chains as structural features of coffee arabinogalactans. *Phytochemistry*, 69, 1573-1585.

[26] Selvendran, R. R. and Ryden, P. (1990). Isolation and analysis of plant cell walls. *Methods in Plant Biochemistry*, 2, 549-570.

[27] Simões, J., Nunes, F. M., Reis, A., Domingues, M. R. M. and Coimbra, M. A. (2010). Structural features of partially acetylated coffee galactomannans presenting immunostimulatory activity. *Carbohydrate Polymers*, 17, 397-402.

[28] Simões, J., Domingues, P., Reis, A., Nunes, F. M., Coimbra, M. A. and Domingues, M. R. M. (2007). Identification of Anomeric Configuration of Underivatized Reducing Glucopyranosyl-glucose. *Analytical Chemistry*, 79, 5896–5905.

[29] Zhou, Z., Ogden, S. and Leary, J. A. (1990). Linkage Position Determination in Oliogosaccharides: MS/MS Study of Lithium Cationized Carbohydrates. *Journal of Organic Chemistry*, 55, 5444-5446.

[30] Hofmeister, G.E., Zhou, Z. and Leary, J. A. (1991). Gas-phase dissociation mechanisms of dilithiated disaccharides: tandem mass spectrometry and semiempirical calculations. *Journal of the American Chemical Society*, 113, 5964-5970.

[31] Asam, M. R. and Glish, G. L. (1997). Tandem Mass Spectrometry of Alkali Cationized Polysaccharides in a Quadrupole Ion Trap. *Journal of the American Society for Mass Spectrometry*, 8, 987-995.

In: Oligosaccharides: Sources, Properties and Applications ISBN: 978-1-61122-193-0
Editor: Nicole S. Gordon

Chapter 8

BIOLOGICALLY ACTIVE OLIGOSACCHARIDE FUNCTIONS IN PLANT CELL: UPDATES AND PROSPECTS

***Olga A. Zabotina*[1] *and Aleksey I. Zabotin*[2]**
[1]Department of Biochemistry, Biophysics and Molecular Biology, Iowa State University, Ames, IA; [2]Institute of Biochemistry and Biophysics, Russian Academy of Sciences, Kazan, Russia

ABSTRACT

Biologically active oligosaccharides, referred to as oligosaccharins, are the specific group of complex carbohydrates that function in plant cells as molecular signals. Indeed, oligosaccharins participate in the regulation of growth, development, and survival in different environmental conditions. Significant research has helped to shed light on the oligosaccharin concept and to make progress in this field. For instance, various oligosaccharins have been structurally characterized, and for some of them, specific receptors have been discovered. Separate signaling events have also been elucidated. Currently, a number of scientific articles and comprehensive reviews present a large body of evidence supporting the idea of oligosaccharin existence and function. This review provides a brief overview of most recent information obtained about oligosaccharin functions, signal perception, their possible origins, and movement within the plant. The discussion is focused only on oligosaccharins that function in plant cells, either originating from the plant cell wall or from the fungal wall. Please note that there are a number of other biologically active carbohydrate-containing molecules, such as fragments of glycoproteins or lipo-oligosaccharides, which are not discussed in this review. Despite gaps in our knowledge of oligosaccharin function, there is no doubt that this group of signaling molecules plays distinct roles in plant cells. The identification of new structures and activities of oligosaccharins in a broader range of organisms, together with the elucidation of their signal transduction pathways, will broaden our understanding of the important roles of these molecules in different aspects of plant life and will yield valuable insight into how cells progress through developmental programs and respond to environmental changes.

Introduction

The concept of "oligosaccharins" evolved almost 40 years ago from a discovery made by Peter Albersheim and his colleagues (Albersheim and Valent, 1978; Albersheim et al., 1983). They demonstrated that specific oligosaccharide fragments released as products of polysaccharide decomposition exert biological activity when reapplied into plant cells. They called these bioactive oligosaccharides "oligosaccharins" and hypothesized that this new group of biologically active molecules may serve as signaling molecules in the regulation of a broad spectra of metabolic processes in plant cells. This discovery opened a new and exciting era for cell wall polysaccharide biology and made a significant contribution toward uncovering the "mystery" of cell wall polysaccharide complexity, particularly their highly dynamic metabolism. Since their original discovery, significant progress has been made elucidating the various functions of oligosaccharins, producing numerous publications at the end of twentieth century. Several reviews (Aldington and Fry, 1993; Cote and Hahn, 1994; Ryan, 1994; Shibuya and Minami, 2001) have documented major achievements in understanding these molecules, and readers are encouraged to review those articles in order to gain a comprehensive overview of the subject. The current review focuses on updating the information in the field of bioactive oligosaccharides that function in plant cells and includes a brief history of their discovery.

Short History of Oligosaccharins

The first research into oligosaccharide bioactivity occurred in the 1970s during plant-microbial interaction studies. Elicitor-active glucans were first detected in the culture media of fungal (*Phytophthora sojae*) cells (Ayers et al., 1976b). These fungal-derived oligosaccharides demonstrated the ability to elicit phytoalexin synthesis in various plants (Ayers et al., 1976a; Cline et al., 1978; Gunia et al., 1991). Later, the fungal glucans were purified, and their structure and functional features were determined. Additional elicitor-active oligosaccharides were derived from the fungal cell wall polysaccharides chitin, and chitosan (Hadwiger and Beckman, 1980; Lesney, 1989a). The greatest amount of research was carried out with investigations of the structural features of bioactive chitin, chitosan, and β-glucan oligosaccharides and their perception by plant cells. The most recent results will be discussed later in this chapter.

Plants are also a source of elicitor-active oligosaccharides. Biologically active oligogalacturonides were prepared by digesting plant pectic polysaccharides with the fungal hydrolytic enzyme endo-polygalacturonase (Albersheim and Darvill, 1985; Ryan and Farmer, 1991). Purification and structural characterization of these oligosaccharides demonstrated that oligogalacturonide fragments composed of 10-15 monomers in length exert the highest elicitor activity.

Another group of bioactive oligosaccharides discovered later in plants includes those involved in the regulation of plant growth and development. For instance, it was found that nonasaccharides released after endo-glucanase treatment of xyloglucan, the most abundant polysaccharide in the primary cell walls of most plants, exert anti-auxin activity (York et al., 1984).

Elicitor-active oligogalacturonides also demonstrate anti-auxin activity in the regulation of rhizogenesis in tobacco explants (Branca et al., 1988). Structurally, the same oligogalacturonides that elicit phytoalexin production can inhibit auxin-initiated root formation when added exogenously to plant explants at a significantly lower concentration (1 μg/ml) than that required for their elicitor activity (100 μg/ml). Further research uncovered other oligosaccharides involved in the regulation of developmental and metabolic processes in plant cells. For example, pectic fragments similar to oligogalacturonides regulate morphogenic responses in tobacco explants (Tran Thanh Van et al., 1985). Neutral oligosaccharides prepared by mild acid hydrolyses of pectin (and presumably composed of arabinogalactan fragments) stimulate root formation in buckwheat explants in the absence of phytohormones in the culture medium (Zabotina et al., 1996a). Another group of neutral oligosaccharides released after the acid hydrolysis of pectin inhibit root formation in buckwheat explants (Zabotina et al., 2002). These oligosaccharides are less structurally characterized and are often not purified to homogeneity due to the elaborate methodology required to purify and analyze their structural complexity.

Studies of cell wall alterations in winter wheat seedlings during their low temperature acclimation revealed the accumulation of endogenous oligosaccharides in the apoplast during the first 6 h of plant exposure to low temperature (2°C). These endogenously formed oligosaccharides stimulated freezing tolerance acquisition in winter plants when added to their growth medium before acclimation (Zabotin et al., 2005). Notably, these observations of freezing tolerance acquisition are currently the only indication that oligosaccharins participate in regulatory processes during plant adaptation.

To conclude this introduction, it is worth noting that oligosaccharins were discovered as products obtained *in vitro* by mild acid or enzymatic hydrolyses of purified or partially purified polysaccharides. These purified oligosaccharides were then used for *in vitro* treatments of plant tissues or cells during studies of their activities. Therefore, their biological relevance was accepted with skepticism until additional information was obtained demonstrating their existence *in vivo*.

Oligosaccharins and Plant Defense Reactions Against Pathogens

The most advanced research on oligosaccharins has been done in the field of plant-microbial interactions and elicitor activity of oligosaccharins during plant defense responses. As previously mentioned, this is the area from which the first piece of evidence about biologically active oligosaccharides was obtained. During evolution, plants developed thickened outer epidermal walls to form a barrier against microbial penetration. In addition, plants also developed a combination of acquired, preformed defenses and an inducible basal defense/immunity that is initiated during a pathogen attack. Although this innate immunity appears to be the only line of defense for plants, various glycoconjugates and oligosaccharides from microbial or plant cell walls have been found to act as elicitors. Glycoconjugates with elicitor activity, such as lipopolysaccharides from gram-negative bacteria and peptidoglycans from gram-negative and gram positive bacteria, were discussed in a recent review (Silipo et al., 2010). Here the focus of our discussion is limited only to elicito-active oligosaccharides. Currently, two groups of oligosaccharins are characterized as being involved in plant defense reactions: 1) oligosaccharides released from fungal polysaccharides by the action of plant hydrolytic enzymes; and 2) oligosaccharides liberated from plant cell wall polysaccharides that are the result of microbial hydrolytic activity. These oligosaccharins of different origins can elicit defense reactions in plant cells by initiating a broad spectrum of intracellular signaling pathways, some of which are similar for different oligosaccharins, while others are specific for each.

Oligosaccharins Originating from Microbial Walls

Fungal cell walls are usually composed of long chains of mannans, glucans, chitin, galactomannans, glucomannans, rhamnomannans, and phosphomannans. In mannans, the mannose residues have α-linkages in the polysaccharide backbone. In glucans, the glucose residues are β-linked, with the most common being β-1,3 and β-1,6. Chitin is a polymer of β-1,4-linked acetylgucosamine (GlcNAc) with a polymer chain length that typically exceeds 1,000 residues.

B-Glucans as Elicitor Active Oligosaccharins

Elicitor-active β-glucans derived from fungal and oomycete cell walls are microbe-specific molecules and are related to microbe-associated molecular patterns (MAMPs) (Shibuya and Minami, 2001). These molecular patterns or structural motifs are common to an entire class of microbes and are often referred to as general elicitors. As an elicitor, MAMPs generate a distress signal in the plant that then responds by activating of a set of defense reactions. Hepta-β-glucoside, the most highly characterized MAMP, was obtained from pathogenic oomycetes (Sharp et al., 1984) and from commercially available yeast extracts (Hahn and Albersheim, 1978). Similar elicitor-active glucans have been released by chemical or enzymatic digestion of mycelia walls of *P. sojae* (Ayers et al., 1976b). These oligosaccharides are presumably generated at the site of host–pathogen interaction by plant endo-β-glucanases (Okinaka et al., 1995), and it is known that hepta-β-glucan elicits the synthesis of glyceollin in soybean cells (Sharp et al., 1984). Detailed studies of the structure-function relationships of sixteen chemically synthesized and structurally related β-glucans have demonstrated that hexa-β-glucoside (Fig. 1) is the minimum structure recognized by host cells. Further, the presence of all three non-reducing terminal glucosyl residues and the spacing between them are the specific features necessary for the complete elicitor-active structure of oligosaccharin (Cheong et al., 1991; Cote and Hahn, 1994).

Glucan oligosaccharides have been shown to induce phytoalexin synthesis in legumes, including chickpea (Gunia et al., 1991), bean (Cline et al., 1978), pea (Bhandal and Paxton, 1991), and in the solanaceous species (Bhandal and Paxton, 1991; Cline et al., 1978). This indicates that similar oligosaccharin-recognition sites are present in these taxonomically related plants. Significant research has been done to reveal the specific plasma membrane-localized binding receptors for elicitor-active β-glucan (Hahn, 1996), and a line of evidence supported the existence of such specific plant receptors for elicitor-active oligosaccharins.

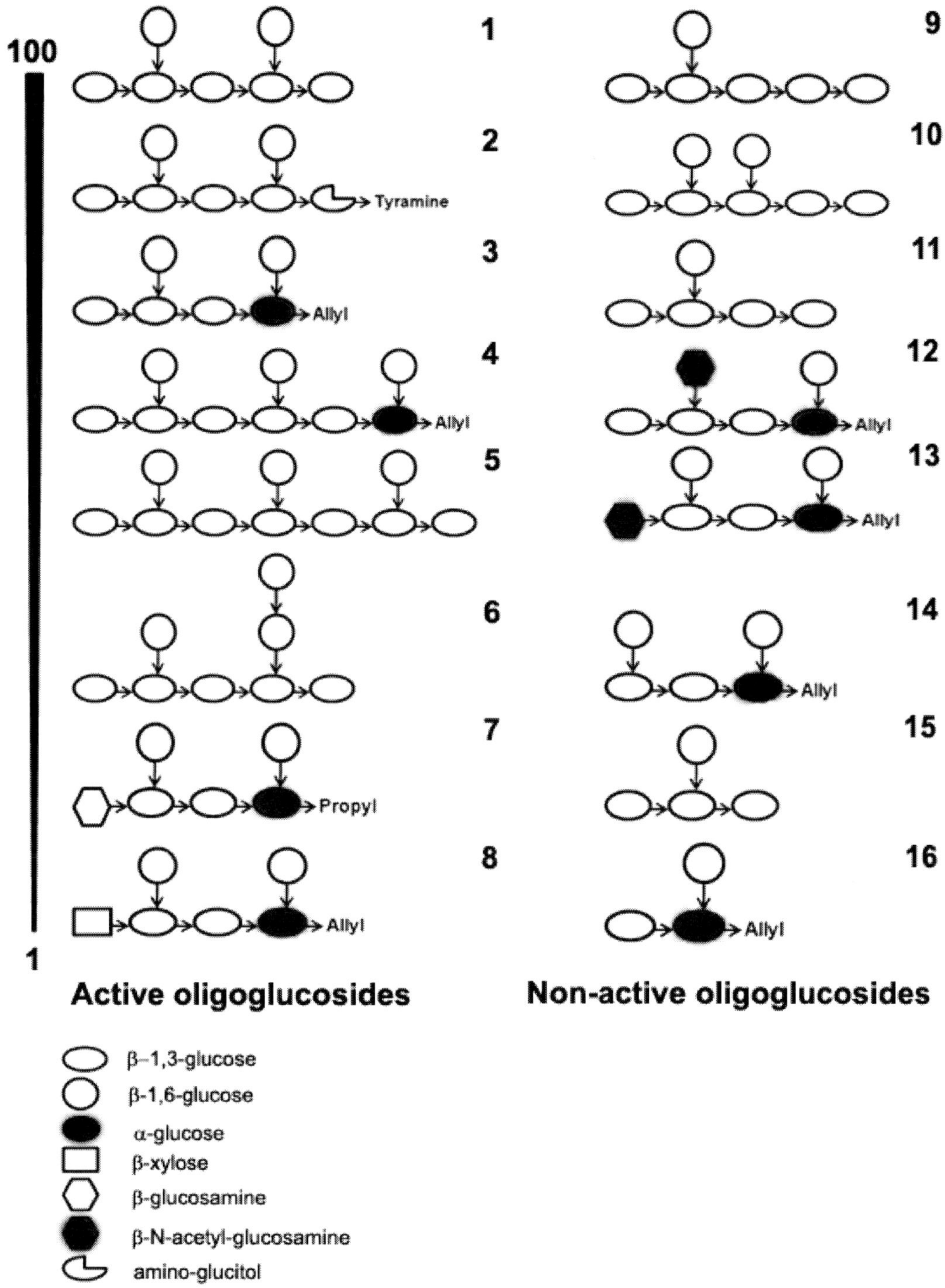

Figure 1. Structures of fungal and synthetic oligoglucosides used in biological activity studies. Active oligoglucosides are shown in the order of diminishing elicitor activity (Cote and Hahn 1994). Fungal hepta-β-glucan (1) demonstrated highest activity assigned as 100 and synthetic hexa- β-glucan with glucose at non-reducing end substituted by xylose (8) showed 100 times lower activity. Eight other synthetic oligoglucosides (9-16) were not active.

For example, the short apparent lag phase in inducing the host responses to the elicitor, the dose-response and stringent structure-function relationships for elicitor activity, and the high sensitivity and selectivity of various host cells to elicitors (Ebel, 1998; Ebel and Cosio, 1994) all point to the existence of one or more receptors. Photoaffinity labeling experiments using ^{125}I-cojugated glucans identified a 70 kDa protein that binds hepta-β-glucan with high-affinity in soybean tissues (Cheong et al., 1993; Cosio et al., 1992; Cosio et al., 1996). The binding receptor was detected in membrane fractions prepared from all soybean seedling organs and soybean cells, but it was not detected in the soluble protein fractions. The binding was saturable, reversible, and β-glucan-specific with a high affinity (Cheong et al., 1993; Ebel, 1998). The binding receptor in the soybean plasma membrane behaved as a 240 kDa oligomeric protein complex (Cheong et al., 1993; Mithofer et al., 1996a; Mithofer et al., 1996b). This complex has a single class of binding sites with an apparent K_d = 1-3 nM for the pure hepta-β-glucoside and K_d = 10-40 nM for the fungal β-glucan fraction (Ebel, 1998). Electron microscopic analysis of soybean roots using antibodies against cloned binding proteins clearly suggests that binding sites are restricted to the cytoplasmic site of the cell walls (Fliegmann et al., 2004). Further, the affinity of the binding site for a glucan ligand correlates with its ability to elicit phytoalexin production (Cheong and Hahn, 1991; Cosio et al., 1990; Cosio et al., 1996). The results of other studies indicate that the binding protein has β-1,3-glucanase activity, which enables the sensing of elicitor and generation of glucan oligosaccharides (Fliegmann et al., 2004; Fliegmann et al., 2005). However, the protein does not have a typical intracellular domain for signaling and likely requires the presence of additional factors, thereby forming a high-molecular-mass protein complex as mentioned above.

Studies of binding sites for hepta-β-glucan in other legume species demonstrated a similar range of affinity for membranes prepared from pea, alfalfa, and lupin. Lower affinities were shown for membranes prepared from *Phaseolus vulgaris* and *Cicer arietinum*. Very little specific binding was detected in *Vicia fab*, while membranes from *Arabidopsis* and tobacco did not bind the glucan elicitor. These observations indicate the existence of similar receptor mechanisms for this fungal glucan by taxonomically related plants. Indeed, the search using bioinformatic approach found genes encoding proteins related to receptor proteins that were identified in multiple plant species (Fliegmann et al., 2004).

Another elicitor active β-glucan was isolated from the rice blast fungus, *Magnaporthe grisea*, and demonstrated the ability to induce phytoalexin biosynthesis in rice suspension-cultured cells (Yamaguchi et al., 2000). This penta-glucoside was distinct from the hepta-β-glucan structure discussed above, with a polysaccharide backbone composed of 1,3-linked glucoside and branches at the 6-position composed of glucose residues. Structure-function studies demonstrated a strict recognition of the active penta-glucoside in rice cells. A similar structure was not active in soybean, indicating a difference in the specificity of binding sites in rice and soybean.

Only fungal glucans are active as elicitors. In contrast, bacterial cyclic (1,3-1,6)-β-glucans from *Bradyrhyzobium japonicum* prevent phytoalexin production induced by fungal glucan elicitors by potentially inhibiting their binding to the membrane-localized receptor (Bhagwat et al., 1999; Mithofer et al., 2001).

In addition, an algal 1,3-β-glucan with some 1,6 linkages (laminarin) (Fig. 2) is known to elicit plant defense responses in different plants, e.g., in tobacco and grapevine (Aziz et al., 2003; Kaczynski et al., 2007; Shinya et al., 2006). A significant overlap between the sets of genes induced by algal and fungal glucans in tobacco cells suggests that the algal oligosaccharides with 1,6-linkages can mimic the action of the fungal β-glucan (Shinya et al., 2007).

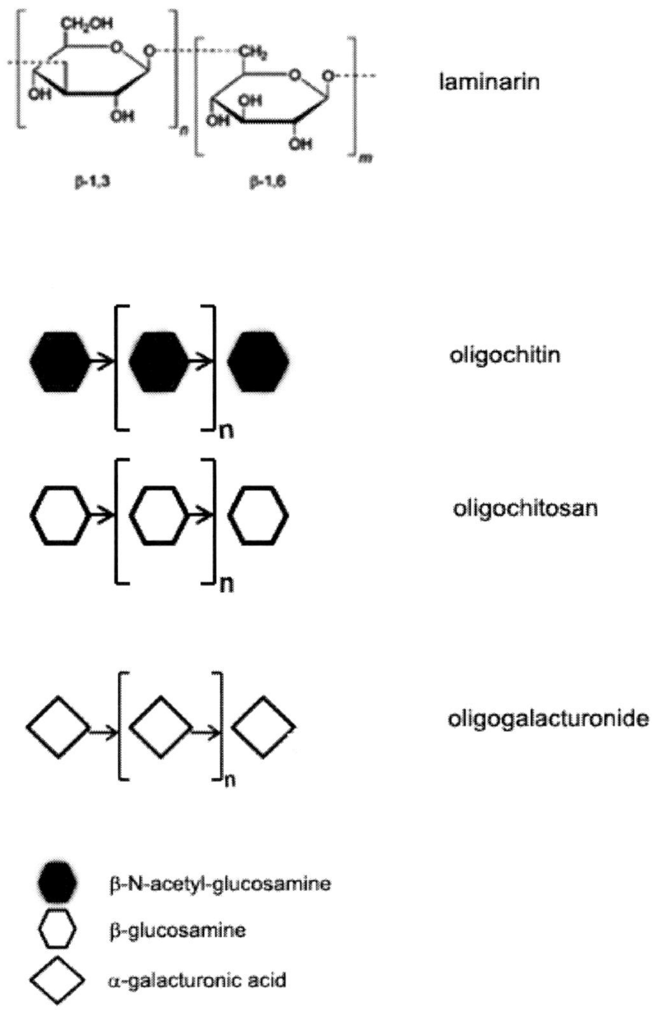

Figure 2. Structures of four different types of elicitor-active oligosaccharins. Laminarin, chitin and chitosan oligosaccharins are of microbial origin. Oligogalacturonides are formed in plant tissues.

Chitin- and Chitosan-Derived Oligosaccharins

Another common type of fungal cell wall polysaccharide is chitin, i.e., 1,4-β-N-acetylglucosamine. Chitin fragments, N-acetylchitooligosaccharides, (Fig. 2) act as a potent MAMP in different plants (Cote and Hahn, 1994; Hadwiger et al., 1994; Lesney, 1989b; Shibuya and Minami, 2001). Oligochitin and its de-acetylated analog, oligochitosan, are liberated by plant chitinases that are activated or induced during plant-microbial interactions.

Different plant systems require chitin or chitosan fragments of different sizes, which indicates variation in receptor specificity. In general, active oligosaccharins must be four to eight monomers long to induce a biological response, and a stronger response can be induced by larger fragments (Barber et al., 1989; Felix et al., 1993; Yamada et al., 1993). The receptor system for oligochitin is very sensitive, and frequently, a response is generated at less than nano-molar concentrations. In the plasma membrane of several plant species, a high affinity binding site for oligochitin was identified (Baureithel and Boller, 1995; Day et al., 2001; Shibuya et al., 1996), and binding affinities were determined for the corresponding 75 kDa proteins (Ito et al., 1997; Okada et al., 2002). Assays with radiolabeled ligands demonstrate the presence of a single class site with high-affinity binding for active oligochitin in membranes prepared from rice cells (apparent K_d =5 nM) (Shibuya et al., 1993) and from tomato cells (apparent K_d=23 nM) (Baureithel et al., 1994). The binding protein, CEBiP, was purified from rice (Kaku et al., 2006) and has two extracellular lysin motifs (LysM) and a transmembrane domain but lacks an intracellular domain for signal transduction. Consequently, the chitin binding site requires additional components for signal transduction through the plasma membrane inside the cell, in a manner similar to the glucan binding receptor discussed previously. Recently, a receptor-like kinase, CERK1 (Chitin Elicitor Receptor Kinase) was identified in *Arabidopsis* and demonstrated to be essential for elicitor signaling (Miya et al., 2007; Wan et al., 2008). This protein has three extracellular LysM domains, a transmembrane domain and an intracellular Ser/Thr kinase domain. Because CERK1 and CEBiP were identified in different plant species, their simultaneous presence in the same binding protein complex remains to be confirmed.

Plant Oligosaccharins Involved in Defense Reactions

Pectins are major polysaccharides found in primary cell walls in non-graminaceous plants. In grasses, primary cell walls also contain small amount of pectins. The middle lamella in all plants is constructed from a combination of pectins and callose. Pectins are composed of several major polymeric blocks. Homogalacturonan is a linear homopolymer composed of 1,4-β-galacturonides. Some stretches of polygalacturonide can be xylosylated forming xylogalacturonide. Rhamnogalacturonan I (RGI) is a branched polymer comprised of a regular sequence of rhamnose-galacturonic acid disaccharides, where rhamnoses carry side chains of arabinans, galactans or arabinogalactans of various lengths. Finally, rhamnogalacturonan II (RGII) is the most diverse branched polysaccharide, with side chains composed of aldo- and keto-sugar-containing oligosaccharides (Albersheim et al., 2010).

Oligogalacturonides as Elicitors

Distinct elicitor-active oligosaccharins are produced in plants as a result of microbial hydrolytic action. Thus, 1,4-α-oligogalacturonides (OGs) (Fig. 2) are released from the pectic polysaccharides of plant cell walls by microbial endo-polygalacturonase (PGase) during host-pathogen interactions. These elicitor-active OGs are host-associated molecular patterns (HAMPs) that activate defense responses in many different plant systems. It is known that many microbial pathogen species produce PGase and pectate lyase to degrade plant cell walls during their invasion of host cells (De Lorenzo et al., 1994), thereby releasing OGs that initiate plant defense reactions. Concomitantly, plants synthesize polygalacturonase-inhibiting proteins (PGIPs) that accumulate in the extracellular matrix to inhibit microbial PGase degradation of plant cell walls and prevent the degradation of elicitor-active OGs (De Lorenzo and Ferrari, 2002). Indeed, overexpression of PGIP enhances plant resistance to pathogenic infection (De Lorenzo and Ferrari, 2002; Powell et al., 2000). PGIPs contain leucine-rich repeat domains (LRR) that are characteristic of resistance gene products (De Lorenzo et al., 2000; Gomez-Gomez and Boller, 2000). Further, plants expressing microbial PGase (Capodicasa et al., 2004) are more resistant to pathogens and acquire constitutively activated defense responses (Ferrari et al., 2008), which is potentially due to the accumulation of elicitor-active OGs produced by introduced PGase activity.

Initially, it was suggested that OGs act by changing physical properties of the plasma membrane rather than via a specific receptor (Aldington et al., 1991). This assumption was based on observations that OGs require a higher concentration (μM) to act as elicitor in comparison with the other oligosaccharins discussed above. The ability of differently sized oligogalacturonides (from 10 to 15 monomers) to exert activity indicates less stringency in their structure-function relationships in comparison with fungal elicitors. However, proteomic analyses of differential expression of apoplastic proteins in response to treatment of *Arabidopsis* seedlings with OGs demonstrated the induction and up-regulation of two receptor-like proteins: At3g22060, a protein related to receptor-like kinase comprising DUF26 domains; and At1g33590, an LRR disease resistance-related protein (Casasoli et al., 2008). These findings suggest the possibility that OGs also act via a receptor, similar to chitin and glucan oligosaccharins, and the two proteins that are up-regulated by OGs are involved in OG signal reception. Two lectin-like proteins (At3g15356 and At1g78830) also accumulate in response to OGs (Casasoli et al., 2008). The recently developed domain swap approach (Brutus et al., 2010) demonstrated that wall-associated kinase WAK1 functions as a receptor for OGs to initiate plant immune responses when OGs mimic damage-associated molecular patterns. WAKs have the Ser/Thr kinase signature and extracytoplasmic domain (composed of several EGF-like repeats) typical of eukaryotic cells (Verica et al., 2003). In *Arabidopsis,* this kinase family includes five tightly clustered genes: *WAK1, WAK2, WAK3, WAK4,* and *WAK5*. WAK1 has been demonstrated to bind to OGs and pectin *in vitro* under the conditions required for the formation of "egg-box" structures induced by calcium (Cabrera et al., 2008; Decreux and Messiaen, 2005). Using protoplasts, WAK2 was shown to be required for activation of a number of genes by pectin, including that for a vacuolar invertase (Kohorn et al., 2009). Further, WAKs have also been suggested to be involved in the regulation of cell expansion (Wagner and Kohorn, 2001). It will be intriguing to determine if other members of the WAK family can sense the OG signal.

It is interesting to note that the sensing of oligosaccharins originating from microbial cell walls (e.g., β-glucan, chitin, and chitosan) occurs via protein receptors that lack intracellular domains for signal transduction and therefore require the presence of other associated proteins. In contrast, OGs interact directly with a kinase (WAK1), which has a specific domain and theoretically does not require any assistance in initiating signal transduction. However, the structure of the OG receptor may be more complex, and participants in the signal transduction pathway remain to be identified.

Reactions Elicited by Oligosaccharides

Many defense reactions induced by oligosaccharins are similar to those observed when plant cells are treated with other elicitors or pathogens, with the exception being the hypersensitive reaction that leads to cell death (Hahn, 1996; Shibuya and Minami, 2001). Oligosaccharins that elicit cell death are yet unknown.

Membrane depolarization and ion flux are immediate events (within minutes) that have been characterized as the cellular response to elicitor signal perception (Aldington and Fry, 1993). All studied elicitors induce a rapid change in membrane potential, via Ca^{2+}/H^{+} influx and K^{+} efflux, which results in cytoplasmic acidification and the extracellular alkalinization of plant cells (Ebel et al., 1995; Kikuyama et al., 1997; Kuchitsu et al., 1997; Mathieu et al., 1991). These rapid alterations in the ionic status of the cytoplasm likely modify enzymatic activities in the plant cells, leading to metabolic changes (Shibuya and Minami, 2001). For example, the enhancement of secondary metabolite production was observed in response to elicitors (Roos et al., 1998). It was demonstrated that transient depolarization of the plasma membrane is not affected by anaerobic treatment or by the addition of azide (Zimmermann et al., 1997), indicating that ion channels (but not energy-dependent ion pumps) are involved in this process. Because no Ca^{2+} and anion channel inhibitors can completely block depolarization, the involvement of ion channels with broad specificity was suggested.

Using inhibitor analyses, it was shown that the processes of protein phosphorylation/ dephosphorylation are involved in the regulation of proton flux across the plasma membrane during elicitor action (Felix et al., 1993; He et al., 1998; Mathieu et al., 1996). Serine/threonine kinase inhibitors block ion flux, protein phosphorylation, and defenses activated by elicitors, while protein phosphatase inhibitors stimulate an inducible defense in the absence of elicitors (Nurnberger and Scheel, 2001).

Another rapid response of plant cells to elicitors or pathogen attack is the oxidative burst, which usually reaches its highest magnitude within minutes (Aldington and Fry, 1993). Reactive oxygen species (ROS) produced during this reaction can stimulate lignification and cross-linking of the cell wall (Barcelo and Laura, 2009), transcription of defense response genes, and phytoalexin biosynthesis (Ryan, 1988). At times, different initial signaling intermediates generated during the oxidative burst are induced by different elicitors. The pathway initiated by OGs involves G-proteins (Legendre et al., 1992) and phospholipase C (Legendre et al., 1993), whereas another cell wall elicitor from *Vericillium dahlia* functions via phospholipase A (Chandra et al., 1996; Legendre et al., 1993). Despite differences in the initial signaling events, the different oxidative burst pathways proceed via a common downstream pathway that involves an influx of Ca^{2+} and the activation of kinases (Chandra and Low, 1997; Jabs et al., 1997). In addition, a flavin-dependent oxidase is involved in many elicitor-stimulated burst pathways as a final component (Romeis et al., 1999). Two mitogen-activated protein (MAP) kinases (44 and 47 kDa) are involved in the oxidative burst pathway initiated by different elicitors (Taylor et al., 2001), suggesting that common MAP kinase intermediates can be shared by different initial pathways, leading to the oxidative burst. The following sequence of events leading to the oxidative burst was proposed: The elicitor via receptor activates MEK (a tyrosine-specific kinase), which in turn activates the 47 kDa MAP kinase. The latter induces anion flux and Ca^{2+} release from internal storage, and this activates the 44 kDa MAP kinase, leading to the oxidative burst. Althour, other unknown constituents of this proposed pathway may also be required.

The oxidative burst-independent pathway is considered to control the transcriptional activation of genes involved in phytoalexin synthesis. The β-glucan-mediated signal transduction pathway leading to phytoalexin production also involves MAP kinase cascade initiation. Two MAP kinases (GmMPK3 and GmMPK6) and one MAP kinase kinase (GmMAPKK1), which function as signaling elements of the defense reaction initiated by elicitor-active glucan, have been identified in soybean (Daxberger et al., 2007).

The reactions described above are the portion of the signal transduction pathways initiated after an elicitor-active oligosaccharide binds to a specific receptor, as described in the previous section. However, some slower downstream cell responses, such as metabolic changes, have also been shown to occur. Currently, limited information does not allow the construction of a complete chain of sequential events from the oligosaccharide's signal perception to the resultant metabolic changes, but some relations can be established. For example, cell wall lignification induced by elicitor-active oligosaccharides results from the rapid activation of phenylpropanoid pathway enzymes, such as PAL, *0*-diphenolmethyl transferase, and chalcone synthase (Aldington and Fry, 1993). These events are likely caused by changes in the ionic status of the cytosol due to ion fluxes, one of earliest events in elicitor action. Changes in mRNA synthesis of these enzymes have also been reported (Aldington and Fry, 1993), whih can also suggest the transcriptional activation of enzymes by oligosaccharins. Elicitor stimulation of the mRNA of another enzyme involved in lignification, alcohol dehydrogenase (CAD), is even faster (within minutes) than for PAL and CHS. Additionally, an increase of hydroxyproline-rich glycoproteins, such as extensins, in plant cell walls (Aldington and Fry, 1993) may be caused by the activation of synthetic enzymes stimulated by changes in ionic fluxes. The rapidity of these enzyme activations suggests that a small number of events occur between elicitor signal sensing and the activation of defense-related reactions (Templeton and Lamb, 1988).

In Summary

Multi-component receptor complexes in plants recognize pathogen- or host-derived elicitors and trigger transient changes in the plasma membrane and cytosolic status through increased ion fluxes, inducing the formation of secondary messengers such as Ca^{2+}, ROS, and NO. This signal sensing system includes the interactions of complementary pairs of pathogen-encoded avirulence genes and plant resistance genes (Nurnberger and Scheel, 2001). Further, the activation of MAP kinase cascades is a key constituent of elicitor-initiated signaling networks. Finally, the transcriptional activation and initiation of enzymatic activities involved in secondary metabolism take place.

Oligosaccharin Involvement in Plant Reactions to Wounding

Many structurally different molecules serve as regulators in wound signaling, and oligosaccharides released from the damaged cell wall are among these signaling molecules (Bishop et al., 1981). The induction of the plant wound response requires the simultaneous action of different signals and regulators, and the combination of these molecules very frequently depends on the plant species (Leon et al., 2001). Many plants can respond to localized injury via the systemic induction of protease inhibitors that help to prevent the digestion of plant proteins by invading microorganisms or insects. The messenger that evokes the synthesis of protease inhibitors in neighboring uninjured cells is termed the "protease inhibitor inducing factor" (PIIF). Systemin, an 18 amino acid peptide generated from prosystemin during the wounding response, (Mcgurl and Ryan, 1992; Mcgurl et al., 1992) and jasmonic acid (JA), synthesized at a wound site, (Li et al., 2002) are thought to function as long-distance signals for wound-activated gene expression in the Solanaceae family (where the systemic signaling response has been intensely studied) and in *Arabidopsis* (Leon et al., 2001). Upon injury, oligogalacturonides are released from damaged plant cell walls (Benhamou et al., 1990). Because there is evidence that even small oligogalacturonides are unable to migrate out of injured tissues (Baydoun and Fry, 1985), they are considered to be local signals involved in wound-inducible responses. OGs and chitosan oligosaccharides both activate proteinase inhibitor (*Pin*) gene expression in solanaceous plants (Doares et al., 1995; Ryan et al., 1986). Wound- and systemin-inducible polygalacturonase, which may be responsible for releasing oligogalacturonides from the plant cell wall, has also been described (Bergey et al., 1999). Considering that oligogalacturonides also elicit an oxidative burst in tomato (Stennis et al., 1998), the sequence of events for wound-signaling (minimally in tomato) is suggested to be wound-systemin-OGs-ROS (Leon et al., 2001).

Wound signal molecules, including OGs, promote rapid membrane depolarization followed by proton influxes (Thain et al., 1995). The sequence of events also includes an increase of the intracellular Ca^{2+} content and changes in differential protein phosphorylation in tomato and *Arabidopsis* (Leon et al., 2001). The presence of two parallel signaling pathways, JA-dependent and OG-dependent/JA-independent, has been proposed (Leon et al., 2001; Titarenko et al., 1997). Both are regulated by the mobilization of Ca^{2+} from intracellular stores, as well as through calmodulin-related activity (Bergey et al., 1999; Leon et al., 1998). Additionally, reversible protein phosphorylation via protein kinase/phosphatase activation is another key event during wound-inducible signaling. For the OG-dependent/JA-independent pathway, the induction of wound-responsive genes by a protein kinase activated by OGs has been proposed (Leon et al., 2001). This proposition is supported by the recent demonstration of OG signal reception by WAK1 (as described above). OG-mediated repression of the JA-dependent pathway through the production and sensing of ethylene is locally present in damaged tissues (Rojo et al., 1999) where OG-responsive genes are expressed.

In Summary

Mechanical wounding releases OGs from damaged cell walls, which bind to WAK1 and activate proton influx, Ca^{2+} accumulation and ROS production. This induces a downstream protein kinase cascade (MEK, MPK, and MEKK) leading to the activation of wound-responsive genes (*Pin*). In parallel, the induction of ethylene suppresses the local JA-dependent pathway, potentially redirecting JA to systemic tissues.

Oligosaccharins and Plant Development

Another important but less advanced area of biologically active oligosaccharide studies is the regulation of plant growth and development. The ability of undifferentiated totipotent cells to differentiate and develop into specialized organs is a unique feature of plants and is regulated by a combination of endogenous and exogenous factors (e.g., phytohormones and light). Phytohormones are polytrophic regulators, each of which has the ability to regulate a broad range of cellular processes. Therefore, the involvement of other effectors, for example oligosaccharins, that would tune-in the phytohormone's broader mode of action down to a more specific regulation of a single process or pathway is a very intriguing hypothesis proposed by Albersheim when he first introduced oligosaccharins (Albersheim and Darvill, 1985).

OGs generated by fungal PGase that stimulate defense reaction in plants, as discussed above, can also induce flower formation and inhibit rhizogenesis in tobacco thin-layer explants (TLCs) or leaf explants (Altamura et al., 1998; Bellincampi et al., 1993). Tissue explant- or organ segment-based assays are very common model systems in studies of the effects of phytohormones and other growth effectors on plant morphogenesis (Mohnen et al., 1990). OGs can change the morphological response of tobacco TLCs incubated in media containing a particular ratio of two hormones: auxin and cytokinin. The concentrations of OGs required for stimulation of morphological responses are approximately 100-fold lower (≤ 1 μg/ml) than their concentrations required for elicitor activity in plant defense reactions. The ability of OGs to inhibit root formation in TLCs or leaf explants from tobacco or *Arabidopsis* is the result of the OG's anti-auxin activity (Bellincampi et al., 1996). Only auxin-dependent root formation can be blocked by the addition of OGs to the media of explants. Additionally, OGs inhibit auxin-induced elongation of pea stem segments (Branca et al., 1988), presumably by interacting with auxin-binding sites (Filippini et al., 1992). For example, OGs were demonstrated to depress the expression of auxin-inducible promoters, *rolβ, rolD,* and *Nt114*, when inhibiting the development of root primordia in leaf explants (Mauro et al., 2002). The ability of OGs to induce flowering in tobacco explants may also be explained by anti-auxin activity. When OGs are added to explants incubated in media with a certain ratio of auxin and cytokinin, they shift the hormonal balance toward cytokinin (Eberhard et al., 1989) by depressing auxin's action.

Structure-function studies of biologically active oligogalacturonides involved in developmental processes have demonstrated that the lengths of the most active fragments are between 10 and 14 residues, and the reducing end of the OGs is important for signaling function (Spiro et al., 1998). It was reported that alginate-derived trisaccharide with *O*-4-deoxy-l-*erythro*-hex-4-enopyranosyluronic acid on the non-reducing end (obtained from *Alteromonas macleodii* by lyase digestion) promotes root growth in barley (Natsume et al., 1994), but no further work has been done with this fragment.

In contrast to the ability of acidic pectic oligosaccharides to inhibit root formation, neutral oligosaccharides released from pectins activate root formation in TLCs from buckwheat hypocotyls grown in the dark and incubated without the addition of phytohormones (Lozovaya et al., 1993). The ability of buckwheat TLCs to form roots in media without exogenous phytohormones is potentially due to sufficient levels of endogenous auxin present in etiolated hypocotyls. A mixture of oligosaccharides added to buckwheat TLCs at a concentration of 1 μg/ml doubled the number of roots formed on each explant.

Xyloglucan is another highly abundant polysaccharide in primary cell walls in non-graminaceous plants; in grasses, xyloglucan is present in lower amounts. This hemicellulosic polysaccharide is composed of a backbone of 1,4-linked glucosyl residues, such that two or three of four backbone residues are substituted with α-xylosyl residues. Approximately half of the xylosyl residues are further substituted with either single α-galactosyls or fucose-galactosyl dimers. The xyloglucan nonasaccharide XXFG (Fig. 3), purified from a mixture of oligosaccharides prepared by the treatment of xyloglucan with endo-β-1,4-glucanase from *Trichoderma reesei,* inhibits 2,4-dichlorophenoacetic acid (2,4-D)- and gibberellic acid-induced (Warneck and Seitz, 1993) elongation of pea stem segments similarly to OGs. Structure-function studies have demonstrated that the terminal fucosyl residue of XXFG is essential for anti-auxin activity, while the terminal xylosyl closest to non-reducing end and the reducing glucosyl are not required for this activity (Fig. 3). The oligosaccharide with two terminal fucosyls, XFFG, exerts higher activity than XXFG, and decasaccharide XLFG was reported to be inactive (Fig.3) (Mcdougall and Fry, 1989). The fucose-containing pentasaccharide (Fig. 3) is less active than the nonasaccharide (Fry, 1989), and its synthetic analog inhibits fusicoccin-stimulated elongation of pumpkin cotyledons (Pavlova et al., 1992). The growth-affecting activities of xyloglucan oligosaccharides are observed at nanomolar concentrations. This, together with highly stringent structure-function requirements, suggests that these oligosaccharins signal via specific receptors. However, binding sites for xyloglucan oligosaccharides have not been determined.

When XXFG nonasaccharide is added at higher concentrations (1 μM) to pea segments in the absence of 2,4-D, it promotes their elongation (Aldington and Fry, 1993). Structure-function studies have demonstrated that the presence of a cellotetraose core with at least one xylosyl is required for the growth promoting activity of XXFG (Fig. 3), but it does not depend on the presence of the terminal fucose (Mcdougall and Fry, 1989; McDougall and Fry, 1990). Thus, it appears that XXFG oligosaccharide, when present at different concentrations, is able to exert the opposite effects on plant growth. At low concentrations, XXFG has a growth inhibitory effect, while at higher concentrations, it begins to promote the growth. Overall, the growth promoting effect of XXFG dominates.

Galactoglucomannans are other hemicellulosic polysaccharides that are predominantly present in vegetative tissues and seeds of legumes, and in wood tissues of trees. Oligogalactoglucomannans (GGMOs) (degree of polymerization 4-8) extracted from the wood of poplar inhibit the 2,4-D-stimulated elongation of pea and spruce stem segments (Auxtová et al., 1995) and differentially affect the activities of extracellular, intracellular, and cell wall glycosidases (Bilisics et al., 2004). GGMOs also promote the differentiation of bud primordia on spruce cotyledons in the presence of IAA and zeatin in the growth media, as well as increase the viability of spruce embryos (Lišková et al., 1995). GGMOs were shown to inhibit adventitious root formation on mung bean hypocotyl cuttings, and their effect depends on the type of auxin present in the growth media (Kollarova et al., 2005). Structure-function analyses have demonstrated that a reduction in the amount of galactosyl residues linked to the glucomannan backbone decreases the inhibitory activity of GGMOs, while a modification of the reducing end does not affect their activity (Kollarova et al., 2006). It was reported that the GGMO anti-auxin activity observed during the inhibition of elongation is accompanied by changes of peroxidase activities that are localized in the cell wall. Therefore, their effect on cell wall rigidity has been suggested (Kollarova et al., 2009).

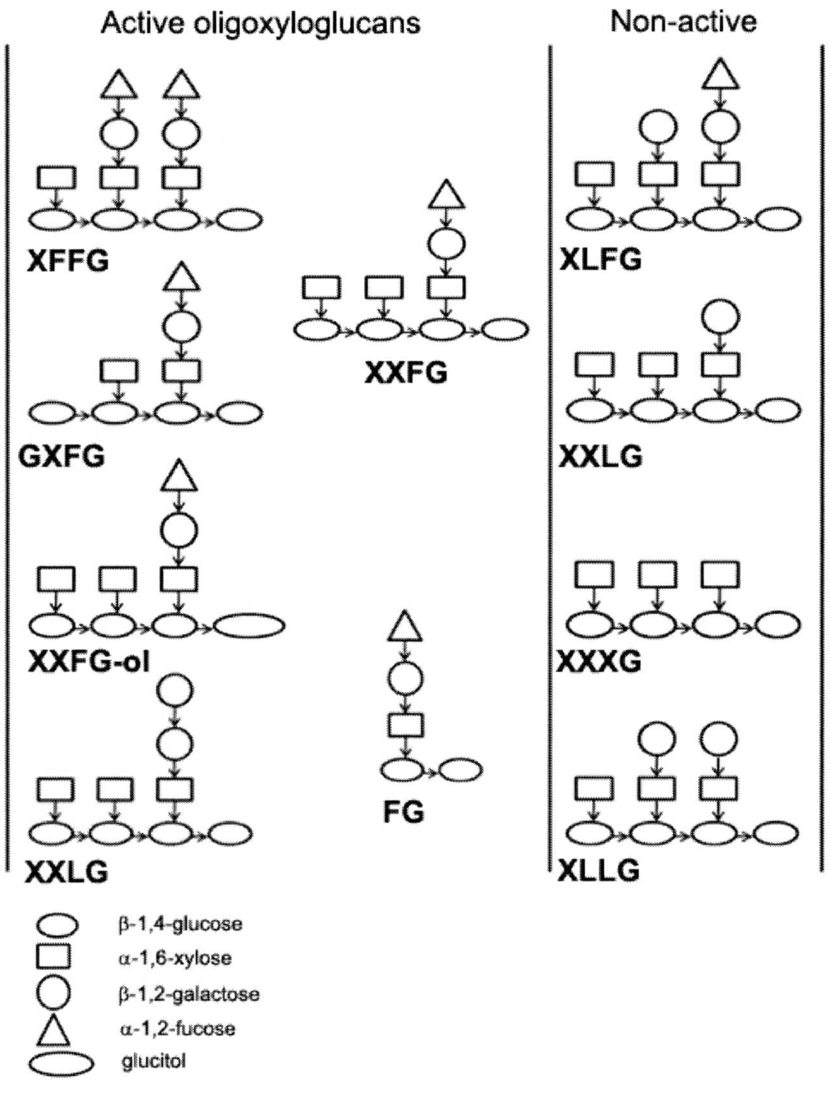

Figure 3. Structures of xyloglucan oligosaccharides used for growth regulation studies. Oligosaccharides XXXG, XXFG, XLFG, XFFG, and FG were prepared as the products of xyloglucan digestion by endo-glucanase. XXLG and GXFG were generated by enzymatic treatment of XXXG. XXFG-ol was obtained after reduction of reducing end in XXFG, and XXLG with substituted fucose by galactose was prepared from *mur3* plants grown on the media with galactose (McDougall and Fry, 1989; Augur et al., 1992; Zablackis et al., 1996).

All of the aforementioned oligosaccharins were obtained *in vitro* by enzymatic or acidic degradation of isolated cell walls or purified polysaccharides and therefore represent, to some extent, artificial prototypes of oligosaccharins that may be released in plant tissues *in vivo*. A legitimate question as to whether oligosaccharins actually are present *in vivo* has been the central obstruction in all of these previous studies. However, some indirect observations have been reported as initial evidence that bioactive oligosaccharides can be generated *in vivo*. Thus, the presence of xyloglucan oligosaccharides structurally similar to those that exert biological activities was demonstrated in the growth medium of suspension-cultured cells (Mcdougall and Fry, 1991). Oligogalacturonides purified from the extracellular medium of *Lithospermum erythrorhizon* cell culture were capable of inducing naphtoquinone pigment synthesis (Tani et al., 1992). Xylosyl- and mannosyl-containing acetylglucosamine oligomers found in the extracellular medium of *S. alba* culture stimulate tomato fruit ripening (Priem and Gross, 1992; Priem et al., 1994). Further, several uncharacterized, putatively bioactive acidic oligosaccharides that are capable of promoting the viability of cells in low density cultures were separated from tobacco and carrot cell culture media (Schroder and Knoop, 1995). It was also found that the growth medium of *Zinnia elegans* cell-suspension cultures contain heat and protease stable, low molecular elements that are presumably oligosaccharides. This medium was added to freshly prepared Zinnia cells and affected cell expansion, early formation of large tracheary elements, and the synchrony of cell differentiation (Roberts et al., 1997). Additionally, elicitor-active glucans are released into the extracellular medium by germinating *Phytophthora sojae* spores cultured without any plant tissues (Cote and Hahn, 1994).

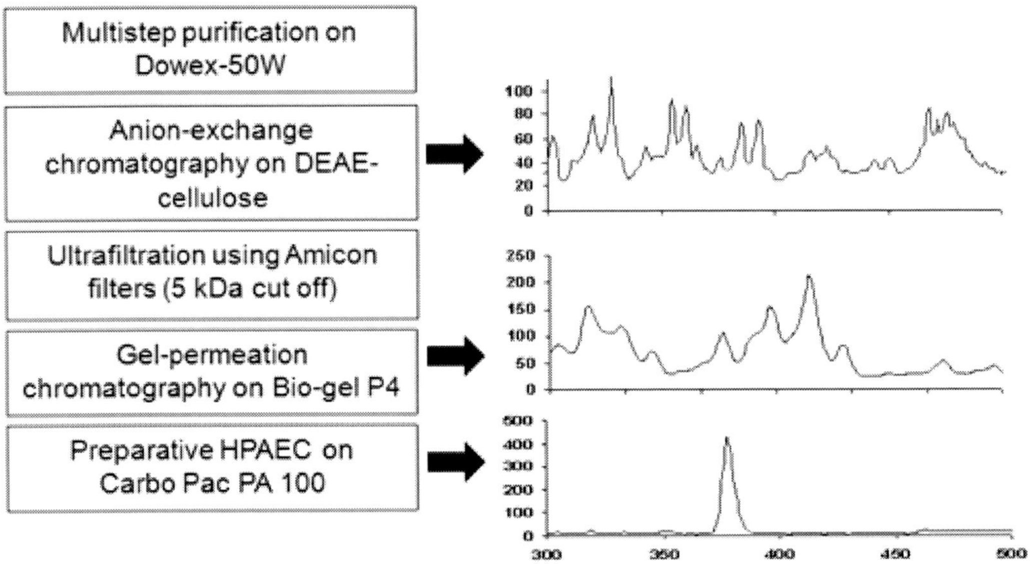

Figure 4. Scheme of endogenous bioactive oligosaccharide purification from plant tissues. Obtained fractions were analyzed by the analytical HPAEC and chromatograms demonstrate the complexity of the active fraction on each step of purification. At the end of purification the fraction with single peak was obtained indicating the high purity of root-inducing oligosaccharin (GAG).

The first direct demonstration of oligosaccharin's presence in plant tissue *in vivo* was the purification of biologically active oligosaccharides from the tissue sap of young pea seedlings (Zabotina et al., 1996b). Sap was prepared, avoiding contamination with cell wall material and the activation of hydrolytic enzymes present in plant tissue. Neutral oligosaccharides were then purified using various chromatographic techniques (Fig. 4) to high homogeneity, as confirmed by high performance anion-exchange chromatography (HPAEC). This highly purified oligosaccharide fraction stimulates root formation in TLC explants from buckwheat hypocotyls (Fig. 5) in the absence of phytohormones (Zabotina et al., 1998a). It also increases auxin-dependent root formation in tobacco and *Arabidopsis* leaf explants, as well as auxin-stimulated lateral root formation in maize root segments (Zabotina, unpublished results). The size of the active oligosaccharide (abbreviated as GAG) was estimated as 10-11 monomers with a composition of Gal:Ara:Glc (60:30:10 mol%). The ability of GAG to stimulate root formation in buckwheat TLCs is completely abolished after treatment with Driselase but is not altered after boiling for 20 min or after treatment with proteases, indicating that the molecule responsible for the observed activity is an oligosaccharide. GAG stimulates root formation by increasing the number of root primordia but does not affect root growth. In agreement with this, GAG has no effect on the elongation of pea segments in the presence or absence of 2,4-D. Promoting an auxin effect on root formation, GAG synergistically works with auxin to increase the expression of a *rolβ-GUS* reporter gene, stimulating an auxin-inducible promoter (Zabotina, unpublished results). It will be of great interest to determine the exact structure of this endogenous root-promoting oligosaccharide and to evaluate the dynamics of its endogenous content.

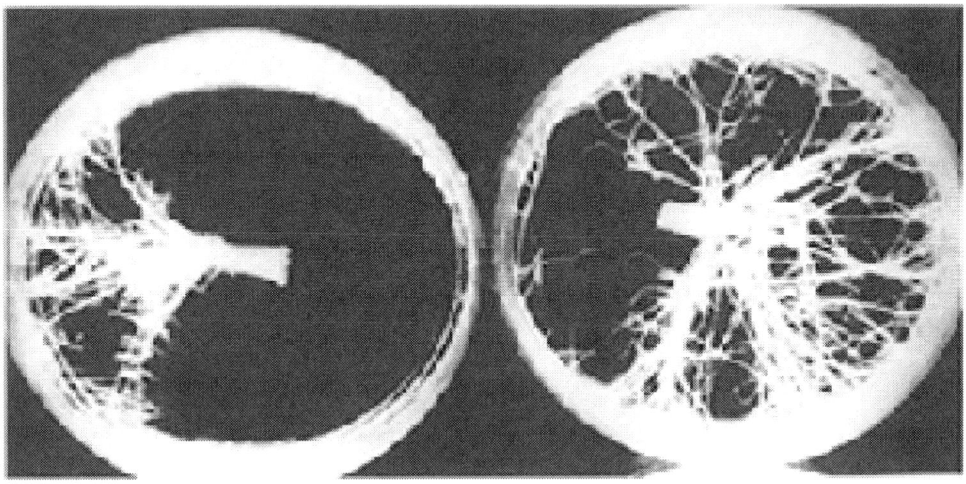

Figure 5. Increase of root systems in Buckwheat thin-layer explants incubated with root-stimulating oligosaccharin (1μg). Pictures were taken after 4 weeks of incubation.

Another type of endogenous biologically active oligosaccharide was purified from the water soluble extract prepared from the perennial medicinal plant *Paris polyphylla.* Linear hepta- and octasaccharides composed of 1,6-linked glucoses with 1,4-linked mannose on the reducing end were demonstrated to stimulate shoot formation, hairy root growth, and saponin production in these roots in *P. polyphylla* culture (Zhou et al., 2003).

Some pectic oligosaccharides that are released and accumulate as the products of cell wall degradation during fruit ripening can further promote ripening by inducing ethylene synthesis when added to unripe fruits (Brecht and Huber, 1988; Dumville and Fry, 2000; Melotto et al., 1994). However, few studies have been performed to investigate the structures of naturally occurring oligosaccharides, and there are no data available on the structure-function relationships of bioactive oligosaccharides involved in ripening. Regardless, it is known that homo-oligogalacturonides and oligosaccharides with neutral side chains can both be naturally produced in tomato fruits during ripening (Dumville and Fry, 2000).

Oligosaccharins and Plant Adaptation

During their life cycle, plants have to defend themselves against pathogenic microorganisms and pests, as well as survive in constantly changing environments while coping with different abiotic stresses. Most plants have acquired complex systems that help them to adapt to environmental changes. It was demonstrated that plant cell walls are the first barrier of defense, not only against pathogenic microorganisms but also acting as effective protection from other environmental stresses, such as temperature, wind, and other forms of mechanical damage, drought, salinity, and flood. Changes in cell wall composition have been demonstrated in response to almost all abiotic stresses, frequently leading to cell wall fortification (Albersheim et al., 2010). Considering that cell wall rearrangements directed towards the fortification of cell walls in response to pathogen attacks result in the release of elicitor-active oligogalacturonides (that in turn protect plants), it is plausible to expect similar results from cell wall rearrangements in response to abiotic stresses.

The first promising evidence of such a possibility is the demonstration of endogenous oligosaccharide accumulation during the first hours of winter wheat hardening at 2°C. In nature, the adaptation process, referred to as fall frost hardening, typically proceeds over the course of the fall and early winter and depends on the rate of decrease in the average daily temperature. The gradual natural process of fall frost hardening can be reproduced in the laboratory over a much shorter period of time (3-4 weeks) by using a two-phase acclimation procedure. This involves: 1) exposing the plants to low, non-freezing temperatures (~2°C) for 5-7 days, followed by 2) exposure of the plants to sub-freezing temperatures that are above the ice nucleation temperature in the tissue. The first phase of adaptation is essential for the full expression of freeze tolerance because plant metabolism undergoes the most dramatic changes during this phase. Rapid transient activation of cell wall glycosidases accompanied by transient increases in cell wall polysaccharide catabolism are observed during the first hours of winter wheat first stage acclimation (Zabotin et al., 1998). This polysaccharide degradation, initiated by low positive temperatures, results in the transient accumulation of soluble, low molecular weight oligosaccharides in plant tissues by 6-10 h after the initiation of hardening (Zabotina et al., 1998b). A freezing tolerance-inducing oligosaccharide was purified from these endogenously accumulated oligosaccharides and partially characterized (Zabotin et al., 2005). The monosaccharide composition of this oligosaccharin, abbreviated GXAG, is Glc:Xyl:Gal:Ara (5:5:1:1), suggesting a xyloglucan origin, although with a nontraditional type of subunit. However, further detailed structural characterizations are required to confirm its structure. GXAG stimulates freezing tolerance in winter varieties of wheat and rye but does not affect spring varieties of these species (Zabotin et al., 2004). This suggests its specificity to the process of freezing tolerance acquisition in temperate species. GXAG is active at concentrations as low as 10 nM and stimulates freezing tolerance acquisition in winter plants even at 20°C. The ability of oligosaccharin to initiate freezing tolerance at room temperatures is similar to abscisic acid (ABA), a phytohormone well known to be involved in this process (Gilmour and Thomashow, 1991; Gusta et al., 2005). Studies of the relationship between GXAG and ABA during the initiation of freezing tolerance acquisition demonstrated that oligosaccharin synergistically potentiates ABA activity by increasing the cell's sensitivity to the phytohormone (Zabotin et al., 2009). Only pretreatment with GXAG before the addition of ABA demonstrated this synergistic effect, while the opposite order of treatments or their simultaneous addition to winter wheat seedlings showed an additive effect (Fig. 6). The mixture of oligosaccharides prepared from sycamore xyloglucan digested with endo-β-1,4-glucanase, purified XXLG oligosaccharide from Tamarind seeds, and hepta-β-1,6-glucan were tested in the same assay and were not able to initiate freezing tolerance and potentiate ABA activity, demonstrating GXAG specificity for the stimulation of freezing tolerance in winter species (Zabotin et al., 2009).

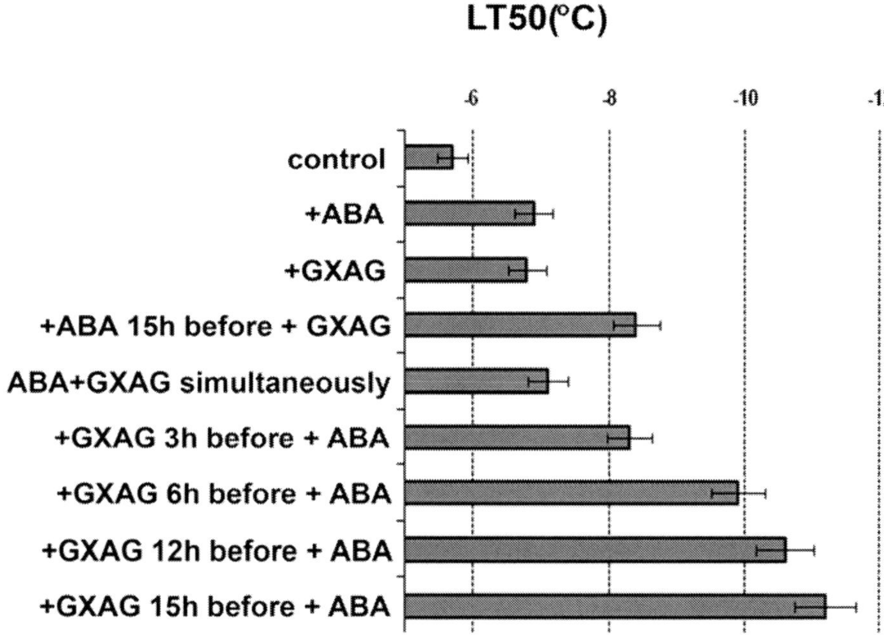

Figure 6. Effect of ABA and GXAG on the acquisition of freezing tolerance in winter wheat seedlings. Seedlings were treated with ABA (50 μM) and GXAG (5 μg/ml) in different order and time period, and their LT_{50} values were determined after growth for 7 days at 20°C. Control represents LT_{50} of non-treated seedlings grown at the same conditions.

In Summary

Currently, there is no information about the specific receptors capable of binding oligosaccharins involved in the regulation of growth and development. The low concentrations and high structural stringency of most oligosaccharins required for their signaling activities suggests that signal perception occurs via specific receptors that remain to be identified. One plausible possibility is the involvement of WAKs in sensing oligosaccharin signals during the initiation of developmental programs. Similarly, because WAK1 was demonstrated to perceive OG signals during the wound-induced reaction, it is tempting to imagine that other members of this protein family may function as specific receptors for other types of oligosaccharins. For example, it was suggested that WAK2 is involved in the regulation of cell expansion (Wagner and Kohorn, 2001) and therefore may serve as a receptor for growth-regulating oligosaccharins. Simultaneously, most of the oligosaccharins participating in the regulation of growth and development act together with phytohormones, either as antagonists or as promoters of their effects. Therefore, it is possible that oligosaccharins can interact directly with hormone-specific receptors, at least with those localized in the plasma membrane, modifying their binding to the ligand. The fact that receptors were discovered for all major known elicitor-active oligosaccharins supports the hypothesis of the requirement for a specific receptor for each type of oligosaccharin and, possibly, for each specific signaling pathway initiated by the same oligosaccharin. Thus, in the case of OGs, WAK1 may be the specific receptor for their elicitor signaling during the initiation of the defense reaction, while other receptors may be required for their anti-auxin activity. This receptor must have a higher affinity due to the lower concentrations required for OG anti-auxin activity. The discovery of new receptors will undoubtedly stimulate a new wave of studies in oligosaccharin signal transduction pathways, more firmly confirming their significance for plant cell functions.

FORMATION OF OLIGOSACCHARINS *IN VIVO*

Theoretically, oligosaccharins can be produced *in vivo* by two different means: 1) synthesis in the Golgi as intermediate products of polysaccharide biosynthesis or 2) being liberated as a result of partial degradation of polysaccharides localized in the cell wall. The machineries required for both pathways of oligosaccharin formation are present in the plant cell. Thus, the Golgi houses a complete set of enzymes required for the synthesis of cell wall polysaccharides. These enzymes possess flexible and dynamic synthetic machinery capable of supporting plants through their entire lifecycle. This machinery synthesizes a wide variety of polysaccharides and other cell wall components, supporting growth and developmental processes and quickly reacting to changes in environmental conditions. Plants also have a diverse set of hydrolytic enzymes localized in the apoplast, representing other dynamic and complex machinery. These degrading enzymes also play a key role in the growth, development, and protection of plant cells that are constantly remodeling the cell wall. Recent results demonstrate that oligosaccharins are the products of partial degradation of the cell wall polysaccharides during rapid plant growth or in response to environmental changes. Microorganisms have also developed sophisticated hydrolytic machinery enabling their penetration through plant cell walls during invasion. As discussed above, oligosaccharins can be liberated during host-pathogen interaction as products of microbial hydrolytic activity.

Experiments using radiolabel substrates have demonstrated that polysaccharides are synthesized in the Golgi and delivered in vesicles to the cell wall surface as large polymeric molecules, possibly as preformed structural complexes (e.g., between xyloglucan and RGI) (Popper and Fry, 2008). Similar to other signal molecules, oligosaccharins are produced transiently and usually in response to certain changes in conditions. It is unlikely that their *de novo* synthesis and delivery in vesicles to the receptor location is fast enough to support quick responses. Therefore, the production of oligosaccharins directly in the apoplast where they can quickly accumulate to sufficient concentrations and have direct access to specific receptors localized in the plasma membrane is the most logical event.

The middle lamella is rich in pectins and is the first polysaccharide layer in the plant barrier against pathogen attack. Therefore, pathogenic microorganisms produce a variety of enzymes to degrade pectic polysaccharides (de Vries and Visser, 2001). The initial indirect information about elicitor-active pectic oligosaccharide formation was obtained using hydrolytic microbial enzymes. For example, pectinases from *Rhizopus stolonifer* and from *Aspergillus niger* induce phytoalexin synthesis and hypersensitivity in castor beans (Bruce and West, 1989; Cervone et al., 1987), and endo-pectate lyase from *Erwinia carotovora* induces phytoalexin synthesis in soybeans (Davis et al., 1984). Pectic enzymes are secreted by *Psedomonas syringae* (Bashan et al., 1985) and by mutualistic mycorrhizal fungi (Garcia-Romera et al., 1991). The overexpression of microbial endo-polygalacturonase in *Arabidopsis* and tobacco increases the resistance of transgenic plants to pathogen (Ferrari et al., 2008). Pectin methyl esterase may also be involved in releasing elicitor-active oligosaccharides, facilitating the action of pectinase (Miller and Macmilla.Jd, 1971). Other enzymes may be activated later, after the progressive degradation of the cell wall during microbial penetration. For example, pectinase, pectate lyase, cellulase, protease, and xylanase are activated sequentially after inoculation of potato with *Erwinia* (Pagel and Heitefuss, 1990). Pectinases catalyze the mid-chain hydrolyses of polygalacturonic molecules within the portions of pectin homogalacturonan that consist of several contiguous, non-esterified residues. This releases the OGs and, thus, amplifies their signal in the host cell. At the same time, the plant cell produces inhibitors of microbial hydrolytic enzymes that protect the plant cell walls and elicitor-active OGs from further degradation. PGIPs and pectin methyl-esterase inhibitors have been found in the cell walls of a variety of plants (De Lorenzo and Ferrari, 2002; Mattei et al., 2002; Toubart et al., 1992), and engineering plants with reduced or increased amounts of these proteins can correspondingly depress or increase plant resistance to pathogens (Ferrari et al., 2006; Lionetti et al., 2007).

Pectinases are activated during the ripening of many fruits in parallel with the accumulation of ethylene (Dumville and Fry, 2000). Because pectic enzymes and pectic oligosaccharides can both induce ethylene biosynthesis, it has been suggested that the initial activation of some pectic enzymes at the beginning of ripening releases pectic oligosaccharins that stimulate ethylene biosynthesis. An increased amount of ethylene further stimulates enzyme activity, increasing polysaccharide degradation, which leads to fruit softening (Aldington and Fry, 1993).

Many plant tissues constitutively produce chitinases and also activate their expression in response to infection, abiotic stress, or hormonal treatments (Aldington and Fry, 1993). The action of chitinase on the fungal wall generates chitin or chitosan oligosaccharides that elicit defense reactions in plants. It is also likely that endo-β-glucanases that are present in many plants can liberate elicitor-active oligoglucans from the β-1,3;1,6-glucans of fungi cell walls, similar to what was shown for soybean endo-glucanase releasing elicitor-active hepta-β-glucan from the walls of *Phytophthora megasperma* (Yoshikawa et al., 1981). The prolongation of chitinase and endo-glucanase activities can further degrade elicitor-active oligosaccharins to shorter fragments, thereby inactivating their post-signaling activity.

Partial degradation of xyloglucan has been demonstrated in the stems of many plants during auxin-promoted elongation or in response to treatment with H^+ (Aldington and Fry,

1993). This degradation can be catalyzed by cellulases, a small amount of which is constitutively present in the plant cell wall, while larger quantities are synthesized in response to auxin treatment (Verma et al., 1975). Although plant cellulases seem unable to convert xyloglucan directly to short fragments, even stochastic cleavage would release a certain amount of XG9 that would be enough to reach the 1 nM concentration to exert biological activity (Aldington and Fry, 1993). However, it is more plausible to believe that xyloglucan endotransglysosylase (XET), which cuts the backbone of xyloglucan and re-connects the newly formed reducing end to the non-reducing terminus of a different xyloglucan molecule (Fry et al., 1992), is more effective in producing bioactive oligoxyloglucans (Fry et al., 1993). These proteins are highly abundant in plants and generally encoded by large multi-gene families (Maris et al., 2009). These enzymes are also implicated in many different processes, such as cell wall extension during cell elongation, root growth, fruit ripening, and responses to stress (Fry, 1995), which suggests their potentially diverse physiological functions. Some XETs can act only as endo-glycosylases, breaking down xyloglucan molecules somewhere in the middle and were named XHTs. Hydrolyses of xyloglucan molecules by activated XET/XHT enzymes during growth or stress reaction may be the source of oligosaccharins. Released oligoxyloglucans accumulate and either interact with their corresponding receptor in the same cell or move to other cells. After the signal role of oligosaccharins is accomplished, they are then recombined with cell wall xyloglucan polysaccharides. An example of such sequence of events could be the transient accumulation of freezing tolerance-inducing oligosaccharin GAXG (Zabotin et al., 2009; Zabotin et al., 2005).

Movement of Oligosaccharins in Plants

Studies using radiolabeled bioactive oligosaccharides have demonstrated that they can move acropetally via the xylem of the plants (Faugeron et al., 1999; Warneck et al., 1998). Transport by the phloem was only observed in the case of N-glycans, but it was unclear if the structure of the oligosaccharide remained intact or if only smaller degradation products moved (Faugeron et al., 1999). Because pectic oligosaccharides cannot move basipetally in tomato plants (Baydoun and Fry, 1985), their action during the wounding response is considered to occur locally, near the site of injury where they are released. Conversely, during other defense responses, such as the elicitation of phytoalexin synthesis or triggering peroxidase accumulation, oligosaccharins are mobile enough to serve as signaling molecules in other cells. Another example is the formation of GAXG during low temperature acclimation. When this oligosaccharin is applied to media containing the immerged roots of wheat seedlings, it stimulates freezing tolerance in the entire seedling, as determined by measuring the electrolyte efflux from leaf tissues (Zabotin et al., 2009). These results suggest that exogenously applied GXAG is quickly picked up by roots and distributed to the upper parts of seedlings. Because GXAG is detected endogenously in all parts of the seedlings, it has also been suggested that all cells may form this oligosaccharide in response to chilling temperatures.

During movement, oligosaccharins (OGs and XG9) can be modified or degraded (Aldington and Fry, 1993; Macdougall et al., 1992; Warneck et al., 1998). However, some oligoxyloglucan molecules applied to pea segments moved for 6-10 mm without any modification and maintained their intact structure for 24 h after application (Warneck et al., 1998). This suggests that oligosaccharins are maintained in low steady-state concentrations and can be transported to the site of their action.

There are two possible major mechanisms for reducing the concentration of oligosaccharins when they fulfill their signaling role. The first is degradation by hydrolytic enzymes, i.e., glycosyl hydrolases. Thus, the presence of fucosidase (Augur et al., 1993), xylosidase (Sampedro et al., 2001), galactosidase, and glucosidase (Iglesias et al., 2006), enzymes potentially capable of cleaving xyloglucan molecules, was demonstrated in plant tissues. However, the demonstration of XG9 stability in pea tissues up to 24 h, despite the presence of α-fucosidase and α-xylosidase, still leaves open questions about the mechanism of oligosaccharin inactivation by degradation. Polygalacturonidase localized in plant cell walls may be responsible for degradation of OGs and other pectic oligosaccharides into smaller inactive fragments. The analysis of degradation products of N-linked glucans during their movement (Faugeron et al., 1999) also suggests the involvement of α-mannosidase.

The second mechanism is modification via incorporation into larger polymeric molecules (as was demonstrated for oligoxyloglucans) or by binding to other small molecules (as was shown in the case of OGs bound to unidentified alcohol) (Macdougall et al., 1992). It has been demonstrated that xyloglucan xylosyl transferase from barley can also bind to β-1,3,1,4-glucan (Hrmova et al., 2007), suggesting the potential existence of such a mechanism of incorporation for other types of oligosaccharides besides xyloglucan.

In Summary

Clearly, strong evidence exists demonstrating that there is established machinery for producing and inactivating oligosaccharins. However, less is known about their transport, and their spatial distribution is still uncertain. Do endogenously produced oligosaccharins act in the cell *in situ* or move through the apoplast to other cells as other types of secondary messengers? Although, exogenously applied, labeled oligosaccharins seem to be transported (minimally) via xylem, the potential ability of each cell to form bioactive oligosaccharides may eliminate the necessity to transport them out of the origin site.

Conclusion

We have summarized the currently available information on the diversity of oligosaccharins, their structure-function characteristics, their routes of synthesis, distribution, and inactivation, and their possible mode of action. However, there are still many gaps in our knowledge concerning each of these topics.

Despite all of these gaps, the oligosaccharins are a new group of signaling molecules that clearly exist and have a wide range of physiological functions. The oligosaccharins identified to date likely represent only the tip of the iceberg, the main body of which remains to be seen. Due to the complexity of the plant cell wall, additional plant-derived oligosaccharins with novel activities will undoubtedly be identified. Theoretically, the plant cell wall contains a vast number of hydrolytic enzymes large enough to produce oligosaccharins from all types of polysaccharides. The same can be said about microbial hydrolytic enzymes. Detailed biochemical investigations of the cellular signaling pathways initiated by oligosaccharins require the isolation/purification of structurally homogeneous oligosaccharides. The development of more efficient approaches for oligosaccharin detection directly in plant tissues and their purification may help to discover new structures with novel functions. Revealing new types of oligosaccharins in a wider range of organisms will broaden our understanding of their important role as biological signals and uncover the mechanisms of the cell's perception of environmental changes. One approach to searching for novel oligosaccharins might be to challenge the plant with environmental stresses. The example of GAXG produced in response to low temperature suggests that plant responses to stress, when the plant cell undergoes significant metabolic changes, potentially involve a range of signaling molecules. For example, temperature, drought, salinity, and drugs may be stimuli for oligosaccharin production and their function as a signal during plant responses to these stresses. Another approach is to increase the hydrolytic activity in the plant apoplast by overexpressing glycosyl hydrolases or even by introducing new types of hydrolytic enzymes. The overexpression of microbial polygalacturonase in plants increases their resistance to systemic pathogens (see in the text), likely by releasing higher amounts of OGs. In a similar way, other microbial hydrolytic enzymes with diverse specific activities can be introduced and localized into plant apoplast. They will release structurally variable oligosaccharides, some of which may be similar to naturally existing ones or those produced under certain conditions. Physiological and biochemical studies of these transgenic plants may reveal yet unknown active oligosaccharides and their functions.

Recently, plant cell wall research has been significantly intensified due to the demands for bio-renewable fuel. Progress has been made in understanding cell wall biosynthesis and metabolism, though there is still a long way to go. Along this path, there will be new discoveries in the field of cell wall component signaling functions. It will be interesting to see if oligosaccharins regain their attraction from researchers and assume an equally deserved position among other regulatory molecules. We believe the most intriguing questions in the area of biologically active oligosaccharides and their roles in the plant cell remain to be investigated.

REFERENCES

Albersheim, P., and B.S. Valent. 1978. Host-Pathogen Interactions in Plants - Plants, When Exposed to Oligosaccharides of Fungal Origin, Defend Themselves by Accumulating Antibiotics. Journal of Cell Biology 78:627-643.

Albersheim, P., and A.G. Darvill. 1985. Oligosaccharins. Scientific American 253:58-64.

Albersheim, P., A. Darvill, K. Roberts, R. Sederoff, and A. Staehelin. 2010. Plant Cell Walls Garland Science, Taylor&Francis Group, LLC, New York.

Albersheim, P., A. Darvill, M. McNeil, B. Valent, J. Sharp, E. Nothnagel, K. Davis, N. Yamazaki, D. Gollin, W.S. York, W. Dudman, J. Darvill, and A. Dell. 1983. Oligosaccharins: Naturally occurring carbohydrates with biological regulatory function., p. 293-311, *In* O. Ciferri and L. Dure, eds. Structure and function of plant genomes. Plenum Press.

Aldington, S., and S.C. Fry. 1993. Oligosaccharins. Advances in Botanical Research 19:1-101.

Aldington, S., G.J. McDougall, and S.C. Fry. 1991. Structure-activity relationships of biologically active oligosaccharides. Plant, Cell and Environment 14:625-636.

Altamura, M.M., D. Zaghi, G. Salvi, G. De Lorenzo, and D. Bellincampi. 1998. Oligogalacturonides stimulate pericycle cell wall thickening and cell divisions leading to stoma formation in tobacco leaf explants. Planta 204:429-436.

Augur, C., N. Benhamou, A. Darvill, and P. Albersheim. 1993. Purification, characterization, and cell wall localization of an à-fucosidase that inactivates a xyloglucan oligosaccharin. The Plant Journal 3:415-426.

Auxtová, O., D. Lišková, D. Kákoniová, M. Kubačková, Š. Karácsonyi, and L. Bilisics. 1995. Effect of galactoglucomannan-derived oligosaccharides on elongation growth of pea and spruce stem segments stimulated by auxin. Planta 196:420-424.

Ayers, A.R., J. Ebel, B. Valent, and P. Albersheim. 1976a. Host-Pathogen Interactions .10. Fractionation and Biological-Activity of an Elicitor Isolated from Mycelial Walls of Phytophthora-Megasperma Var Sojae. Plant Physiology 57:760-765.

Ayers, A.R., J. Ebel, F. Finelli, N. Berger, and P. Albersheim. 1976b. Host-Pathogen Interactions .9. Quantitative Assays of Elicitor Activity and Characterization of Elicitor Present in Extracellular Medium of Cultures of Phytophthora-Mega-Sperma Var Sojae. Plant Physiology 57:751-759.

Aziz, A., B. Poinssot, X. Daire, M. Adrian, A. Bezier, B. Lambert, J.M. Joubert, and A. Pugin. 2003. Laminarin elicits defense responses in grapevine and induces protection against Botrytis cinerea and Plasmopara viticola. Molecular Plant-Microbe Interactions 16:1118-1128.

Barber, M.S., R.E. Bertram, and J.P. Ride. 1989. Chitin oligosaccharides elicit lignification in wounded wheat leaves. Physiological and Molecular Plant Pathology 34:3-12.

Barcelo, A., and V.G. Laura. 2009. Reactive Oxygen Species in Plant Cell Walls, p. 73-94, *In* Luis A. del Rio and A. Puppo, eds. Reactive Oxygen Species in Plant Signaling. Springer, Heidelberg.

Bashan, Y., Y. Okon, and Y. Henis. 1985. Detection of Cutinases and Pectic Enzymes during Infection of Tomato by Pseudomonas-Syringae Pv Tomato. Phytopathology 75:940-945.

Baureithel, K., and T. Boller. 1995. Characterization and solubilization of a specific binding site for chitin fragments in suspension-cultured tomato cells and microsomal membranes. Journal of Cellular Biochemistry Supplement 19B:151.

Baureithel, K., G. Felix, and T. Boller. 1994. Specific, High-Affinity Binding of Chitin Fragments to Tomato Cells and Membranes - Competitive-Inhibition of Binding by Derivatives of Chitooligosaccharides and a Nod Factor of Rhizobium. Journal of Biological Chemistry 269:17931-17938.

Baydoun, E.A.H., and S.C. Fry. 1985. The Immobility of Pectic Substances in Injured Tomato Leaves and Its Bearing on the Identity of the Wound Hormone. Planta 165:269-276.

Bellincampi, D., G. Salvi, G. De Lorenzo, F. Cervone, V. Marf..., S. Eberhard, A. Darvill, and P. Albersheim. 1993. Oligogalacturonides inhibit the formation of roots on tobacco explants. The Plant Journal 4:207-213.

Bellincampi, D., M. Cardarelli, D. Zaghi, G. Serino, G. Salvi, C. Gatz, F. Cervone, M.M. Altamura, P. Costantino, and G. De Lorenzo. 1996. Oligogalacturonides prevent rhizogenesis in rolB-transformed tobacco explants by inhibiting auxin-induced expression of the rolB gene. The Plant Cell 8:477-487.

Benhamou, N., H. Chamberland, and F.J. Pauze. 1990. Implication of Pectic Components in Cell-Surface Interactions between Tomato Root-Cells and Fusarium-Oxysporum F Sp Radicis-

Lycopersici - a Cytochemical Study by Means of a Lectin with Polygalacturonic Acid-Binding Specificity. Plant Physiology 92:995-1003.

Bergey, D.R., M. Orozco-Cardenas, D.S. de Moura, and C.A. Ryan. 1999. A wound- and systemin-inducible polygalacturonase in tomato leaves. Proceedings of the National Academy of Sciences of the United States of America 96:1756-1760.

Bhagwat, A.A., A. Mithofer, P.E. Pfeffer, C. Kraus, N. Spickers, A. Hotchkiss, J. Ebel, and D.L. Keister. 1999. Further studies of the role of cyclic beta-glucans in symbiosis. An ndvC mutant of Bradyrhizobium japonicum synthesizes cyclodecakis-(1 -> 3)-beta-glucosyl. Plant Physiology 119:1057-1064.

Bhandal, I.S., and J.D. Paxton. 1991. Phytoalexin Biosynthesis Induced by the Fungal Glucan Polytran L in Soybean, Pea, and Sweet-Pepper Tissues. Journal of Agricultural and Food Chemistry 39:2156-2157.

Bilisics, L., J. Vojtassak, P. Capek, K. Kollarova, and D. Liskova. 2004. Changes in glycosidase activities during galactoglucomannan oligosaccharide inhibition of auxin induced growth. Phytochemistry 65:1903-1909.

Bishop, P.D., D.J. Makus, G. Pearce, and C.A. Ryan. 1981. Proteinase Inhibitor-Inducing Factor Activity in Tomato Leaves Resides in Oligosaccharides Enzymically Released from Cell-Walls. Proceedings of the National Academy of Sciences of the United States of America-Biological Sciences 78:3536-3540.

Branca, C., G. De Lorenzo, and F. Cervone. 1988. Competitive inhibition of the auxin-induced elongation by à-D-oligogalacturonides in pea stem segments. Physiologia Plantarum 72:499-504.

Brecht, J.K., and D.J. Huber. 1988. Products released from enzymically active cell wall stimulate ethylene production and ripening in preclimacteric tomato (Lycopersicon esculentum Mill.) fruit. Plant Physiology 88:1037-1041.

Bruce, R.J., and C.A. West. 1989. Elicitation of Lignin Biosynthesis and Isoperoxidase Activity by Pectic Fragments in Suspension-Cultures of Castor Bean. Plant Physiology 91:889-897.

Brutus, A., F. Sicilia, A. Macone, F. Cervone, and G. De Lorenzo. 2010. A domain swap approach reveals a role of the plant wall-associated kinase 1 (WAK1) as a receptor of oligogalacturonides. Proceedings of the National Academy of Sciences of the United States of America 107:9452-9457.

Cabrera, J.C., A. Boland, J. Messiaen, P. Cambier, and P. Van Cutsem. 2008. Egg box conformation of oligogalacturonides: The time-dependent stabilization of the elicitor-active conformation increases its biological activity. Glycobiology 18:473-482.

Capodicasa, C., D. Vairo, O. Zabotina, L. McCartney, C. Caprari, B. Mattei, C. Manfredini, B. Aracri, J. Benen, J.P. Knox, G. De Lorenzo, and F. Cervone. 2004. Targeted modification of homogalacturonan by transgenic expression of a fungal polygalacturonase alters plant growth. Plant Physiology 135:1294-1304.

Casasoli, M., S. Spadoni, K.S. Lilley, F. Cervone, G. De Lorenzo, and B. Mattei. 2008. Identification by 2-D DIGE of apoplastic proteins regulated by oligogalacturonides in Arabidopsis thaliana. Proteomics 8:1042-54.

Cervone, F., G. Delorenzo, L. Degra, and G. Salvi. 1987. Elicitation of Necrosis in Vigna-Unguiculata Walp by Homogeneous Aspergillus-Niger Endo-Polygalacturonase and by Alpha-D-Galacturonate Oligomers. Plant Physiology 85:626-630.

Chandra, S., and P.S. Low. 1997. Measurement of Ca2+ fluxes during elicitation of the oxidative burst in aequorin-transformed tobacco cells. Journal of Biological Chemistry 272:28274-28280.

Chandra, S., P.F. Heinstein, and P.S. Low. 1996. Activation of phospholipase a by plant defense elicitors. Plant Physiology 110:979-986.

Cheong, J.J., and M.G. Hahn. 1991. A specific, high-affinity binding site for the hepta-beta-glucoside elicitor exists in soybean membranes. Plant Cell 3:137-47.

Cheong, J.J., R. Alba, F. Cote, J. Enkerli, and M.G. Hahn. 1993. Solubilization of functional plasma membrane-localized hepta-beta-glucoside elicitor-binding proteins from soybean. Plant Physiol 103:1173-82.

Cheong, J.J., W. Birberg, P. Fugedi, A. Pilotti, P.J. Garegg, N. Hong, T. Ogawa, and M.G. Hahn. 1991. Structure-activity relationships of oligo-beta-glucoside elicitors of phytoalexin accumulation in soybean. Plant Cell 3:127-36.

Cline, K., M. Wade, and P. Albersheim. 1978. Host-Pathogen Interactions .15. Fungal Glucans Which Elicit Phytoalexin Accumulation in Soybean Also Elicit Accumulation of Phytoalexins in Other Plants. Plant Physiology 62:918-921.

Cosio, E.G., T. Frey, and J. Ebel. 1990. Solubilization of Soybean Membrane-Binding Sites for Fungal Beta-Glucans That Elicit Phytoalexin Accumulation. Febs Letters 264:235-238.

Cosio, E.G., T. Frey, and J. Ebel. 1992. Identification of a High-Affinity Binding-Protein for a Hepta-Beta-Glucoside Phytoalexin Elicitor in Soybean. European Journal of Biochemistry 204:1115-1123.

Cosio, E.G., M. Feger, C.J. Miller, L. Antelo, and J. Ebel. 1996. High-affinity binding of fungal á-glucan elicitors to cell membranes of species of the plant family Fabaceae. Planta 200:92-99.

Cote, F., and M.G. Hahn. 1994. Oligosaccharins: structures and signal transduction. Plant Molecular Biology 26:1375-1411.

Davis, K.R., G.D. Lyon, A.G. Darvill, and P. Albersheim. 1984. Host-Pathogen Interactions .25. Endopolygalacturonic Acid Lyase from Erwinia-Carotovora Elicits Phytoalexin Accumulation by Releasing Plant-Cell Wall Fragments. Plant Physiology 74:52-60.

Daxberger, A., A. Nemak, A. Mithofer, J. Fliegmann, W. Ligterink, H. Hirt, and J. Ebel. 2007. Activation of members of a MAPK module in beta-glucan elicitor-mediated non-host resistance of soybean. Planta 225:1559-1571.

Day, R.B., M. Okada, Y. Ito, K. Tsukada, H. Zaghouani, N. Shibuya, and G. Stacey. 2001. Binding site for chitin oligosaccharides in the soybean plasma membrane. Plant Physiology 126:1162-1173.

De Lorenzo, G., and S. Ferrari. 2002. Polygalacturonase-inhibiting proteins in defense against phytopathogenic fungi. Current Opinion in Plant Biology 5:295-299.

De Lorenzo, G., F. Cervone, D. Bellincampi, C. Caprari, A.J. Clark, A. Desiderio, A. Devoto, R. Forrest, F. Leckie, L. Nuss, and et al. 1994. Polygalacturonase, PGIP and oligogalacturonides in cell-cell communication. Biochem Soc Trans 22:394-7.

De Lorenzo, G., F. Cervone, D. Bellincampi, C. Capodicasa, C. Caprari, L. Federici, S. Ferrari, P. Giuli, M. Laurenzi, B. Mattei, A. Raiola, G. Salvi, F. Sicilia, D. Vairo, and O. Zabotina. 2000. Structure-function studies on the leucine-rich repeat polygalacturonase-inhibiting protein (PGIP). Biology of Plant-Microbe Interactions, Vol 2:126-130

454.

de Vries, R.P., and J. Visser. 2001. Aspergillus enzymes involved in degradation of plant cell wall polysaccharides. Microbiology and Molecular Biology Reviews 65:497-+.

Decreux, A., and J. Messiaen. 2005. Wall-associated kinase WAK1 interacts with cell wall pectins in a calcium-induced conformation. Plant and Cell Physiology 46:268-278.

Doares, S.H., T. Syrovets, E.W. Weiler, and C.A. Ryan. 1995. Oligogalacturonides and chitosan activate plant defensive genes through the octadecanoid pathway. Proceedings of the National Academy of Sciences of the United States of America 92:4095-4098.

Dumville, J.C., and S.C. Fry. 2000. Uronic acid-containing oligosaccharins: Their biosynthesis, degradation and signalling roles in non-diseased plant tissues. Plant Physiology and Biochemistry 38:125-140.

Ebel, J. 1998. Oligoglucoside elicitor-mediated activation of plant defense. BioEssays 20:569-576.

Ebel, J., and E.G. Cosio. 1994. Elicitors of Plant Defense Responses. International Review of Cytology - a Survey of Cell Biology, Vol 148 148:1-36.

Ebel, J., A.A. Bhagwat, E.G. Cosio, M. Feger, U. Kissel, A. Mithofer, and T. Waldmuller. 1995. Elicitor Binding-Proteins and Signal-Transduction in the Activation of a Phytoalexin Defense Response. Canadian Journal of Botany-Revue Canadienne De Botanique 73:S506-S510.

Eberhard, S., N. Doubrava, V. Marfa, D. Mohnen, A. Southwick, A. Darvill, and P. Albersheim. 1989. Pectic Cell-Wall Fragments Regulate Tobacco Thin-Cell-Layer Explant Morphogenesis. Plant Cell 1:747-755.

Faugeron, C., S. Sakr, S. Lhernould, J.C. Michalski, S. Delrot, and H. Morvan. 1999. Long-distance transport and metabolism of unconjugated N-glycans in tomato plants. Journal of Experimental Botany 50:1669-1675.

Felix, G., M. Regenass, and T. Boller. 1993. Specific perception of subnanomolar concentrations of chitin fragments by tomato cells: Induction of extracellular alkalinization, changes in protein phosphorylation, and establishment of a refractory state. The Plant Journal 4:307-316.

Ferrari, S., R. Galletti, D. Vairo, F. Cervone, and G. De Lorenzo. 2006. Antisense expression of the Arabidopsis thaliana AtPGIP1 gene reduces polygalacturonase-inhibiting protein accumulation and enhances susceptibility to Botrytis cinerea. Molecular Plant-Microbe Interactions 19:931-936.

Ferrari, S., R. Galletti, D. Pontiggia, C. Manfredini, V. Lionetti, D. Bellincampi, F. Cervone, and G. De Lorenzo. 2008. Transgenic expression of a fungal endo-polygalacturonase increases plant resistance to pathogens and reduces auxin sensitivity. Plant Physiology 146:669-681.

Filippini, F., F. Loschiavo, M. Terzi, C. Branca, D. Bellincampi, G. Salvi, A. Desiderio, G. Delorenzo, and F. Cervone. 1992. Phytoalexin Elicitor-Active Alpha-1,4-D-Oligogalacturonides Reduce Auxin Perception by Plant-Cells and Tissues. Progress in Plant Growth Regulation 13:517-521

963.

Fliegmann, J., A. Mithofer, G. Wanner, and J. Ebel. 2004. An ancient enzyme domain hidden in the putative beta-glucan elicitor receptor of soybean may play an active part in the perception of pathogen-associated molecular patterns during broad host resistance. Journal of Biological Chemistry 279:1132-1140.

Fliegmann, J., E. Montel, A. Djulic, S. Cottaz, H. Driguez, and M.G. Ebel. 2005. Catalytic properties of the bifunctional soybean á-glucan-binding protein, a member of family 81 glycoside hydrolases. FEBS Letters 579:6647-6652.

Fry, S.C. 1989. The structure and functions of xyloglucan. Journal of Experimental Botany 40:1-11.

Fry, S.C. 1995. Polysaccharide-Modifying Enzymes in the Plant-Cell Wall. Annual Review of Plant Physiology and Plant Molecular Biology 46:497-520.

Fry, S.C., S. Aldington, P.R. Hetherington, and J. Aitken. 1993. Oligosaccharides as signals and substrates in the plant cell wall. Plant Physiol 103:1-5.

Fry, S.C., R.C. Smith, K.F. Renwick, D.J. Martin, S.K. Hodge, and K.J. Matthews. 1992. Xyloglucan Endotransglycosylase, a New Wall-Loosening Enzyme-Activity from Plants. Biochemical Journal 282:821-828.

Garcia-Romera, I., J.M. Garcia-Garrido, E. Martinez-Molina, and J.A. Ocampo. 1991. Production of Pectolytic Enzymes in Lettuce Root Colonized by Glomus-Mosseae. Soil Biology & Biochemistry 23:597-601.

Gilmour, S.J., and M.F. Thomashow. 1991. Cold acclimation and cold-regulated gene expression in ABA mutants of Arabidopsis thaliana. Plant Mol Biol 17:1233-40.

Gomez-Gomez, L., and T. Boller. 2000. FLS2: An LRR receptor-like kinase involved in the perception of the bacterial elicitor flagellin in Arabidopsis. Molecular Cell 5:1003-1011.

Gunia, W., W. Hinderer, U. Wittkampf, and W. Barz. 1991. Elicitor Induction of Cytochrome-P-450 Monooxygenases in Cell-Suspension Cultures of Chickpea (Cicer-Arietinum L) and Their Involvement in Pterocarpan Phytoalexin Biosynthesis. Zeitschrift Fur Naturforschung C-a Journal of Biosciences 46:58-66.

Gusta, L.V., R. Trischuk, and C.J. Weiser. 2005. Plant cold acclimation: The role of abscisic acid. Journal of Plant Growth Regulation 24:308-318.

Hadwiger, L.A., and J.M. Beckman. 1980. Chitosan as a Component of Pea-Fusarium-Solani Interactions. Plant Physiology 66:205-211.

Hadwiger, L.A., T. Ogawa, and H. Kuyama. 1994. Chitosan polymer sizes effective in inducing phytoalexin accumulation and fungal suppression are verified with synthesized oligomers. Molecular Plant-Microbe Interactions 7:531-533.

Hahn, M.G. 1996. Microbial elicitors and their receptors in plants. Annu Rev Phytopathol 34:387-412.

Hahn, M.G., and P. Albersheim. 1978. Host-Pathogen Interactions: XIV. Isolation and Partial Characterization of an Elicitor from Yeast Extract. Plant Physiol 62:107-111.

He, D.Y., Y. Yazaki, Y. Nishizawa, R. Takai, K. Yamada, K. Sakano, N. Shibuya, and E. Minami. 1998. Gene activation by cytoplasmic acidification in suspension-cultured rice cells in response to the potent elicitor, N - acetylchitoheptaose. Molecular Plant-Microbe Interactions 11:1167-1174.

Hrmova, M., V. Farkas, J. Lahnstein, and G.B. Fincher. 2007. A barley xyloglucan xyloglucosyl transferase covalently links xyloglucan, cellulosic substrates, and (1,3;1,4)-beta-D-glucans. Journal of Biological Chemistry 282:12951-12962.

Iglesias, N., J.A. Abelenda, M. Rodino, J. Sampedro, G. Revilla, and I. Zarra. 2006. Apoplastic glycosidases active against xyloglucan oligosaccharides of Arabidopsis thaliana. Plant and Cell Physiology 47:55-63.

Ito, Y., H. Kaku, and N. Shibuya. 1997. Identification of a high-affinity binding protein for N - acetylchitooligosaccharide elicitor in the plasma membrane of suspension-cultured rice cells by affinity labeling. The Plant Journal 12:347-356.

Jabs, T., M. Tschope, C. Colling, K. Hahlbrock, and D. Scheel. 1997. Elicitor-stimulated ion fluxes and O-2(-) from the oxidative burst are essential components in triggering defense gene activation and phytoalexin synthesis in parsley. Proceedings of the National Academy of Sciences of the United States of America 94:4800-4805.

Kaczynski, Z., S. Braun, B. Lindner, K. Niehaus, and O. Holst. 2007. Investigation of the chemical structure and biological activity of oligosaccharides isolated from rough-type Xanthomonas campestris pv. campestris B100 lipopolysaccharide. Journal of Endotoxin Research 13:101-108.

Kaku, H., Y. Nishizawa, N. Ishii-Minami, C. Akimoto-Tomiyama, N. Dohmae, K. Takio, E. Minami, and N. Shibuya. 2006. Plant cells recognize chitin fragments for defense signaling through a plasma membrane receptor. Proceedings of the National Academy of Sciences of the United States of America 103:11086-11091.

Kikuyama, M., K. Kuchitsu, and N. Shibuya. 1997. Membrane depolarization induced by N - acetylchitooligosaccharide elicitor in suspension-cultured rice cells. Plant and Cell Physiology 38:902-909.

Kohorn, B.D., S. Johansen, A. Shishido, T. Todorova, R. Martinez, E. Defeo, and P. Obregon. 2009. Pectin activation of MAP kinase and gene expression is WAK2 dependent. Plant Journal 60:974-982.

Kollarova, K., M. Henselova, and D. Liskova. 2005. Effect of auxins and plant oligosaccharides on root formation and elongation growth of mung bean hypocotyls. Plant Growth Regulation 46:1-9.

Kollarova, K., D. Liskova, and P. Capek. 2006. Further biological characteristics of galactoglucomannan oligosaccharides. Biologia Plantarum 50:232-238.

Kollarova, K., D. Richterova, I. Slovakova, M. Henselova, P. Capek, and D. Liskova. 2009. Impact of galactoglucomannan oligosaccharides on elongation growth in intact mung bean plants. Plant Science 177:324-330.

Kuchitsu, K., Y. Yazaki, K. Sakano, and N. Shibuya. 1997. Transient cytoplasmic pH change and ion fluxes through the plasma membrane in suspension-cultured rice cells triggered by N - acetylchitooligosaccharide elicitor. Plant and Cell Physiology 38:1012-1018.

Legendre, L., P.F. Heinstein, and P.S. Low. 1992. Evidence for participation of GTP-binding proteins in elicitation of the rapid oxidative burst in cultured soybean cells. Journal of Biological Chemistry 267:20140-20147.

Legendre, L., Y.G. Yueh, R. Crain, N. Haddock, P.F. Heinstein, and P.S. Low. 1993. Phospholipase-C Activation during Elicitation of the Oxidative Burst in Cultured Plant-Cells. Journal of Biological Chemistry 268:24559-24563.

Leon, J., E. Rojo, and J.J. Sanchez-Serrano. 2001. Wound signalling in plants. Journal of Experimental Botany 52:1-9.

Leon, J., E. Rojo, E. Titarenko, and J.J. Sanchez-Serrano. 1998. Jasmonic acid-dependent and - independent wound signal transduction pathways are differentially regulated by Ca2+/calmodulin in Arabidopsis thaliana. Molecular and General Genetics 258:412-419.

Lesney, M.S. 1989a. Growth-Responses and Lignin Production in Cell-Suspensions of Pinus-Elliottii Elicited by Chitin, Chitosan or Mycelium of Cronartium-Quercum F-Sp Fusiforme. Plant Cell Tissue and Organ Culture 19:23-31.

Lesney, M.S. 1989b. Growth responses and lignin production in cell suspensions of Pinus elliottii `elicited' by chitin, chitosan or mycelium of Cronartium quercum f.sp. fusiforme Plant Cell, Tissue and Organ Culture 19:23-31.

Li, L., C.Y. Li, G.I. Lee, and G.A. Howe. 2002. Distinct roles for jasmonate synthesis and action in the systemic wound response of tomato. Proceedings of the National Academy of Sciences of the United States of America 99:6416-6421.

Lionetti, V., A. Raiola, L. Camardella, A. Giovane, N. Obel, M. Pauly, F. Favaron, F. Cervone, and D. Bellincampi. 2007. Overexpression of pectin methylesterase inhibitors in Arabidopsis restricts fungal infection by Botrytis cinerea. Plant Physiol 143:1871-80.

Lišková, D., O. Auxtová, D. Kákoniová, M. Kubačková, Š. Karácsonyi, and L. Bilisics. 1995. Biological activity of galactoglucomannan-derived oligosaccharides. Planta 196:425-429.

Lozovaya, V., O. Zabotina, N. Rumyantseva, R. Malihov, and M. Zihareva. 1993. Stimulation of root development on buckwheat thin cell-layer explants by pectic fragments from pea stem cell walls. Plant Cell Reports 12:530-533.

Macdougall, A.J., N.M. Rigby, P.W. Needs, and R.R. Selvendran. 1992. Movement and Metabolism of Oligogalacturonide Elicitors in Tomato Shoots. Planta 188:566-574.

Maris, A., D. Suslov, S.C. Fry, J.P. Verbelen, and K. Vissenberg. 2009. Enzymic characterization of two recombinant xyloglucan endotransglucosylase/hydrolase (XTH) proteins of Arabidopsis and their effect on root growth and cell wall extension. Journal of Experimental Botany 60:3959-3972.

Mathieu, Y., D. Lapous, S. Thomine, C. LauriŠre, and J. Guern. 1996. Cytoplasmic acidification as an early phosphorylation-dependent response of tobacco cells to elicitors. Planta 199:416-424.

Mathieu, Y., A. Kurkdijan, H. Xia, J. Guern, A. Koller, M. Spiro, M. O'Neill, P. Albersheim, and A. Darvill. 1991. Membrane responses induced by oligogalacturonides in suspension-cultured tobacco cells. The Plant Journal 1:333-343.

Mattei, B., A. Raiola, C. Caprari, L. Federici, D. Bellincampi, G. De Lorenzo, F. Cervone, A. Giovane, and L. Camardella. 2002. Studies on plant inhibitors of pectin modifying enzymes: Polygalacturonase-inhibiting protein (PGIP) and pectin methylesterase inhibitor (PMEI). Carbohydrate Bioengineering: Interdisciplinary Approaches:160-167

195.

Mauro, M.L., G. De Lorenzo, P. Costantino, and D. Bellincampi. 2002. Oligogalacturonides inhibit the induction of late but not of early auxin-responsive genes in tobacco. Planta 215:494-501.

Mcdougall, G.J., and S.C. Fry. 1989. Structure-Activity-Relationships for Xyloglucan Oligosaccharides with Antiauxin Activity. Plant Physiology 89:883-887.

McDougall, G.J., and S.C. Fry. 1990. Xyloglucan Oligosaccharides Promote Growth and Activate Cellulase: Evidence for a Role of Cellulase in Cell Expansion. Plant Physiol. 93:1042-1048.

Mcdougall, G.J., and S.C. Fry. 1991. Xyloglucan Nonasaccharide, a Naturally-Occurring Oligosaccharin, Arises Invivo by Polysaccharide Breakdown. Journal of Plant Physiology 137:332-336.

Mcgurl, B., and C.A. Ryan. 1992. The Organization of the Prosystemin Gene. Plant Molecular Biology 20:405-409.

Mcgurl, B., G. Pearce, M. Orozcocardenas, and C.A. Ryan. 1992. Structure, Expression, and Antisense Inhibition of the Systemin Precursor Gene. Science 255:1570-1573.

Melotto, E., L.C. Greve, and J.M. Labavitch. 1994. Cell wall metabolism in ripening fruit. VII. Biologically-active pectin oligomers in ripening tomato (Lycopersicon esculentum Mill.) fruits. Plant Physiology 106:575-581.

Miller, L., and Macmilla.Jd. 1971. Purification and Pattern of Action of Pectinesterase from Fusarium-Oxysporum F Sp Vasinfectum. Biochemistry 10:570-&.

Mithofer, A., F. Lottspeich, and J. Ebel. 1996a. One-step purification of the beta-glucan elicitor-binding protein from soybean (Glycine max L) roots and characterization of an anti-peptide antiserum. Febs Letters 381:203-207.

Mithofer, A., A.A. Bhagwat, M. Feger, and J. Ebel. 1996b. Suppression of fungal beta-glucan-induced plant defence in soybean (Glycine max L) by cyclic 1,3-1,6-beta-glucans from the symbiont Bradyrhizobium japonicum. Planta 199:270-275.

Mithofer, A., A.A. Bhagwat, D.L. Keister, and J. Ebel. 2001. Bradyrhizobium japonicum mutants defective in cyclic beta-glucan synthesis show enhanced sensitivity to plant defense responses. Zeitschrift Fur Naturforschung C-a Journal of Biosciences 56:581-584.

Miya, A., P. Albert, T. Shinya, Y. Desaki, K. Ichimura, K. Shirasu, Y. Narusaka, N. Kawakami, H. Kaku, and N. Shibuya. 2007. CERK1, a LysM receptor kinase, is essential for chitin elicitor signaling in Arabidopsis. Proceedings of the National Academy of Sciences of the United States of America 104:19613-19618.

Mohnen, D., S. Eberhard, V. Marfa, N. Doubrava, P. Toubart, D.J. Gollin, T.A. Gruber, W. Nuri, P. Albersheim, and A. Darvill. 1990. The Control of Root, Vegetative Shoot and Flower Morphogenesis in Tobacco Thin Cell-Layer Explants (Tcls). Development 108:191-&.

Natsume, M., Y. Kamo, M. Hirayama, and T. Adachi. 1994. Isolation and characterization of alginate-derived oligosaccharides with root growth-promoting activities. Carbohydrate Research 258:187-197.

Nurnberger, T., and D. Scheel. 2001. Signal transmission in the plant immune response. Trends in Plant Science 6:372-379.

Okada, M., M. Matsumura, Y. Ito, and N. Shibuya. 2002. High-affinity binding proteins for N-acetylchitooligosaccharide elicitor in the plasma membranes from wheat, barley and carrot cells: Conserved presence and correlation with the responsiveness to the elicitor. Plant and Cell Physiology 43:505-512.

Okinaka, Y., K. Mimori, K. Takeo, S. Kitamura, Y. Takeuchi, N. Yamaoka, and M. Yoshikawa. 1995. A Structural Model for the Mechanisms of Elicitor Release from Fungal Cell-Walls by Plant Beta-1,3-Endoglucanase. Plant Physiology 109:839-845.

Pagel, W., and R. Heitefuss. 1990. Enzyme-Activities in Soft Rot Pathogenesis of Potato-Tubers - Effects of Calcium, Ph, and Degree of Pectin Esterification on the Activities of Polygalacturonase and Pectate Lyase. Physiological and Molecular Plant Pathology 37:9-25.

Pavlova, Z.N., O.A. Ash, V.A. Vnuchkova, A.V. Babakov, V.I. Torgov, O.A. Nechaev, A.I. Usov, and V.N. Shibaev. 1992. Biological activity of a synthetic pentasaccharide fragment of xyloglucan. Plant Science 85:131-134.

Popper, Z.A., and S.C. Fry. 2008. Xyloglucan-pectin linkages are formed intra-protoplasmically, contribute to wall-assembly, and remain stable in the cell wall. Planta 227:781-794.

Powell, A.L.T., J. van Kan, A. ten Have, J. Visser, L.C. Greve, A.B. Bennett, and J.M. Labavitch. 2000. Transgenic Expression of Pear PGIP in Tomato Limits Fungal Colonization. Molecular Plant-Microbe Interactions 13:942-950.

Priem, B., and K.C. Gross. 1992. Mannosyl- and xylosyl-containing glycans promote tomato (Lycopersicon esculentum Mill.) fruit ripening. Plant Physiology 98:399-401.

Priem, B., H. Morvan, and K.C. Gross. 1994. Unconjugated N -glycans as a new class of plant oligosaccharins. Biochemical Society Transactions 22:398-402.

Roberts, A.W., S.G. Donovan, and C.H. Haigler. 1997. A secreted factor induces cell expansion and formation of metaxylem-like tracheary elements in xylogenic suspension cultures of Zinnia. Plant Physiology 115:683-692.

Rojo, E., J. Leon, and J.J. Sanchez-Serrano. 1999. Cross-talk between wound signalling pathways determines local versus systemic gene expression in Arabidopsis thaliana. Plant Journal 20:135-142.

Romeis, T., P. Piedras, S.Q. Zhang, D.F. Klessig, H. Hirt, and J.D.G. Jones. 1999. Rapid Avr9- and Cf-9-dependent activation of MAP kinases in tobacco cell cultures and leaves: Convergence of resistance gene, elicitor, wound, and salicylate responses. Plant Cell 11:273-287.

Roos, W., S. Evers, M. Hieke, M. Tschope, and B. Schumann. 1998. Shifts of intracellular pH distribution as a part of the signal mechanism leading to the elicitation of benzophenanthridine alkaloids - Phytoalexin biosynthesis in cultured cells of Eschscholtzia californica. Plant Physiology 118:349-364.

Ryan, C.A. 1988. Oligosaccharides as recognition signals for the expression of defensive genes in plants. Biochemistry 27:8879-8883.

Ryan, C.A. 1994. Oligosaccharide Signals - from Plant Defense to Parasite Offense. Proceedings of the National Academy of Sciences of the United States of America 91:1-2.

Ryan, C.A., and E.E. Farmer. 1991. Oligosaccharide signals in plants: A current assessment. Annual Review of Plant Physiology and Plant Molecular Biology 42:651-674.

Ryan, C.A., P.D. Bishop, J.S. Graham, R.M. Broadway, and S.S. Duffey. 1986. Plant and Fungal Cell-Wall Fragments Activate Expression of Proteinase-Inhibitor Genes for Plant Defense. Journal of Chemical Ecology 12:1025-1036.

Sampedro, J., C. Sieiro, G. Revilla, T. Gonzalez-Villa, and I. Zarra. 2001. Cloning and expression pattern of a gene encoding an alpha-xylosidase active against xyloglucan oligosaccharides from Arabidopsis. Plant Physiology 126:910-920.

Schroder, R., and B. Knoop. 1995. An Oligosaccharide Growth-Factor in Plant Suspension-Cultures - a Proposed Structure. Journal of Plant Physiology 146:139-147.

Sharp, J.K., M. McNeil, and P. Albersheim. 1984. The primary structures of one elicitor-active and seven elicitor-inactive hexa(á-D-glucopyranosyl)-D-glucitols isolated from the mycelial walls of Phytophthora megasperma f. sp. glycinea Journal of Biological Chemistry 259:11321-11336.

Shibuya, N., and E. Minami. 2001. Oligosaccharide signaling for defense responses in plants. Physiological and Molecular Plant Pathology.

Shibuya, N., H. Kaku, K. Kuchitsu, and M.J. Maliarik. 1993. Identification of a novel high-affinity binding site for N - acetylchitooligosaccharide elicitor in the membrane fraction from suspension-cultured rice cells. FEBS Letters 329:75-78.

Shibuya, N., N. Ebisu, Y. Kamada, H. Kaku, J. Cohn, and Y. Ito. 1996. Localization and binding characteristics of a high-affinity binding site for N -acetylchitooligosaccharide elicitor in the plasma membrane from suspension-cultured rice cells suggest a role as a receptor for the elicitor signal at the cell surface. Plant and Cell Physiology 37:894-898.

Shinya, T., R. Menard, I. Kozone, H. Matsuoka, N. Shibuya, S. Kauffmann, K. Matsuoka, and M. Saito. 2006. Novel beta-1,3-, 1,6-oligoglucan elicitor from Alternaria alternata 102 for defense responses in tobacco. Febs Journal 273:2421-2431.

Shinya, T., I. Galis, T. Narisawa, M. Sasaki, H. Fukuda, H. Matsuoka, M. Saito, and K. Matsuoka. 2007. Comprehensive analysis of glucan elicitor-regulated gene expression in tobacco BY-2 cells reveals a novel MYB transcription factor involved in the regulation of phenylpropanoid metabolism. Plant and Cell Physiology 48:1404-1413.

Silipo, A., G. Erbs, T. Shinya, J.M. Dow, M. Parrilli, R. Lanzetta, N. Shibuya, M.-A. Newman, and A. Molinaro. 2010. Glycoconjugates as elicitors or suppressors of plant innate immunity. Glycobiology 20:406-419.

Spiro, M.D., B.L. Ridley, S. Eberhard, K.A. Kates, Y. Mathieu, M.A. O'Neill, D. Mohnen, J. Guern, A. Darvill, and P. Albersheim. 1998. Biological activity of reducing-end-derivatized oligogalacturonides in tobacco tissue cultures. Plant Physiology 116:1289-1298.

Stennis, M.J., S. Chandra, C.A. Ryan, and P.S. Low. 1998. Systemin potentiates the oxidative burst in cultured tomato cells. Plant Physiology 117:1031-1036.

Tani, M., H. Fukui, M. Shimomura, and M. Tabata. 1992. Structure of endogenous oligogalacturonides inducing shikonin biosynthesis in Lithospermum cell cultures. Phytochemistry 31:2719-2723.

Taylor, A.T.S., J. Kim, and P.S. Low. 2001. Involvement of mitogen-activated protein kinase activation in the signal-transduction pathways of the soya bean oxidative burst. Biochemical Journal 355:795-803.

Templeton, M.D., and C.J. Lamb. 1988. Elicitors and Defense Gene Activation. Plant Cell and Environment 11:395-401.

Thain, J.F., I.R. Gubb, and D.C. Wildon. 1995. Depolarization of tomato leaf cells by oligogalacturonide elicitors. Plant, Cell and Environment 18:211-214.

Titarenko, E., E. Rojo, J. Leon, and J.J. SanchezSerrano. 1997. Jasmonic acid-dependent and -independent signaling pathways control wound-induced gene activation in Arabidopsis thaliana. Plant Physiology 115:817-826.

Toubart, P., A. Desiderio, G. Salvi, F. Cervone, L. Daroda, G. Delorenzo, C. Bergmann, A.G. Darvill, and P. Albersheim. 1992. Cloning and Characterization of the Gene Encoding the Endopolygalacturonase-Inhibiting Protein (Pgip) of Phaseolus-Vulgaris L. Plant Journal 2:367-373.

Tran Thanh Van, K., P. Toubart, A. Cousson, A.G. Darvill, D.J. Gollin, P. Chelf, and P. Albersheim. 1985. Manipulation of the morphogenetic pathways of tobacco explants by oligosaccharins. Nature 314:615-617.

Verica, J.A., L. Chae, H. Tong, P. Ingmire, and Z.-H. He. 2003. Tissue-Specific and Developmentally Regulated Expression of a Cluster of Tandemly Arrayed Cell Wall-Associated Kinase-Like Kinase Genes in Arabidopsis. Plant Physiol. 133:1732-1746.

Verma, D.P.S., G.A. Maclachlan, H. Byrne, and D. Ewings. 1975. Regulation and Invitro Translation of Messenger-Ribonucleic-Acid for Cellulase from Auxin-Treated Pea Epicotyls. Journal of Biological Chemistry 250:1019-1026.

Wagner, T.A., and B.D. Kohorn. 2001. Wall-associated kinases are expressed throughout plant development and are required for cell expansion. Plant Cell 13:303-318.

Wan, J.R., X.C. Zhang, D. Neece, K.M. Ramonell, S. Clough, S.Y. Kim, M.G. Stacey, and G. Stacey. 2008. A LysM receptor-like kinase plays a critical role in chitin signaling and fungal resistance in Arabidopsis. Plant Cell 20:471-481.

Warneck, H., and H.U. Seitz. 1993. Inhibition of Gibberellic Acid-Induced Elongation-Growth of Pea Epicotyls by Xyloglucan Oligosaccharides. Journal of Experimental Botany 44:1105-1109.

Warneck, H.M., D.C. Fulton, H.U. Seitz, and S.C. Fry. 1998. Transport, degradation and cell wall-integration of XXFGol, a growth-regulating nonasaccharide of xyloglucan, in pea stems. Planta 204:78-85.

Yamada, A., N. Shibuya, O. Kodama, and T. Akatsuka. 1993. Induction of phytoalexin formation in suspension-cultured rice cells by N -acetylchitooligosaccharides. Bioscience, Biotechnology and Biochemistry 57:405-409.

Yamaguchi, T., A. Yamada, N. Hong, T. Ogawa, T. Ishii, and N. Shibuya. 2000. Differences in the recognition of glucan elicitor signals between rice and soybean: Characterization of á-glucan fragments from rice blast disease fungus, Pyricularia oryzae , which elicit phytoalexin biosynthesis in suspension-cultured rice cells. The Plant Cell.

York, W.S., A.G. Darvill, and P. Albersheim. 1984. Inhibition of 2,4-Dichlorophenoxyacetic Acid-Stimulated Elongation of Pea Stem Segments by a Xyloglucan Oligosaccharide. Plant Physiol 75:295-297.

Yoshikawa, M., M. Matama, and H. Masago. 1981. Release of a Soluble Phytoalexin Elicitor from Mycelial Walls of Phytophthora-Megasperma Var Sojae by Soybean Tissues. Plant Physiology 67:1032-1035.

Zabotin, A., T.S. Barisheva, I.A. Larskaya, T. Toroshina, O.I. Trofimova, M. Hahn, and O.A. Zabotina. 2004. Oligosaccharins-a new systemic factor in the acquisition of freeze tolerance in winter plants. Book of Abstracts of X Cell Wall Meeting., August 29-September 3, Sorrento, Italy.

Zabotin, A.I., T.S. Barisheva, O.I. Trofimova, T.E. Toroschina, I.A. Larskaya, and O.A. Zabotina. 2009. Oligosaccharin and ABA synergistically affect the acquisition of freezing tolerance in winter wheat. Plant Physiology and Biochemistry 47:854-858.

Zabotin, A.I., T.S. Barisheva, O.A. Zabotina, I.A. Larskaya, V.V. Lozovaya, G. Beldman, and A.G.J. Voragen. 1998. Alterations in cell walls of winter wheat roots during low temperature acclimation. Journal of Plant Physiology 152:473-479.

Zabotin, A.I., T.S. Barisheva, I.A. Larskaya, T.E. Toroshina, O.V. Trofimova, M.G. Hahn, and O.A. Zabotina. 2005. Oligosaccharin - a new systemic factor in the acquisition of freeze tolerance in winter plants. Plant Biosystems 139:36-41.

Zabotina, O.A., O.P. Gurjanov, N.N. Ibragimova, D.A. Ayupova, and V.V. Lozovaya. 1998a. Rhizogenesis in buckwheat thin-cell-layer explants: effect of plant oligosaccharides. Plant Science 135:195-201.

Zabotina, O.A., N.N. Ibragimova, A.I. Zabotin, O.I. Trofimova, and A.P. Sitnikov. 2002. Biologically active oligosaccharides from pectins of Pisum sativum L. seedlings affecting root generation. Biochemistry-Moscow 67:227-232.

Zabotina, O.A., O.P. Gurjanov, D.A. Ayupova, G. Beldman, A.G.J. Voragen, and V.V. Lozovaya. 1996a. Biologically-active soluble oligosaccharides from pea stem tissues. Plant Cell Reports 15:954-957.

Zabotina, O.A., O.P. Gurjanov, D.A. Ayupova, V.V. Lozovaya, G. Beldman, and A.G.J. Voragen. 1996b. Soluble bioactive oligosaccharides from pea stems. Plant Cell Rep. 15:954-957.

Zabotina, O.A., D.A. Ayupova, I.A. Larskaya, O.G. Nikolaeva, G.A. Petrovicheva, and A.I. Zabotin. 1998b. Physiologically active oligosaccharides accumulating in the roots of winter wheat during adaptation to low temperature. Russian Journal of Plant Physiology 45:221-226.

Zhou, L.A., C.Z. Yang, J.Q. Li, S.L. Wang, and J.Y. Wu. 2003. Heptasaccharide and octasaccharide isolated from Paris polyphylla var. yunnanensis and their plant growth-regulatory activity. Plant Science 165:571-575.

Zimmermann, S., T. Nurnberger, J.M. Frachisse, W. Wirtz, J. Guern, R. Hedrich, and D. Scheel. 1997. Receptor-mediated activation of a plant Ca2+-permeable ion channel involved in pathogen defense. Proceedings of the National Academy of Sciences of the United States of America 94:2751-2755.

In: Oligosaccharides: Sources, Properties and Applications ISBN 978-1-61122-193-0
Editor: Nicole S. Gordon

Chapter 9

THE MANNAN OLIGOSACCHARIDES IN AQUACULTURE

Huynh Minh Sang*[*] *and Ravi Fotedar
Aquatic Science Research Unit, Curtin University of Technology, Brodie Hall Building, 1 Turner Avenue, Technology Park, Bentley, Western Australia 6102

ABSTRACT

Increased concern over effects of antibiotics usage in aquaculture, on environment and human health, has prompted the search for alternative products. Recently, immunostimulants such as probiotics and prebiotics have shown promising results as preventive and environmentally friendly alternatives to the antibiotics. Among common immunostimulants used, mannan oligosaccharides (MOS) have received heightened attention in aquaculture. Since the first use of MOS in aquaculture, there has been increased number of studies demonstrating their ability to increase the survival, growth performance and control of the potential pathogens in fish and crustacean including, marron (*Cherax tenuimanus*), yabbies (*Cherax destructor*) and tropical rock lobsters (*Panulirus ornatus*). The chapter reviews the role of MOS on the culture of crayfish and fishes while detailing the effects of MOS on the growth performance, physiology and immune response of these aquatic animals. Suggestions for further research on the application of MOS in crayfish aquaculture are also included in the Chapter.

1. INTRODUCTION

Improvement of animal health in commercial production practices is one of the major factors determining the success of aquaculture industry [1]. For the past several decades, antibiotics have been used in animal feeds worldwide at sub-therapeutic concentrations to obtain positive effects on weight gain, feed utilization and survival [2]. However, there has been increased concern of antibiotic usage on the development of antibiotic resistant bacteria

[*] Email: hmsang2000@yahoo.com

in both animals and humans, and associated risks to human health. Restriction or ban of antibiotic usage as feed additives in fish and crustacean diets in many countries, due to the development of antibiotic resistance has prompted the search for alternative products [3].

Prebiotics as immunostimulants have been demonstrated as promising preventive and environmentally friendly alternative to antibiotics in aquaculture of fishes and crustacean [4]. Prebiotics defined as "selectively fermented ingredients that allow specific changes, both in the composition and/or activity in the gastrointestinal microbiota that confers benefits upon host well-being and health" [5], effect the host indirectly. By selectively providing nutrition for one or a limited number of microorganisms, they cause a selective modification of the host's intestinal microflora [6]. Prebiotics can be applied in wide range of organisms in animal husbandry, poultry and aquaculture industry. Among the most common immunostimulant, mannan oligosaccharide (MOS) has been recently receiving heightened application in aquaculture. Since the first use of MOS in aquaculture, there has been increase in the number of studies demonstrating their ability to increase the survival, growth performance and control of the potential pathogens of fish and crustacean. The chapter reviews the role of MOS on the culture of crayfish and fishes while detailing the effects of MOS on the growth performance, physiology and immune response of these aquatic animals.

2. Chemical Structure and Sources of Mannan Oligosaccharide

An oligosaccharide is a saccharide polymer containing a small number (typically three to ten) of component sugars, also known as simple sugars (monosaccharide).

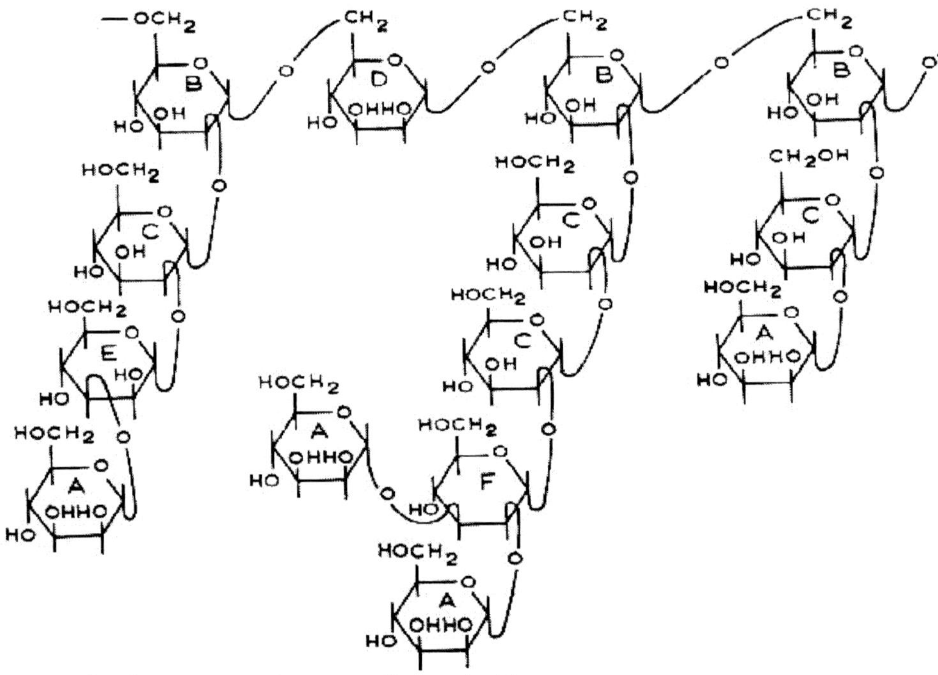

Figure 1. Molecular structure of mannan oligosaccharide.

Examples of oligosaccharides are Fructo-oligosaccharides (FOS) which, are found in many vegetables, consist of short chains of fructose molecules; Galactooligosaccharides (GOS) which, also occur naturally, consist of short chains of galactose molecules. These compounds can be only partially digested by humans. Mannan oligosaccharides (MOS), which are normally obtained from the yeast cell walls of *Saccharomyces cerevisiae,* are glucomannoprotein complex consisting of α -1- 3 MOS and α -1-6 MOS and have a molecular structure as follows:

A commonly used commercial product, which contains about 28% of MOS, is Bio-Mos® (Alltech, USA).

3. Application Methods and Dosage

The methods of MOS application depend on various factors such as culture species, life stage of species, and culture environment. In aquaculture industry, MOS is included in the diets as a non-digestive ingredient [3, 7-10] and/or through live feed enrichment [11, 12]. MOS is supplemented in the diets of marron (*Cherax tennuimanus*), yabbies (*Cherax destructor*) and tropical rock lobster (*Panulirus ornatus*) [13-16]. To date, literature on the effectiveness of different administration methods of MOS in culture of crayfish is not available. During the early larvae development stage of European lobster (*Homarus gammarus)*, MOS was applied to the animal through enriching *Artemia* [11]. Sang and Fotedar [15] reported that inclusion of MOS in the diets of marron did not affect nutrient value of the diet as well as the water quality of culture media.

Dosage of MOS in aquaculture diets is another important criterion to get optimum economical returns. A dosage below minimum threshold levels may result in null effect on the animal while dosage above the maximum requirement can result in the unnecessary high feeding costs and even negative effect on the performance of the animal. In marron, the appropriate dietary inclusion level is 0.2 to 0.4% of MOS [15]. Dosages less than 0.2% of MOS and higher than 0.4% of MOS did not improve the performance of the species [15]. There are no reports on the adverse side effects of higher dietary inclusion of MOS, however the cost of MOS seems to be the driver for the higher inclusion levels in crayfish culture. In tropical rock lobster and yabbies, MOS was applied at 0.4% in the diets resulting in positive performance [13, 16]. Although the optimum dosage of MOS for these species is not yet known, 0.4% dietary inclusion levels is the most common practice in most of the crustacean [3] and fish species [7, 8, 17]. Therefore, further research is required to determine the optimum inclusion levels of MOS in freshwater crayfish and lobsters. In the fish culture, MOS enriched *Artemia* were fed to cobia larvae at 0.2% of the dry basis [12]. MOS at 0.2% to 0.6% were included in the diets of rainbow trout (*Oncorhynchus mykiss*) [1, 18], 0.2 to 0.4% for sea bass [19], 0.2% for channel catfish (*Ictalurus punctatus*) [20] and 0.2 to 0.4% for sea bream (*Sparus aurata*) [21].

4. Effects of MOS on Growth, Survival and Physiological Condition of Aquaculture Species

4.1. Decapods

Dietary inclusion of MOS improved survival of marron, tropical rock lobster, yabbies and European lobster under the optimum culture conditions and even after infection with the bacteria. At the optimum inclusion level, survival of marron increased by 39% after 112 days of rearing (Figure 2) [15]; 22% in tropical rock lobsters [13] and 25% in yabbies [16] (Table 1). The higher survival in turn, increased the total biomass production of the crayfish. When dietary MOS was used, the improved growth was another desired outcome reported for most of the investigated crayfish species, except marron [15]. Growth of yabbies and tropical rock lobsters were improved by 9% and by 22% respectively after 56 days of rearing which MOS supplemented diets were used [13, 16]. On the contrary, MOS's inability to improve the growth of marron could be due to the fact that marron was exposed to dietary MOS for a relatively shorter duration compared to its long growing period.

Some physiological indicators reflecting health condition also improved when MOS was included in the diets of aquatic species. Dietary MOS increased tail muscle index (wet and dry) of marron, lobsters and yabbies (Table 2) [13, 15, 16]. It is quite feasible that MOS can stimulate the storage of energy in the tail muscles and thus reduce the impact of stress [22].

Table 1. Survival and growth of yabbies and tropical rock lobster (mean ± S.E.)

Species	Parameters	Control	0.4% MOS	Reference
Yabbies	Survival (%)	66.67 ± 4.17^a	83.33 ± 4.17^b	[16]
	Weight (g)	42.10 ± 0.74^a	45.71 ± 0.73^b	
	SGR	0.32 ± 0.02^a	0.47 ± 0.07^b	
	AWG (g/week)	0.87 ± 0.07^a	1.32 ± 0.06^b	
Tropical rock lobster	Survival (%)	54.76 ± 2.38^a	66.67 ± 4.76^b	[13]
	Weight (g)	2.35 ± 0.14^a	2.86 ± 0.07^b	
	SGR	1.07 ± 0.11^a	1.44 ± 0.04^b	
	AWG (g/week)*	0.13 ± 0.02^a	0.20 ± 0.01^b	

The mechanisms by which MOS directly stimulate the physiological condition of aquatic species have not yet been investigated though reports have shown that MOS increases digestive enzyme activities in yabbies. For example, protease activity was higher in hepatopancreas; amylase activity was higher in the gut of yabbies, fed MOS [16]. Higher activity of digestive enzyme increases capacity of animals to obtain nutrients from a particular food source [23], break it down and to assimilate specific nutrients [24] thus, improves physiological condition and finally growth [24]. The evidence of MOS influencing the enzyme activity comes from the reports showing that MOS have stimulated the beneficial bacterial community in the intestine which directly stimulated the activity of digestive enzyme [25].

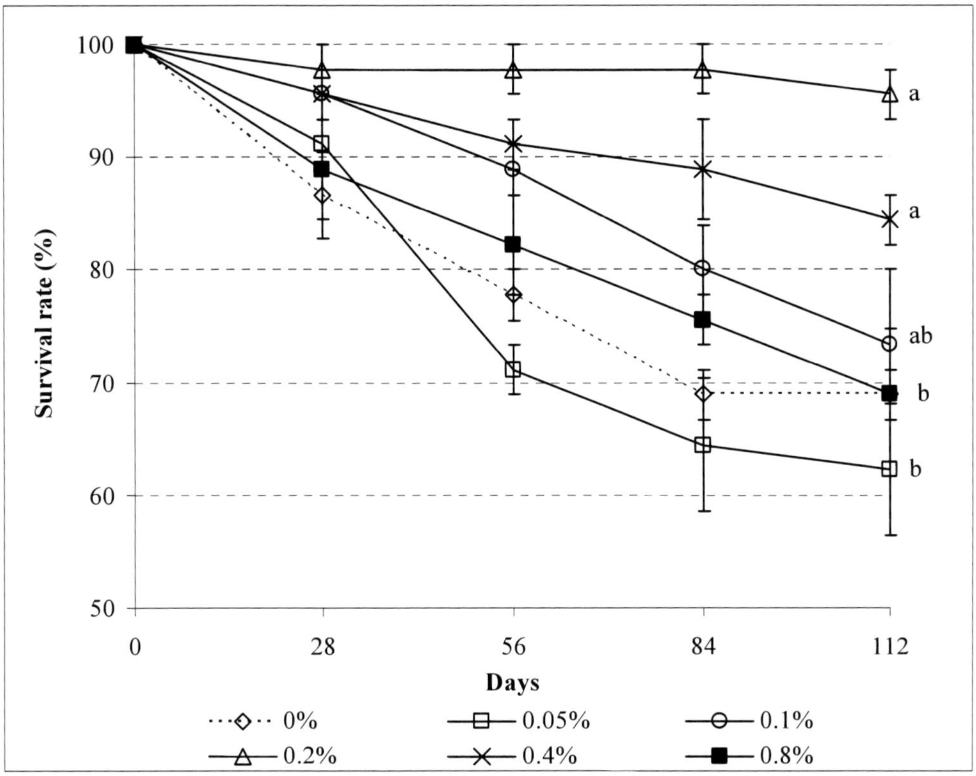

Figure 2. Survival of marron fed different MOS supplemented diets [15].

Table 2. Effect of MOS on physiology of marron and tropical rock lobster

Physiological parameters	Marron [15]	Tropical rock lobster [13]
HSIwet	-	+
HSIdry	-	-
TMIwet	+	+
TMIdry	+	+
TM%	-	-
HM%	-	-
OC	+	N/A

Note: +: Positive effect; -: Null effect.

4.2. Teleosts

MOS has shown positive effects on survival and growth of various fish species. An experiment under controlled conditions was conducted on common carp (*Cyprinus carpio* L)at the University of Trakia - Bulgaria in which 0.2% MOS was incorporated into a standard commercial extruded diet (23.5% protein and 5.4% lipid). Results showed an average weight of 480 g for MOS fed fish and 430 g for the control fish giving an 11.6% higher weight gain in MOS fed fish. Feed conversion ratio (FCR) was improved by 17.6% (1.69 for MOS fed fish and. 2.05 for the control fish). Lower mortalities were also observed in the MOS fed fish

(1.9% vs. 3.6% for the control) [26]. Culjak et al. [27] reported a similar enhancement in weight gain of the carp responding to dietary MOS. MOS included at 0.6% resulted in 24% higher weight gain, FCR was improved from 2.06 to 1.60; and mortality was reduced from 50.0% to 16.7% [27]. Similar trials were conducted on rainbow trout with a 0.2% MOS inclusion rate in the commercial feed, resulting in increased average weights of 13.7% in fish grown from 30 g to just under 100 g. Mortalities and FCR were significantly altered in response to MOS feeding. FCR decreased from 0.91 in controls to 0.83 in trout given MOS. Mortalities decreased from 1.7% in controls to 0.6% in fish fed MOS. For the fish grown from 100 g to approximately 310 g, an improved performance has also been reported in the raceway system. The growth in trout fed MOS was 10% higher than the control and the FCR decreased by 11.2% to 1.07 with MOS addition. Mortalities also decreased from 5 to 3%, a reduction of 41% [18].

Similarly, the addition of MOS to the diets of European catfish (*Silurus glanis)* juveniles has also shown improved growth by 9.7% and higher average body weight. The FCR was also lower by 11.6% for the MOS fed fish and mortality decreased from 28.3 to 16.7% [28]. This data is in agreement with the findings of Hanley *et al.* [29], who also demonstrated that hybrid red tilapia juveniles fed 0.6% MOS in their hatchery diets had a 22.5% improved survival with a 27.2% increase in weight gain [29].

For sub-adult hybrid striped bass (*Morone chrysops* X *M. saxatilis*), an enhanced growth was observed in fish fed diets supplemented with MOS than fish fed the basal diet, with significantly enhanced weight gain observed after 12 weeks of feeding. Fish fed MOS had a significantly enhanced survival (80%) compared to the control treatment (72–73%) at the end of 21 weeks. In commercially run farms, MOS is being used systematically in the feed industry as numerous farmers have reported improved growth, survival and feed conversion rates for the sea bass and sea bream industry [30, 31]. However, the improved growth has not been observed in the controlled experiment [21]. An experiment carried out at the University of Las Palmas of Gran Canaria in Spain with 35 g sea bass (*Dicentrarchus labrax)* juveniles showed the similar effects of dietary MOS (0.2 and 0.4%) on growth, histological and biochemical characteristics. Growth in terms of final body weight, specific growth rate (SGR) and relative growth significantly increased by 10% in sea bass fed MOS [19].

On the contrary, no effect of MOS on the growth of Gulf sturgeon (*Acipenser oxyrinchus desotoi*) [32], turbot larvae (*Psetta maxima*) [33], hybrid tilapia (*Oreochromis niloticus* X *O. aureus*) [34], sea bream [21] and Atlantic salmon (*Salmo sarla*) [35] has been reported under the pilot experiment conditions. The reason for this could be the differences in the experimental condition and life stage of the animals used in the experiment. However, the MOS positively affected the physiological conditions of these fishes. The salmon fed MOS showed 11% lower routine oxygen consumption, 5% lower protein and 3% higher energy concentration in the whole-body and 7% greater energy retention than in the fish fed the basal diet [35]. Condition factor (K) and hepatosomatic index (HSI) of sea bream were significantly lower in the fish fed 0.2% to 0.4% MOS [21]. Craig and McLean [36] found that the incorporation of MOS in Nile tilapia (*Oreochromis niloticus*) diets at levels ranging from 0.25 to 2% resulted in a leaner fillet than in the control animals. However, MOS did not reduce hepatic lipid accumulation in this experiment [36]. In contrast, sea bass fed MOS showed a gradual reduction in the vacuolization, denoting a better utilization of dietary nutrients (Figure 3). Biochemical analysis of the liver revealed a reduced fat content (-5%), less humidity (-3%) and a higher protein content (+2%) with MOS supplementation than the control sea bass group [19].

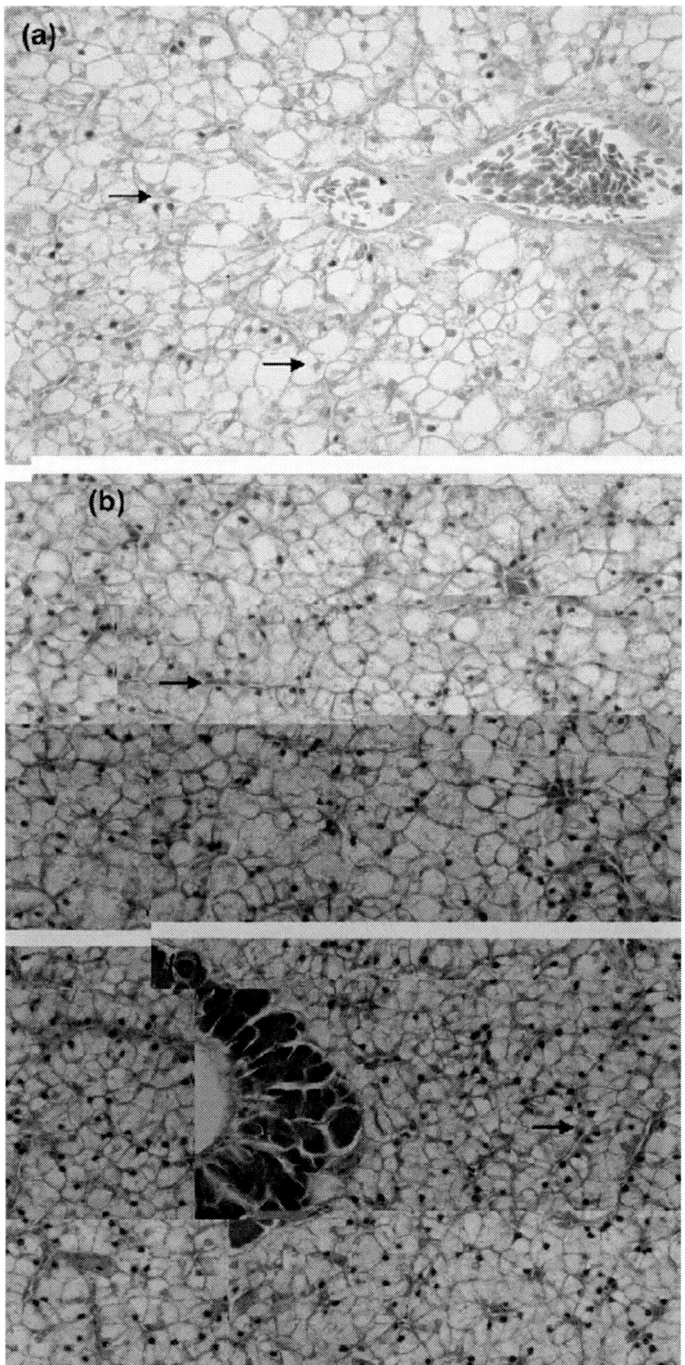

Figure 3. Hepatocytes (H&E. X400) from fish fed (a) Control diet showing foci of swollen hepatoyctes characterized by cytoplasm vacuolization and nuclei displaced to cell periphery; (b) 0.2% MOS supplemented diet with lower number of swollen hepatocytes; and (c) 0.4% MOS supplemented diet with a regular morphology of the hepatocytes located around sinusoidal spaces [19].

5. Effects of MOS on Digestive Tract of Aquatic Species

5.1. Decapods

Digestive tract health of tropical rock lobsters and marron was improved by dietary MOS inclusion [13, 14]. Sang and Fotedar [14] reported ten times increase in the total bacteria and bacteria/Vibrio ratio in intestine microbial of marron when 0.2 and 0.4% MOS were supplemented in their diets. (Table 3). Similarly, positive effect was observed in tropical rock lobster when the diet was supplemented with 0.4% MOS (Table 4) [13]. However, no information is available on the identification of the intestine bacterial strain/species which are stimulated/eliminated by MOS inclusion in the diets of crayfish, unlike some fish species [7, 37].

Table 3. Bacteria in the gut of marron fed different MOS supplemented diets (Mean ± SE)

Diets	Total bacteria (million CFU/g)	*Vibrio* (million CFU/g)	Bacteria/*Vibrio* ratio
Control	$_{a}$974.50 ± 80.66*	$_{a}$0.51 ± 0.09	$_{a}$2 033.34 ± 384.54
0.05% MOS	$_{b}$702.09 ± 71.53	$_{a}$2.78 ± 0.61	$_{bc}$293.51 ± 97.34
0.1% MOS	$_{c}$2 449.51 ± 58.86	$_{b}$9.99 ± 1.56	$_{c}$259.77 ± 47.48
0.2% MOS	$_{d}$1 905.06 ± 84.93	$_{a}$1.46 ± 0.58	$_{a}$1 816.93 ± 670.60
0.4% MOS	$_{e}$3 283.90 ± 72.33	$_{a}$0.82 ± 0.05	$_{d}$4 013.93 ± 244.57
0.8% MOS	$_{a}$1 216.90 ± 124.64	$_{a}$1.34 ± 0.39	$_{abc}$1 144.84 ± 462.75

*Value in the same column of each parameter having different subscript letters are significantly different at $P < 0.05$.

Table 4. Gut bacteria and morphology of the lobster fed different diets

Parameter	Control	0.4% MOS supplementation
Total aerobic bacteria (million cfu/g)	54.00 ± 4.93^{a}	546.67 ± 53.64^{b}
Total *Vibrio* spp. (million cfu/g)	28.00 ± 7.09^{a}	207.66 ± 29.63^{b}
Total bacteria/*Vibrio* spp. ratio	2.09 ± 0.37^{a}	2.68 ± 0.14^{a}
Internal perimeter/External perimeter	2.25 ± 0.10^{a}	3.76 ± 0.12^{b}

Immunostimulants in many organisms have also increased the efficiency of the digestive tract by increasing the regularity, height and integrity of the gut villi [38]. In crustaceans, such effects of MOS were observed in tropical rock lobster (Figure 4) and marron (Figure 5 and Figure 6) [13, 14]. Dietary MOS improved the gut morphology of marron by increasing the number of villi/group and density of villi (Table 5) [14]. This in turn resulted in better ability to irrigate the gut, to protect the cuticle layer and/or in smooth movement of faces and uptake of water [39]. Higher density of gut epidermic cells of marron was also direct consequence of dietary MOS [14] indicating the better ability in the transport, assimilation, storage of nutrients [40] and transport of ions and water [41] through the digestive tract. In tropical rock lobster, dietary MOS increased the absorptive surface area of the gut indicated by the higher internal /external perimeter ratio [13]. Higher absorption area enhances the nutrient and energy assimilated process due to the higher nutrient uptake, consequently higher assimilated energy is stored in the tail muscle in the form of dry tail muscle biomass. The digestive, absorptive and assimilated efficiencies could be improved by the inclusion of MOS in the diets of crayfish. However, evidences of nutrient uptake and transportation, assimilation and digestive efficiency enhanced by dietary MOS in crayfish are still unavailable in the literature. In addition, the changes in digestive enzyme profile of crayfish when manipulated by the dietary MOS and the influence of MOS on digestibility of main dietary ingredients has not yet been investigated for decapods. Those limitations provide great opportunities for conducting further research on understanding the mechanism in which MOS affects the digestive function of the gastrointestinal track of decapods.

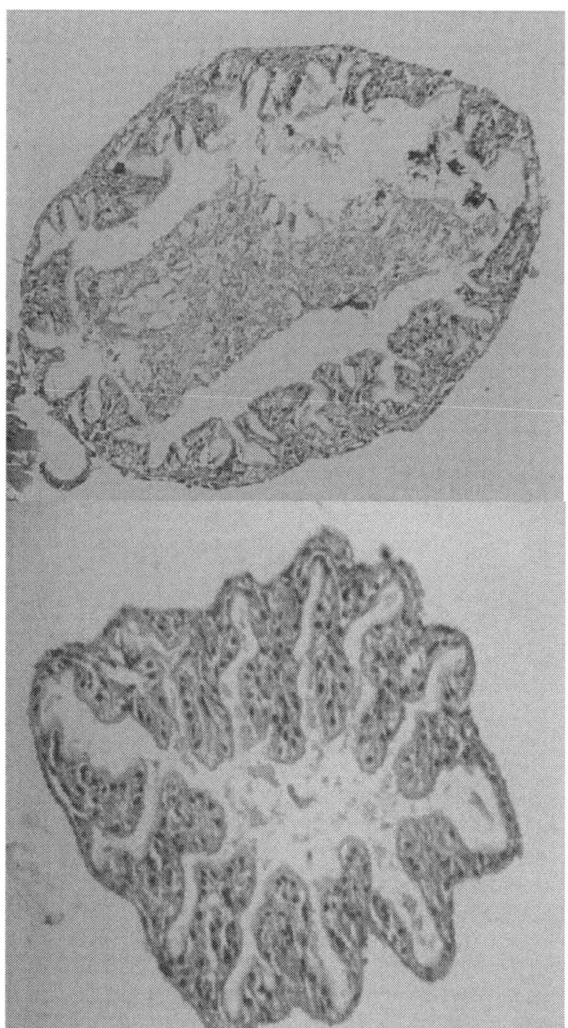

Figure 4. Cross section of the gut of tropical rock lobsters fed the control (A) and 0.4% MOS supplemented diets (B).

Table 5. Villi in the hindgut of marron fed different MOS supplemented diets

Diets	Number of villi/group (Mean ± SE)	Villi density (villi/100 μm²) (Mean ± SE)
Control	$_a$6.13 ± 0.04*	$_a$9.67 ± 0.38
0.05% MOS	$_b$7.44 ± 0.20	$_a$10.93 ± 0.60
0.1% MOS	$_b$7.68 ± 0.20	$_b$13.56 ± 0.48
0.2% MOS	$_c$8.85 ± 0.61	$_b$14.63 ± 0.94
0.4% MOS	$_{bc}$8.22 ± 0.27	$_b$14.04 ± 0.27
0.8% MOS	$_{bc}$8.00 ± 0.26	$_a$10.41 ± 0.72

*Value in the same column of each parameter having different subscript letters are significantly different at $P < 0.05$.

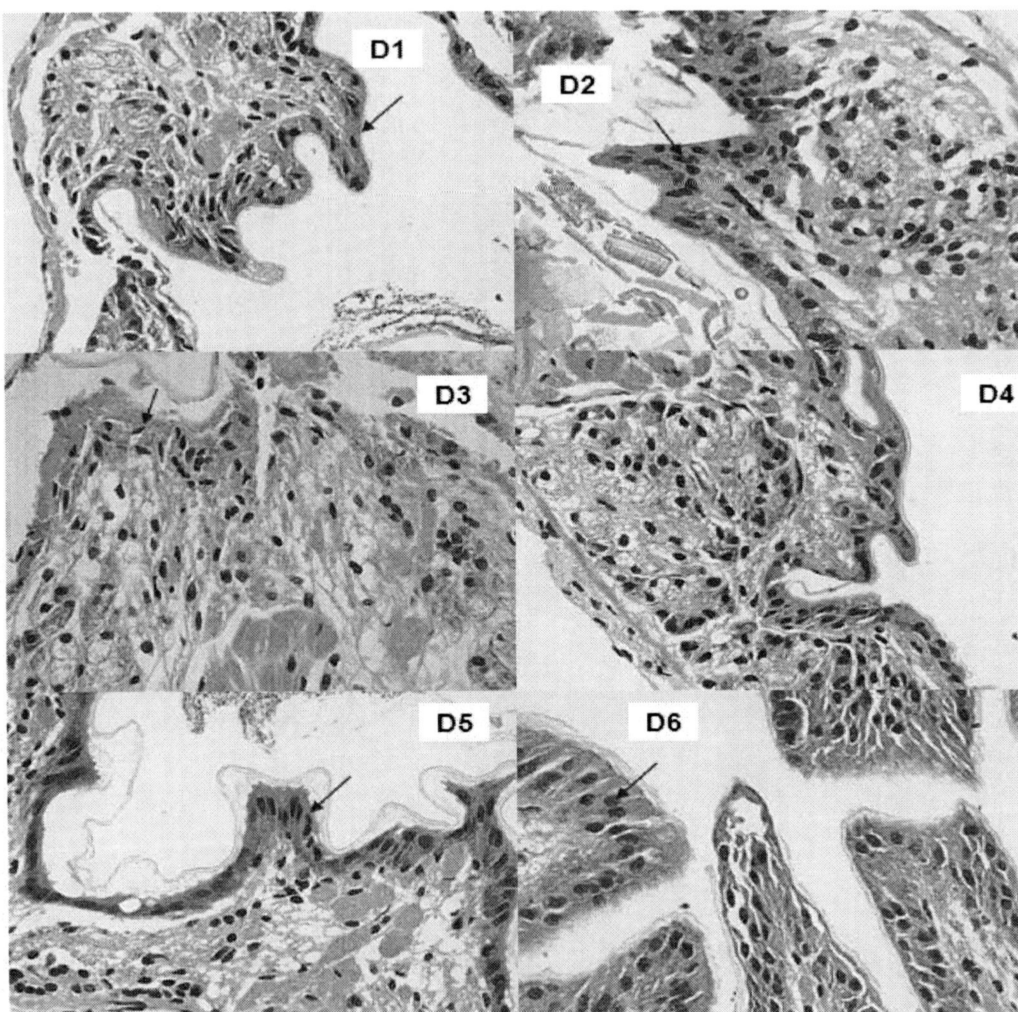

Figure 5. Transverse sections of the gut of marron, fed the different MOS supplemented diets (H&E stained, 400X magnification) (The arrow points to the epidermis cell of the inner gut lining). D1: MOS free; D2: MOS - 0.05%; D3: MOS - 0.1%; D4; MOS - 0.2%; D5; MOS - 0.4%; D6: MOS - 0.8%.

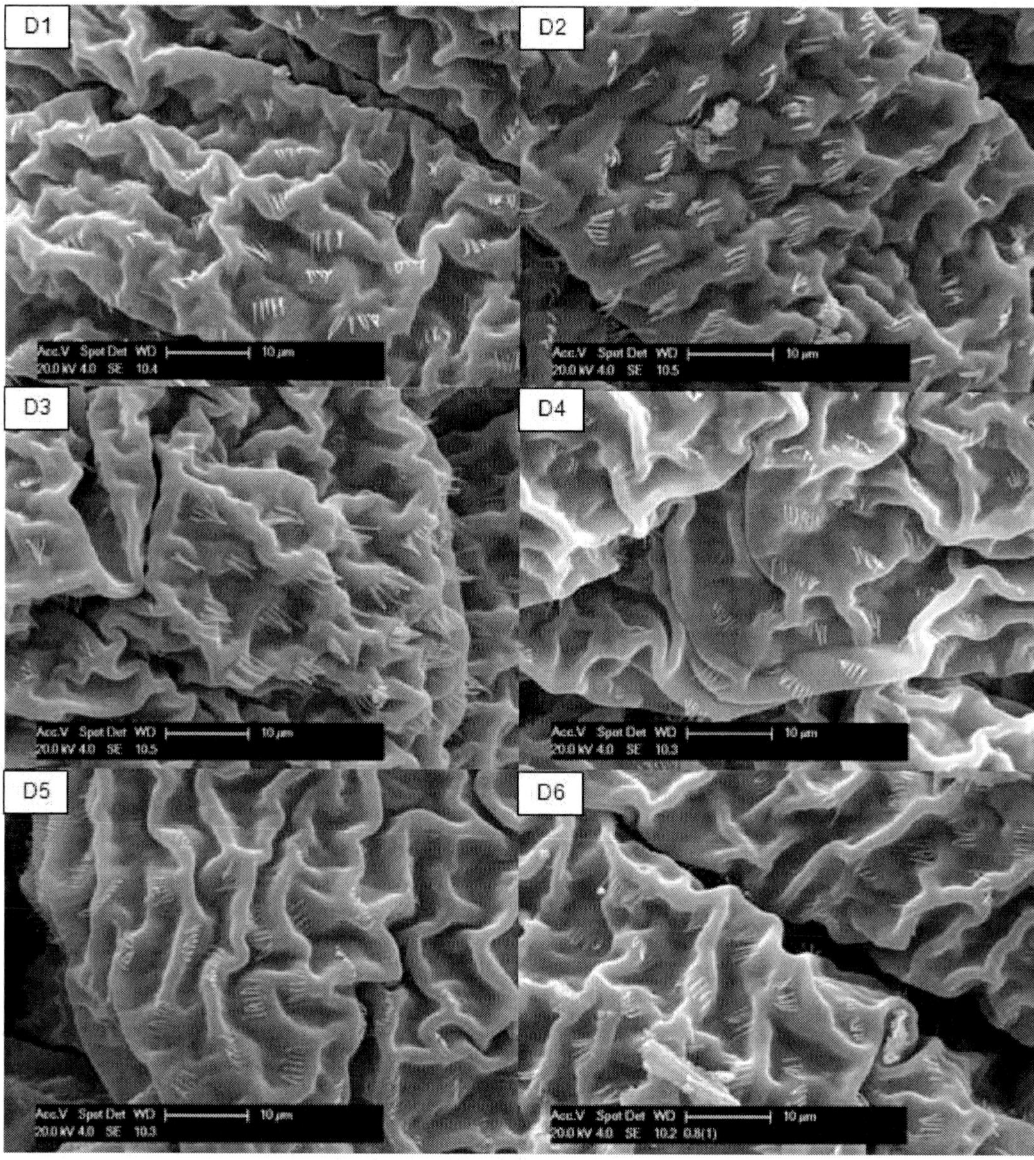

Figure 6. Gut micrographic of marron fed different MOS supplemented diets. D1: MOS free; D2: MOS- 0.05%; D3: MOS - 0.1%; D4; MOS - 0.2%; D5; MOS - 0.4%; D6: MOS - 0.8%.

5.2. Teleosts

Dietary MOS altered the intestinal microbial populations of common carp by reducing the population of *Escherichia coli* and increasing the population of *Bifidobacterium* and *Lactobacillus* [37]. In addition, MOS inclusion also reduced the total aerobically cultivated bacteria load in the gut of rainbow trout. Viable population of microbiota in the intestine of MOS supplemented groups was approximately 2 log scales less than the control group. Changes in the relative abundance of microbiota identified were also observed. The MOS fed fish displayed a reduction in the relative abundance of *Micrococcus spp.* (from 22 to 7% of the total microbiota), *Aeromonas/Vibrio spp.* (from 37 to 9%) and unidentified gram – positive rods (from 25 to 6%). Those changes coincided with the increases of *Enterococcus spp.* (from 3 to 19%) and Enterobacteriaceae (from 5 to 39%) [1, 7]. In sea bream, MOS increased species richness and diversity of intestinal microbiota and reduced similarity between microbial profile of the different MOS supplemented fed group (Table 6) [21].

Table 6. Summary of species richness, Shannon-Weaver diversity index and similarity analysis of the intestinal microbiota of sea bream fed different MOS supplemented diets [21]

MOS supplemented level	Species richness[a]	Shannon-Weaver diversity index[b]	SIMPER similarity[c] (%)
0	15.5 ± 0.7	2.65 ± 0.05	84.79
0.2%	25.0 ± 0.0	3.06 ± 0.02	88.16
0.4%	24.5 ± 0.7	3.1 ± 0.05	81.34

[a] Average number of bands/presumed species.
[b] Shannons diversity index: $H' = -\sum(pi(\ln pi))$.
[c] SIMPER = similarity within group replicates.

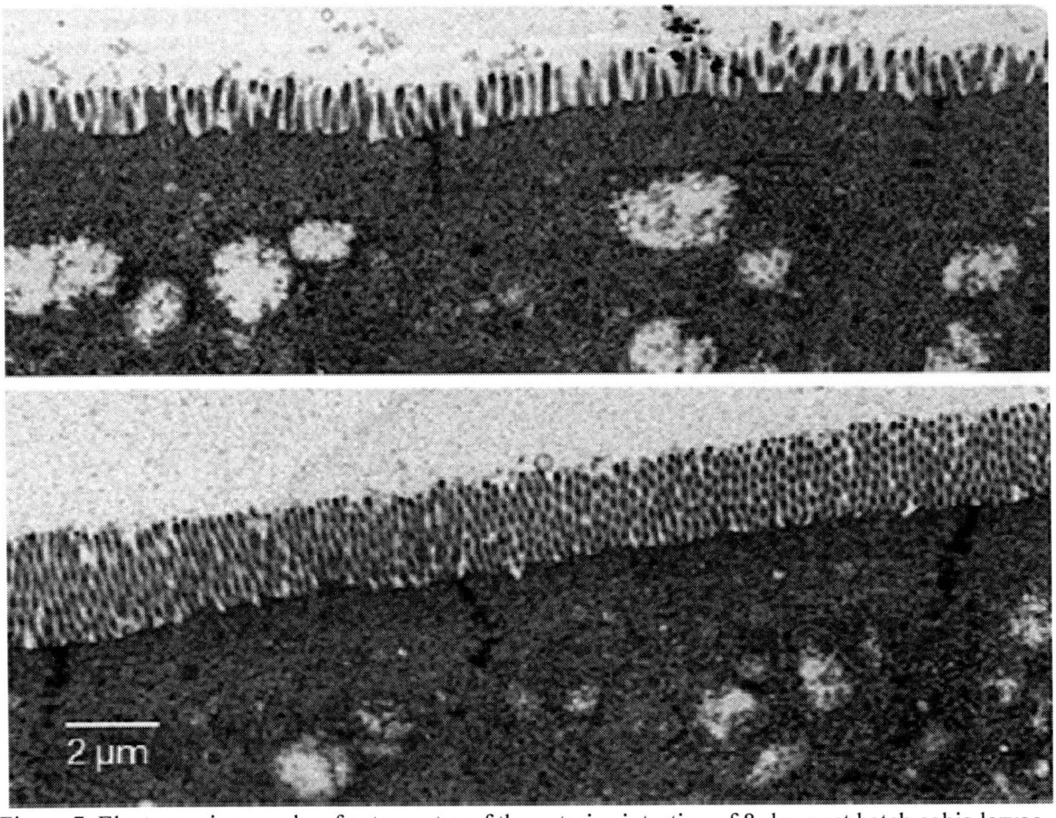

Figure 7. Electron micrographs of enterocytes of the anterior intestine of 8-day post hatch cobia larvae fed control (upper plate) or MOS supplemented diets. MOS supplemented diet resulted in a heightening of absorptive cell microvilli and a reduction in the number and size of vacuoles and vesicles in the supranuclear region of the cell [12].

Dietary MOS was also reported to improve the gut morphology of fish though the increase in the absorptive surface and structure of microvilli. Enterocytes of cobia larvae provided with MOS supplementation expressed more uniform, densely packed, longer and narrower microvilli averaging 2.04±0.02 μm than observed in the identical gut regions of larvae fed control diets (1.18±0.03 μm). Moreover, the number and size of supranuclear (SNV) vacuoles in MOS-treated cobia were lesser and smaller than those observed in control samples from 8 days-post-hatch throughout the length of the intestine. The width of SNV in MOS-treated fish was 0.87±0.32 μm versus 1.69±0.46 μm in control larvae (Figure 7) [12].

Light microscope and scanning electronic microscope observations on the gastrointestinal morphology of salmon, sole (*Solea solea*), rainbow trout and sea bream revealed the better condition in the MOS fed fishes. Dietary MOS increased the intestinal absorption surface of the rainbow trout indicated by bigger internal to external perimeter ratio. For the anterior gut region, the ratio was 4.50±0.30 for MOS-fed fish compared to 3.46±0.48 of the control fish. The ratios in the posterior gut region were 3.03±0.19 for the control fish and 5.29±0.39 for the MOS fed fish. Scanning electron microscopy images showed 0.2% MOS inclusion can significantly affect the posterior gut region by increasing microvilli density (6.54 ± 1.08 of MOS fed group and 2.53 ± 0.75 of control group). Transmission electron microscopy images revealed that in the anterior gut, MOS fed fish (1.9 ± 0.15) had significantly increased microvilli length (μm) in contrast to the controls (1.33 ± 0.72). Increased microvilli length as a result of MOS inclusion was also found in the posterior gut region [7]. Similar results have also been reported for sole [8], sea bream [42] and salmon [43] (Figure 8 and 9).

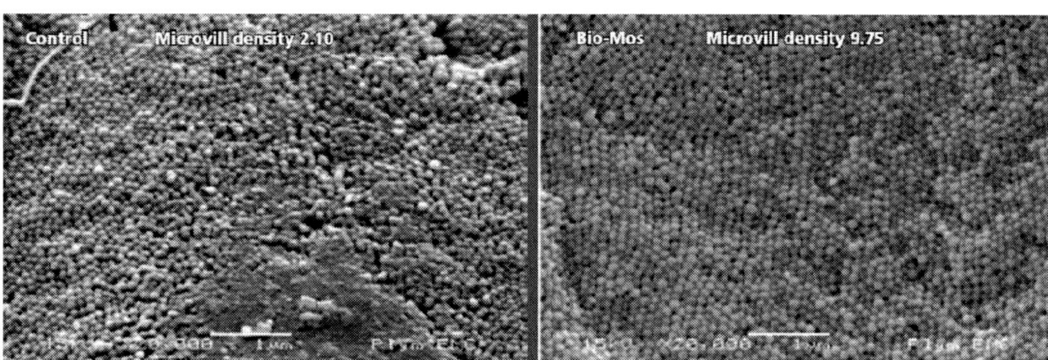

Figure 8. MOS improved the microvilli density of salmon [43].

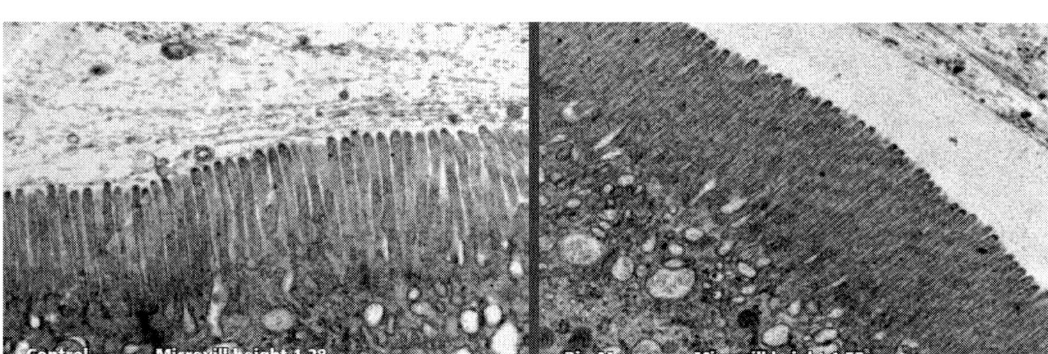

Figure 9. MOS improved the microvilli height of salmon [43].

6. Effect of MOS on Immune System and Health Status

6.1. Decapods

In the haemolymph of crayfish, the hyaline cells are chiefly involved in phagocytosis. The semigranular cells are the cells active in encapsulation which recognize and respond to foreign molecules and particles by degranulation and subsequently attaching and spreading on the foreign surface [44]. Together with granular cells, semigranular cells also participate in cytotoxicity and storage and release of prophenoloxidase activating system [45, 46]. The research has clearly shown that the application of MOS enhances the immune capacity of crayfish at cellular level by altering the hemocyte profile of marron, lobster and yabbies (Table 7). MOS increased total hemocyte count, the proportion of granular and semigranular cells and reduced proportion of hyaline cells of marron and yabbies [15, 16]. Similar effects were observed on the total hemocyte count and granular cells on tropical rock lobster [13]. Sang et al. [13, 15, 16] stated that although the proportion of hyaline cells are reduced, the total number of hyaline cells increased due to the increase in the total hemocyte count when MOS was applied. Haemolymph clotting time of marron and tropical rock lobster also decreased when MOS was provided to the crayfish [13, 15]. In addition, MOS also reduced the bacteria load in the haemolymph of above crayfish [13, 15, 16].

MOS increased ability of marron and lobsters in defending against bacterial infection and environmental stress. Accumulative mortality of MOS-fed marron was reduced by 121% after 96 hours of infection with *Vibrio mimicus* [9]. Similarly, survival of MOS-fed tropical rock lobsters ($77.78 \pm 5.55\%$) was higher comparing to non MOS-fed lobsters ($55.56 \pm 5.55\%$) after 7 days of infection with *Vibrio spp.* [13]. Other immune indicators such as higher total hemocyte count, granular cells and neutral red retention time of lysosome were also observed in MOS-fed marron and tropical rock lobster after challenging them with *Vibrio spp* [9, 13]. Sang et al. [9] also reported the better ability of MOS-fed marron to resist stressors such as NH_3 and air exposure. Total hemocyte count and granular cells were also higher in MOS-fed lobster during the infected period [13]. Higher performances of crayfish to bacterial infection are as a consequence of higher capacity of immune system generated by MOS administration. Unfortunately, very little information is available on the other bacteria-infected crustaceans when fed MOS.

6.2. Teleosts

MOS improved several immune indicators such as antibody titre, bactericidal activity, lysozyme activity, alternative pathway complement activity (APCA) and classical pathway complement activity (CPCA) of rainbow trout. Two trials were conducted in net cages and raceway rearing system which showed that these indicators were positively affected by the dietary MOS (Table 5) [18].

Similar improvements in immune indicators of MOS-fed common carp were also observed (Figure 10). Serum lysozyme activity of MOS-fed carp increased (+13%). and remained significantly higher through out the study Likewise, after receiving MOS in the diet for 60 days, carp had higher blood complement activity for both pathways [10, 26].

Table 7. Effect of MOS on immunological indicators of crayfish

Species	THC	GCs	SGCs	HCs	Haemolymph clotting time	Bacteraemia	Resistance to bacterial infection
Cherax tennuimanus – MOS [1]	+	+	+	+	+	+	+
Panulirus ornatus – MOS[2]	+	+	N/A	N/A	+	+	+
Cherax destructor [3]	+	+	+	+	N/A	+	N/A

[1] [15]; [2] [13]; [3] [16];.

THC: Total hemocyte count; GCs: Granular cells; SGCs: Semigranular cells; HCs: Hyaline cells; - : none effect; +: positive effect; N/A: data not available.

Table 8. Effect of dietary supplementation of MOS on different indicators of immune status in rainbow trout [18]

Culture system – initial fish weight	Immune indicators	Control	MOS	P- value
Net cage – 30 g	Antibody titre	3.37 ± 0.06	6.35 ± 0.02	< 0.01
	Bactericidal activity	2.37 ± 0.05	3.44 ± 0.02	< 0.01
	Lysozyme (mg L^{-1})	7.59 ± 0.09	10.63 ± 0.12	< 0.01
	APCA (CH50)	195.64 ± 0.30	226.12 ± 0.41	< 0.01
	CPCA (CH50)	64.92 ± 0.31	71.86 ± 0.29	< 0.01
Raceway – 101 g	Antibody titre	2.82 ± 0.07	3.45 ± 0.33	> 0.05
	Bactericidal activity	2.40 ± 0.06	2.46 ± 0.16	> 0.05
	Lysozyme (mg L^{-1})	6.10 ± 0.07	7.26 ± 0.40	< 0.05
	APCA (CH50)	180.80 ± 1.03	198.85 ± 5.86	< 0.05
	CPCA (CH50)	52.37 ± 0.44	62.24 ± 3.35	< 0.05

Note: CH50 units correspond to 50% of complement-induced haemolysis of applied erythrocytes.

Torecillas et al. [19] reported the enhanced immune stimulation and improved infection resistance in European sea bass-fed MOS. Immune indicators such as NBT (Nitro Blue Tetrazolium) index (measuring the activity of circulating neutrophils), lysozyme and APC activities and phagocytic index of head kidney macrophages were evaluated following MOS supplemented diets. Phagocytic activity of head kidney macrophages against *Vibrio alginolyticus* significantly increased in fish fed the 0.4% MOS supplemented diet, resulting in a positive correlation between the phagocytic activity and the dietary MOS inclusion levels. No significant differences were found on lysozyme and ACH50, although a positive correlation was observed between the level of MOS inclusions and the level of these parameters in fish serum (Figure 11). Furthermore, the numbers of infected fish were lower in the MOS-fed groups when they were challenged with *V. alginolyticus* transmitted by cohabitation or anal canalization (Figure 12) [10, 26].

7. Current Constrains and Suggestions for Further Research

MOS supplementation shows numerous benefits and can be used as alternative for antibiotic in decapods and teleosts. However, research into the direct roles of MOS is currently limited and, from the available literature, only some species of crayfish have been briefly exposed to MOS. To further maximize the production of aquatic species and obtain the wide effectiveness of MOS application, a number of key research priorities are suggested based on the information reviewed previously.

Under culture conditions the period to reach typical marketable size is about three years for marron; is more than one and a half year for tropical rock lobster and about 1 year for yabbies. However, the effects of MOS have only been examined in those crayfish for relatively short periods and only at juvenile stages. Thus there is a need to validate the effects of MOS in the whole culture cycle of decapods under commercial farming situations. In addition, apparent digestibility of dietary ingredient when MOS is supplemented needs to be

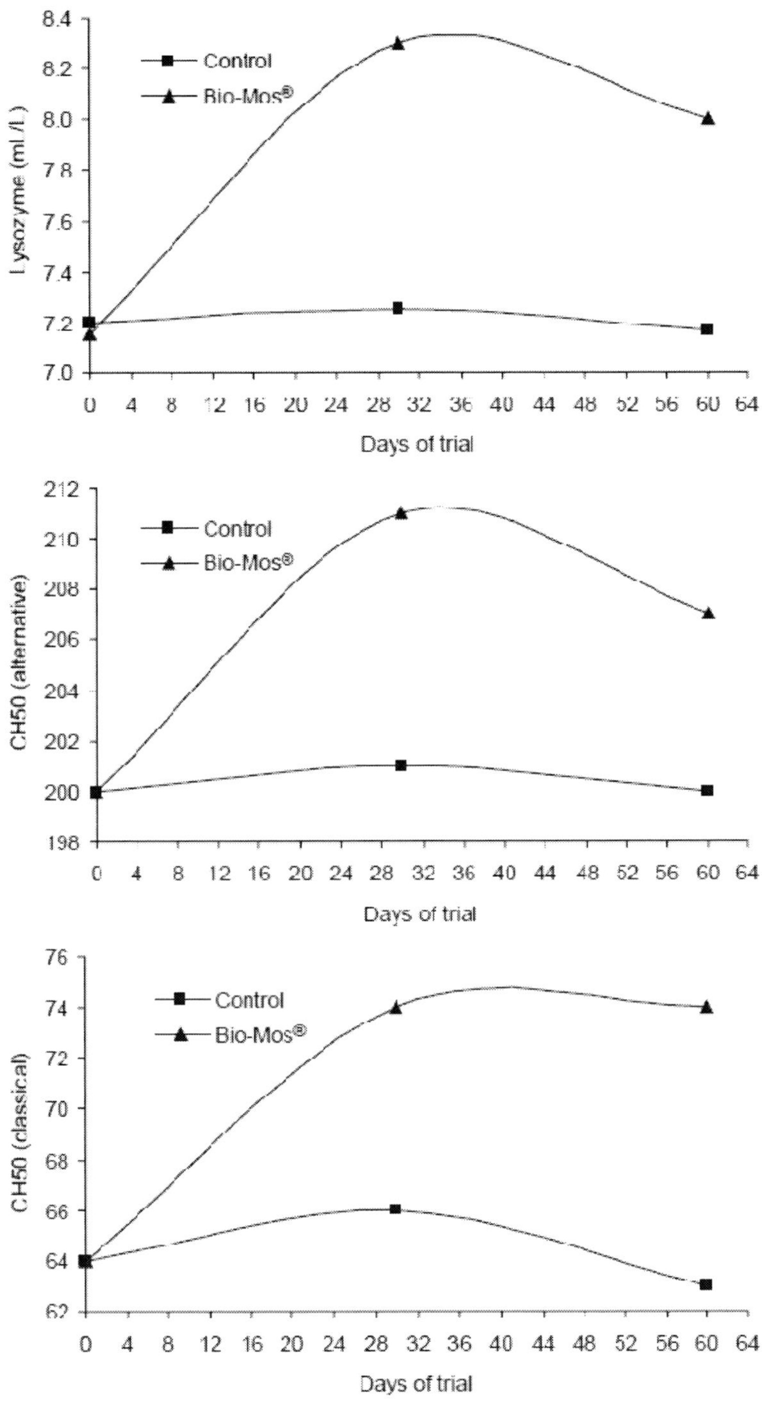

Figure 10. Effect of MOS on serum lysozyme and complement activity in common carp [10, 26].

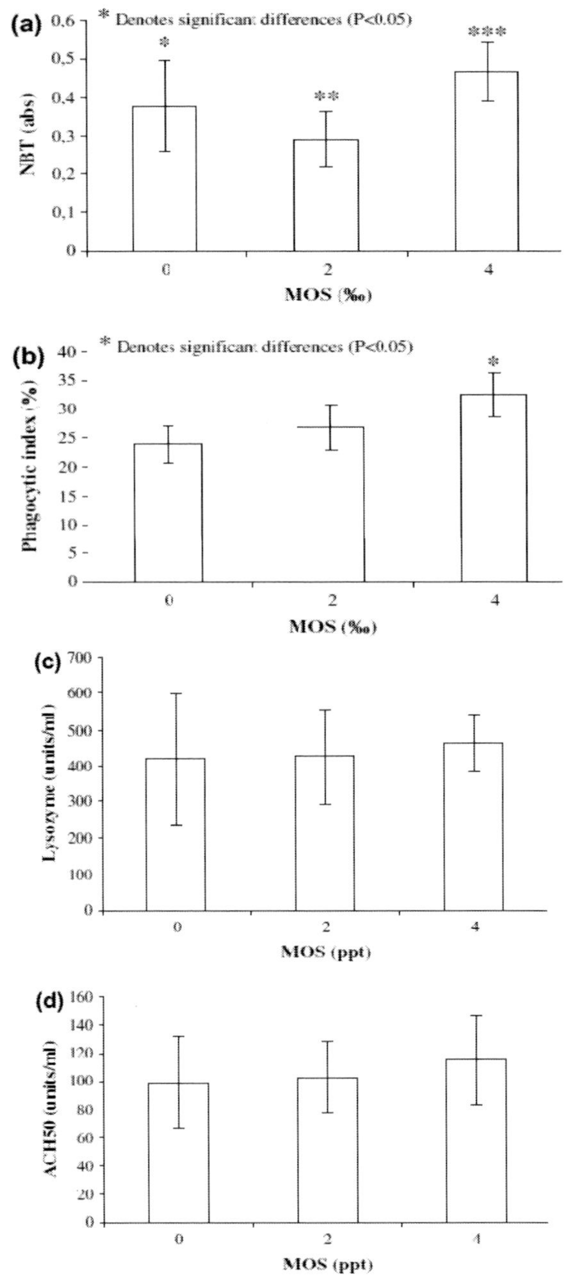

Significant differences ($P < 0.05$) among treatments are indicated by (*).

Figure 11. Influence of MOS on total (a) phagocytic activity of head kidney macrophages; (b) reduction potential of circulating neutrophils; (c) lysozyme activity; and (d) alternative complement pathway activity in European sea bass. Improvement in nonspecific immune system of sea bass by MOS was recently supported by a finding of Terova et al. [47] who evaluated the antimicrobial peptide decentracin mRNA copy number in kidney of the MOS fed fish. After 30 days of feeding fish with supplemented MOS diets at either 0.3 or 0.5% significantly increased the dicentracin mRNA copy number in the head kidney. The mRNA copy number in fish fed 0.3% MOS was significantly higher than that of the group fed 0.5% MOS for the same period of feeding [47]. On the contrary, dietary MOS did not reveal improvements in immune response and resistance to stress and *Edwardsiella ictaluri* infection in channel catfish [20].

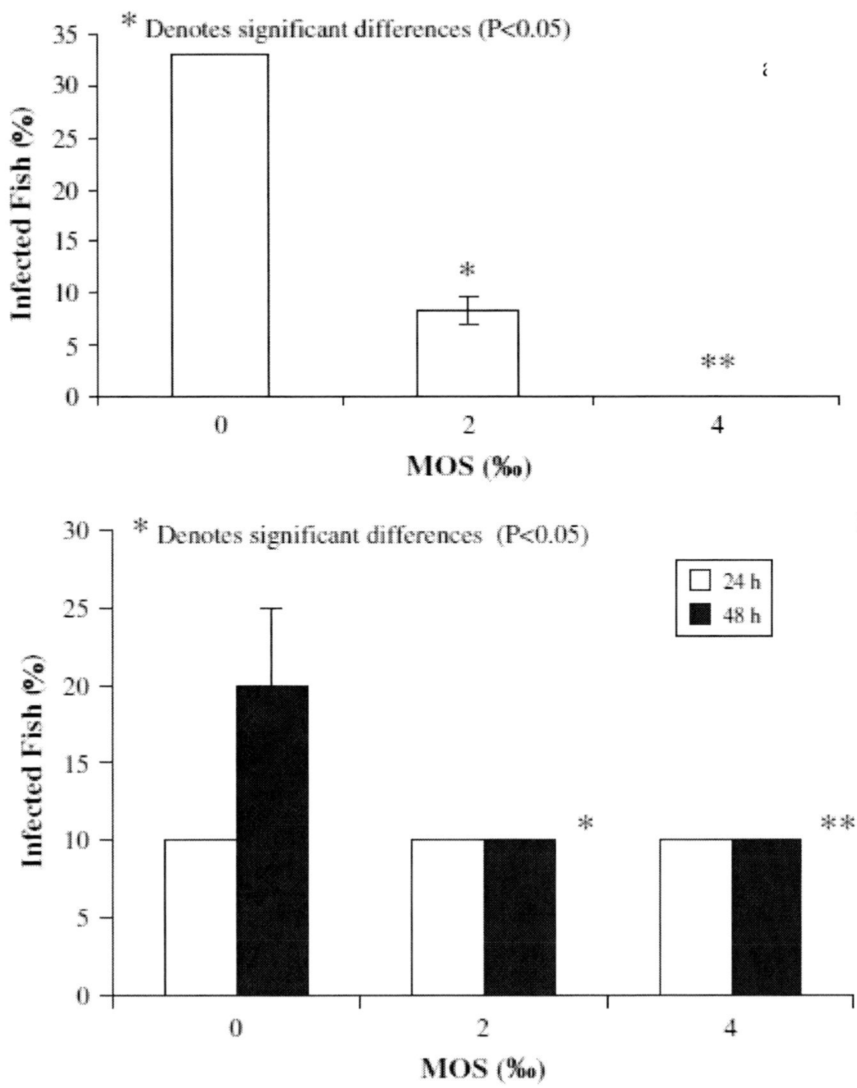

Figure 12. Presence of *Vibrio alginolyticus* on head kidney of European sea bass fed different levels of MOS inclusion and submitted to infection by cohabitation (a) and anal canalization (b) [10, 26].

determined. This would assist in the usage of cheaper protein sources instead of expensive and unreliable fishmeal. In addition, the economical and environmental benefits of MOS applications in decapod crustaceans culture need to be quantified in future research.

To date, the understanding on the benefits of MOS on the digestive tract of decapods within the literature is rather limited. Only morphological and total bacterial dynamic in the gut has been researched. Thus, future research needs to focus on the exact dynamics with proper identification of gastrointestinal microbial communities. Furthermore, roles of MOS in the nutrient transportation, nutrient uptake and digestive enzyme activity profile in the gastrointestinal track need to be investigated.

Although MOS have been reported to improve the immune capacity of aquatic species, whether the MOS stimulate the immune reactions such as prophenoxidase activating system, phagocytosis, cytotoxicity is not clear. Therefore, further research to understand this is warranted. It is also important to understand the underlying mechanism by which MOS inclusions increases the microvilli densities, other anatomical and physiological characteristics of intestinal track of the target species.

References

[1] Dimitroglou, A., Merrifield, D. L., Moate, R., Davies, S. J., Spring, P., Sweetman, J. & Bradley, G. (2009). Dietary mannan oligosaccharide supplementation modulates intestinalmicrobial ecology and improves gut morphology of rainbow trout, Oncorhynchus mykiss (Walbaum). *Journal of Animal Science*, 87, 3226 - 3234.

[2] Rosen, G. D. The nutritional effects of tetracyclines in broiler feeds. In XX World's Poultry Congress, New Delhi, India (WPSA). 141–146; 1996.

[3] Genc, M. A., Aktas, M., Genc, E. & Yilmaz, E. (2007). Effects of dietary mannan oligosaccharide on growth, body composition and hepatopancreas histology of *Penaeus semisulcatus* (de Haan 1844). *Aquaculture Nutrition*, 13, 156 - 161.

[4] Ringo, E., Olsen, R. E., Gifstar, T. O., Dalmo, R. A., Amlund, H., Hemre, G. I. & Bakke, A. M. (2010). Prebiotic in Aquaculture: a review. *Aquaculture Nutrition*, 16, 117 - 136.

[5] Gibson, G. R., Probert, H. M., Van, L. J., Rastall, R. A. & Roberfroid, M. B. (2004). Dietary modulation of the human colonic microbiota: Updating the concept of prebiotics. *Nutrition Research Reviews*, 17, 259–275.

[6] Teitelbaum, J. E. & Walker, W. A. (2002). Nutritional impact of pre- and probiotics as protective gastrointestinal organisms. *Annual Review Nutrition*, 22, 107-138.

[7] Dimitroglou, A., Davies, S. & Sweetman, J. (2008). The effect of dietary mannan oligosaccharides on the intestinal histology of rainbow trout (*Oncorhynchus mykiss*). *Comparative Biochemistry and Physiology - Part A: Molecular & Integrative Physiology*, 150, S63-S63.

[8] Dimitroglou, A., Janssens, T. & Davies, S. Effects of Bio-Mos on Sole (*Solea solea*) gut integrity (Histological perspective). In Proceedings of Alltech's 22st Annual Symposium Lexington, KY.,vol 2006.

[9] Sang, H. M., Ky, T. L. & Fotedar, R. (2009). Dietary supplementation of mannan oligosaccharide improves the immune responses and survival of marron, *Cherax tenuimanus* (Smith, 1912) when challenged with different stressors. *Fish & Shellfish Immunology*, 27, 341 - 348.

[10] Staykov, Y., Spring, P. & Denev, S. Influence of dietary Bio-Mos® on growth, survival and immune status of rainbow trout (*Salmo gairdneri irideus* G.) and common carp (*Cyprinus carpio* L.) In Trakia University, Bulgaria, 2006.

[11] Daniels, C., Boothroyd, D., Davies, S., Pryor, R., Taylor, D. & Wells, C. (2006). Bio-Mos® improves the growth and survival of cultured European lobster. *Shellfish News*, 21, 23 - 25.

[12] Salze, G., McLean, E., Schwarz, M. H. & Craig, S. R. (2008). Dietary mannan oligosaccharide enhances salinity tolerance and gut development of larval cobia. *Aquaculture*, 274, 148 - 152.

[13] Sang, H. M. & Fotedar, R. (2010). Effects of mannan oligosaccharide dietary supplementation on performances of the tropical spiny lobsters juvenile (*Panulirus ornatus*, Fabricius 1798). *Fish & Shellfish Immunology*, 28, 483 - 489.

[14] Sang, H. M. & Fotedar, R. (2010). Dietary mannan oligosaccharide improves health status of the digestive system of marron, *Cherax tenuimanus* (Smith, 1912). *Journal of Applied Aquaculture*, In press,

[15] Sang, H. M., Fotedar, R. & Filer, K. (2010). Effects of dietary Mannan Oligosaccharide on survival, growth, physiological condition and immunological responses of marron, *Cherax tenuimanus* (Smith, 1912). *Journal of the World Aquaculture Society*, In press,

[16] Sang, H. M., Fotedar, R. & Filer, K. (2010). Effects of dietary mannan oligosaccharide on the survival, growth, immunity and digestive enzyme activity of freshwater crayfish, *Cherax destructor* Clark (1936). Aquaculture Nutrition, In press,

[17] D'Abramo, L. R. & Robinson, E. H. (1989). Nutrition of crayfish. *CRC Critical review of aquatic science*, 1, 711 - 728.

[18] Staykov, Y., Spring, P., Denev, E. S. & Sweetman, E. J. (2007). Effect of a mannan oligosaccharide on the growth performance and immune status of rainbow trout (*Oncorhynchus mykiss*). *Aquaculture International*, 15, 153 - 161.

[19] Torrecillas, S., Makol, A., Caballero, M. J., Montero, D., Robaina, L., Real, F., Sweetman, J., Tort, L. & Izquierdo, M. S. (2007). Immune stimulation and improved infection resistance in European sea bass (*Dicentrarchus labrax*) fed mannan oligosaccharides. *Fish & Shellfish Immunology*, 23, 969 - 981.

[20] Welker, T. L., Lim, C., Yildirim-Aksoy, M., Shelby, R. & Klesius, P. H. (2007). Immune response and resistance to stress and *Edwardsiella ictaluri* challenge in channel catfish, *Ictalurus punctatus*, fed diets containing commercial Whole-Cell Yeast or Yeast Subcomponents. *Journal of the World Aquaculture Society*, 38, 24 - 35.

[21] Dimitroglou, A., Merrifield, D. L., Spring, P., Sweetman, J., Moate, R. & Davies, S. J. (2010). Effects of mannan oligosaccharide (MOS) supplementation on growth performance, feed utilisation, intestinal histology and gut microbiota of gilthead sea bream (*Sparus aurata*). *Aquaculture*, 300, 182 - 188.

[22] Fotedar, R., Knott, B. & Evans, L. (1999). Effect of a diet supplemented with cod liver oil and sunflower oil on growth, survival and condition indices of juvenile *Cherax tenuimanus* (Smith). *Freshwater Crayfish*, 12, 478 - 493.

[23] Furne, M., Hidalgo, M. C., Lopez, A., Garcia-Gallego, M., Morales, A. E., Domezain, A., Domezaine, J. & Sanz, A. (2005). Digestive enzyme activities in Adriatic sturgeon *Acipenser naccarii* and rainbow trout *Oncorhynchus mykiss*. A comparative study. *Aquaculture*, 250, 391 - 398.

[24] Houlihan, D. F., Hall, S. J., Gray, C. & Noble, B. S. (1988). Growth rates and protein turnover in Atlantic cod, *Gadus morhua*. *Canadian Journal of Fisheries and Aquatic Sciences*, 45, 951 - 964.

[25] Wang, Y. (2007). Effect of probiotics on growth performance and digestive enzyme activity of the shrimp *Penaeus vannamei*. Aquaculture, 269, 259 - 264.

[26] Staykov, Y., Denev, S. & Spring, P. (2005). Influence of dietary mannan oligosaccharides (Bio-Mos®) on growth rate and immune function of common carp (*Cyprinus carpio* L). *European Aquaculture Society, Special Publication*, 35, 431-432.

[27] Culjak, V., Bogut, G., Has-Schon, E., Milakovic, Z. & Canecki, K. Effect of Bio-Mos on performance and health of juvenile carp. In: Nutrition and biotechnology in the feed

and food industries. In Alltech's 22nd annual symposium (suppl. 1—abstracts of posters presented), Lexington, KY, USA,vol 2006.

[28] Bogut, I., Milakovic, Z., Pavlicevic, J. & Petrovic, D. Effect of Bio-Mos on performance and health of European catfish, *Sillirus glanis*. In Proceedings of Alltech's 22nd Annual Symposium April 23-26, 2006,vol Lexington, KY, USA, 2006.

[29] Hanley, F., Brown, H. & J., C. First observations on the effects of mannan oligosaccharide added to the hatchery diets for warmwater Hybrid Red Tilapia. In: Nutritional Biotechnology in the Feed & Food Industries. Proceedings of Alltech's 11th Annual Symposium (Suppl. 1) (Abstracts of posters presented). Lexington, KY, May,vol 1995.

[30] Hossu, B., Salnur, S. & Gultepe, N. (2005). The effects of yeast derivatives (Bio-Mos®) on growth of Gilthead sea bream, *Sparus aurata*. In: Nutritional Biotechnology in the Feed & Food Industries: Proceedings of Alltech's 21st Annual Symposium (Suppl. 1) (Abstracts of posters presented). Lexington, KY, May 23-25.,

[31] Hossu, B., Salnur, S. & Gultepe, N. (2005). The effects of yeast derivatives (Bio-Mos®) on digestibility of Gilthead sea bream, *Sparus aurata*. In: Nutritional Biotechnology in the Feed & Food Industries: Proceedings of Alltech's 21st Annual Symposium (Suppl. 1) (Abstracts of posters presented). Lexington, KY, May 23-25,

[32] Pryor, G. S., Royes, J. B., Chapman, F. A. & Miles, R. D. (2003). Mannanoligosaccharides in fish nutrition: effects of dietary supplementation on growth and gastrointestinal villi structure in Gulf of Mexico sturgeon. *North American journal of aquaculture*, 65, 106 - 111.

[33] Mahious, A. S., Gatesoupe, F. J., Hervi, M., Metailler, R. & Ollevier, F. (2006). Effect of dietary inulin and oligosaccharides as prebiotics for weaning turbot, *Psetta maxima* (Linnaeus, C. 1758). *Aquaculture International*, 14, 219-229.

[34] Genc, M. A., Yilmaz, E., Genc, E. & Aktas, M. (2007). Effects of dietary mannan oligosaccharides (MOS) on growth, body composition, and intestine and liver histology of the hybrid Tilapia (*Oreochromis niloticus*×*O. aureus*) *Israel Journal of Aquaculture*, 59, 10 - 16.

[35] Grisdale-Helland, B., Helland, S. J. & Gatlin, D. M. (2008). The effects of dietary supplementation with mannanoligosaccharide, fructooligosaccharide or galactooligosaccharide on the growth and feed utilization of Atlantic salmon (*Salmo salar*). *Aquaculture*, 283, 163-167.

[36] Craig, S. R. & McLean, E. (2003). The effect of dietary inclusion of Bio-Mos® upon performance characteristics of Nile tilapia. In: Biotechnology in the Feed Industry: Proceedings of Alltech's 19th Annual Symposium (Suppl. 1) (Abstracts of posters presented). Lexington, KY, May 23-35.,

[37] Zhou, X. Q. & Li, Y. L. The effects of Bio-Mos on intestinal microflora and immune function of juvenile Jain Carp (*Cyprinus carpio* Var. Jian). In Alltech's 20th Annual Symposium,vol Lexington, KY, USA, 2004.

[38] Hooge, D. (2004). Metaanalysis of Broiler chicken pen trials evaluating dietary Mannan Oligosaccharide, 1993 - 2003. *Poultry Science*, 3, 163 - 174.

[39] Chisaka, H., Ueno, M. & Futaesaku, Y. (1999). Spines in the hindgut of the crayfish procambarus clarkii (Decapoda): Their distribution and correlation with hindgut muscles. *Journal of crustacean biology*, 19, 337 - 343.

[40] Ferna´ ndez, I., lvarez, L. H., Pardos, F. & Benito, J. (2002). Gut-Associated Cells of *Derocheilocaris remanei* (Crustacea, Mystacocarida). *Journal of morphology*, 251, 276 - 283.

[41] Mykles, D. L. (1979). Ultrastructure of alimentary epithelia of lobsters,*Homarus americanus* and *H. gammarus*, and crab,*Cancer magister*. *Zoomorphology*, 93, 201 - 215.

[42] Dimitroglou, A., Davies, S., Divanach, P. & Chatzifotis, S. The role of mannan oligosaccharide in gut development of white sea bream, *Diplodus sargus* In Proceedings of Alltech's 21st Annual Symposium vol Lexington, KY, 2005.

[43] Dimitroglou, A., Davies, S., Moate, R., Spring, P. & Sweetman, J. (2007). The beneficial effect of Bio-Mos on gut integrity and enhancement of fish health. Presented at Alltech's Technical Seminar Series held in Dublin, November 2007,

[44] Johansson, M. W. & Söderhäll, K. (1985). Exocytosis of the prophenoloxidase activating system from crayfish haemocytes. *Journal of comparative physiology*, 156, 175-181.

[45] Jackson, C. J. (1994). Effect of intramolt growth and sainity on the larval morphology of *Penaeus semisulcatus* de Haan (Decapode: Penaeoidea). *Journal of crustacean biology*, 14, 463-472.

[46] Johansson, M. W., Keyser, P., Sritunyalucksana, K. & Söderhäll, K. (2000). Crustacean haemocytes and haematopoiesis. *Aquaculture*, 191, 45 - 52.

[47] Terova, G., Forchino, A., Rimoldi, S., Brambilla, F., Antonini, M. & Saroglia, M. (2009). Bio-Mos®: An effective inducer of dicentracin gene expression in European sea bass (*Dicentrarchus labrax*). *Comparative Biochemistry and Physiology, Part B*, 153, 372 - 377.

In: Oligosaccharides: Sources, Properties and Applications ISBN 978-1-61122-193-0
Editor: Nicole S. Gordon

Chapter 10

Facile Synthesis of Unnatural Oligosaccharides by Phosphorylase-Catalyzed Enzymatic Glycosylations Using New Glycosyl Donors

Jun-ichi Kadokawa*
Graduate School of Science and Engineering, Kagoshima University, 1-21-40 Korimoto, Kagoshima 890-0065, Japan

Abstract

Unnatural oligosaccharides can be expected to exhibit new functions and applications in glycoscience such as potential as drug candidates. Enzymatic glycosylation is a useful tool for the preparation of oligosaccharides with well-defined structure. α-1,4-Glucosidic linkages can be prepared by an enzymatic polymerization through the successive phosphorylase-catalyzed glucosylation using α-D-glucose 1-phosphate as a glycosyl donor. Since enzymes often express loose specificity for recognition of substrate structures, extension of the phosphorylase-catalyzed glycosylation using different substrates is useful to obtain new unnatural oligosaccharides. In this chapter, on the basis of above viewpoints, the facile synthesis of unnatural oligosaccharides by the phosphorylase-catalyzed enzymatic glycosylations using new glycosyl donors of glycose 1-phosphates is described. It has been found that α-D-xylose, α-D-mannose, 2-deoxy-α-D-glucopyranose, 3- or 4-deoxy-α-D-glucose, and α-(*N*-formyl)-D-glucosamine 1-phosphates are recognized by phosphorylase as the glycosyl donor, to occur the transfer reaction of the corresponding sugar residues to maltooligosaccharides, giving the unnatural oligosaccharides. Consequently, construction of the new oligosaccharide chains containing the different units by the phosphorylase-catalysis probably leads to development on new applications of unnatural substrates, such as the drug candidate.

---

* Tel: +81-99-285-7743, Fax: +81-99-285-3253,
E-mail: kadokawa@eng.kagoshima-u.ac.jp

1. Introduction

Naturally occurring oligosaccharides in glycoproteins, glycolipids, and other glycoconjugates have important roles and functions in biological processes [1]. For example, differences in oligosaccharide structures are responsible for blood group activities, and are involved in ontogenesis and oncogenesis as differentiation antigens [2]. In addition to such natural oligosaccharides, unnatural oligosaccharide structures can be expected to show new functions and applications in glycoscience such as potential as drug candidates [3]. Particularly, approaches for the efficient synthesis of the unnatural building blocks for the assembly of complex oligosaccharides composed of multiple different monosaccharide units still receive significant attention. In order to supply such substrates with well-defined structures, highly selective glycosylations are in great demand, where a glycosyl donor and a glycosyl acceptor are reacted in the presence of a catalyst to form the glycosidic bond (Scheme 1) [4]. In the reactions, there are two important selectivities that should be controlled to form the desired oligosaccharide structures. Because there are two possible geometric isomers related to the geometry of the anomeric carbon atom of a monosaccharide, namely α-isomer and β-isomer, the control of the formation in such two glycosidic bonds, i.e., stereoselectivity, is one of the two important selectivities in the glycosylation. The other prerequisite selectivity is regioselectivity. Monosaccharides have multiple hydroxy groups that can participate in glycosyl bond formation. Since oligosaccharides can be formed by connecting a hydroxy group at anomeric position in a monosaccharide unit and one of other hydroxy groups of the adjacent monosaccharide unit, there are many possibilities for the regioselective formation of the glycosidic bonds.

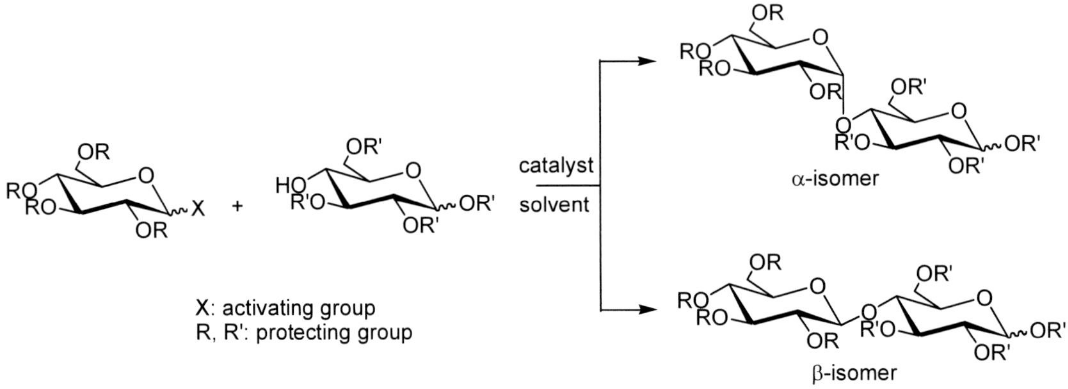

Scheme 1. Principle of glycosylation.

Enzymatic glycosylation is a very powerful tool for the stereo- and regioselective construction of glycosidic bonds under mild conditions, where a glycosyl donor and a glycosyl acceptor can be employed in their unprotected forms, leading to the direct formation of unprotected saccharide chains in aqueous media [5]. Enzymes involved in the formation of glycosidic bonds are categorized into three main classes; hydrolytic enzymes, phosphorolytic enzymes, and synthetic enzymes (Scheme 2). The phosphorolytic enzymes (phosphorylases) catalyze phosphorolytic cleavage of glycosidic bonds in saccharide chains in the presence of inorganic phosphate to give monosaccharide 1-phosphate (glycose 1-phosphate) and the saccharide chains with one smaller degree of polymerization (DP) [6]. Since the bond energy of the produced phosphate is comparable with that of the glycosidic bond, the phosphorylase-catalyzed reactions have reversible nature. Therefore, the phosphorylases can be employed in the practical synthesis of saccharide chains via glycosylations. In such phosphorylase-catalyzed glycosylations, the glycose 1-phsophates are used as glycosyl donors and the glycose unit is transferred from the substrate to a non-reducing end of a glycosyl acceptor to form the stereo- and regio-controlled glycosidic bond accompanied with the production of inorganic phosphate.

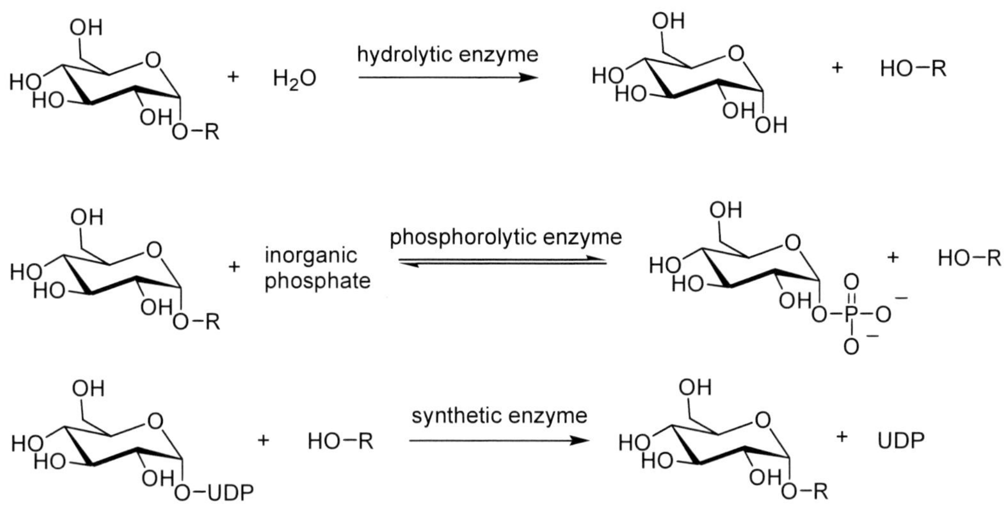

Scheme 2. Enzymes involved in the formation of glycosidic bonds.

Phosphorylases are generally classified by the anomeric forms of the glycosidic bonds in the substrates that are phosphorolyzed or by the anomeric forms of the glycose 1-phosphates that are produced. The other way used to classify the phosphorylases is according to describing them in terms of the anomeric retention or inversion process in the reactions. All of the phosphorylases reported to date catalyze an exowise phosphorolysis at the non-reducing end of the glycosidic bond. The regiospecificities of the phosphorylases are considered to be very strict and they generally phosphorolyze only their specific type of glycosidic linkage. The strict specificities are important in the exploitation of the phosphorylases in the synthesis of oligosaccharides by means of their reverse reactions.

Of the phosphorylases, the phosphorylase EC 2.4.1.1 (glycogen phosphorylase, starch phosphorylase, or α-glucan phosphorylase) is the most extensively studied and is found in animals, plants, and microorganisms [7]. Because this enzyme is often called as simply 'phosphorylase,' the followings in this chapter express it as just 'phosphorylase.' The role of phosphorylase is considered to be in the utilization of storage polysaccharides (glycogen or starch) in the glycolytic pathway. This enzyme catalyzes the reversible phosphorolysis of α-1,4-glucans at the non-reducing end, such as glycogen and starch, in the presence of inorganic phosphate, giving an α-D-glucose 1-phosphate (Glc-1-P) (Scheme 3). By means of the reversibility of the reaction, α-1,4-glucosidic bond can be prepared by the phosphorylase-catalyzed α-glucosylation using Glc-1-P as a glycosyl donor [8]. As a glycosyl acceptor in the glucosylation, which is often called 'primer,' maltooligosaccharides with the DPs higher than the smallest one that is recognized by phosphorylase are used. The smallest glycosyl acceptor for the phosphorylase-catalyzed α-glucosylation is typically maltotetraose (Glc_4), whereas that for the phosphorolysis is maltopentaose (Glc_5). In the glucosylation, a glucose unit is transferred from Glc-1-P to a non-reducing end of the primer to form an α-1,4-glucosydic bond. Then, the successive reactions in the same manner take place as a propagation of polymerization to produce the α-1,4-glucan chain, i.e., amylose. The molecular weight of the amylose thus produced has a narrow distribution ($M_w/M_n < 1.2$) and can be controlled by the Glc-1-P/primer molar ratio [9].

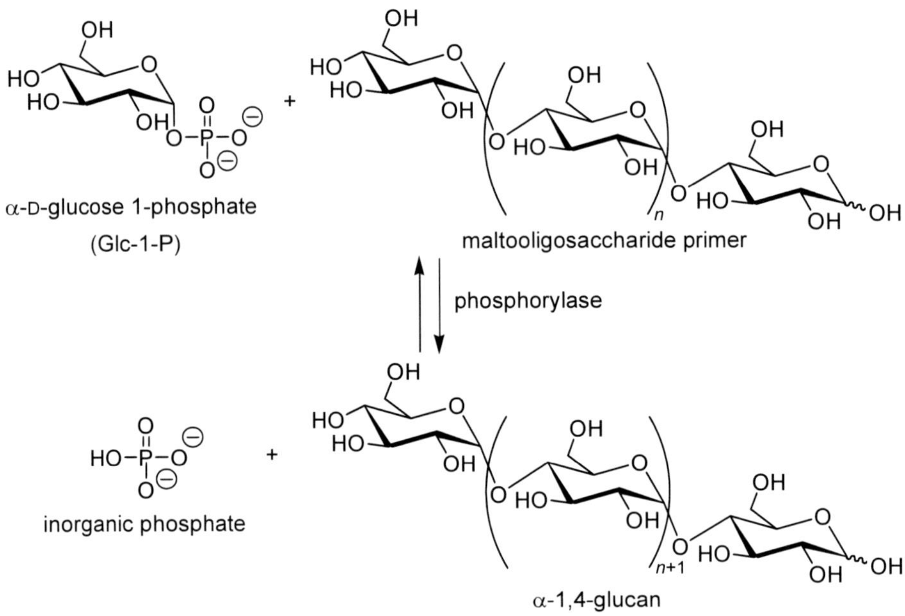

Scheme 3. Phosphorylase-catalyzed phosphorolysis and glucosylation.

Because enzymes often have loose specificity for recognition of the substrate structure, extension of the enzymatic glycosylation using different substrates is useful to obtain unnatural saccharide chains. On the basis of this viewpoint, in this chapter, the synthesis of unnatural oligosaccharides by the phosphorylase-catalyzed glycosylations using different glycose 1-phosphates from a native substrate of Glc-1-P is reviewed.

3. Synthesis of α-D-Mannosylated, α-D-Xylosylated, and α-Deoxy-D-Glucosylated Maltooligosaccharides by Phosphorylase Catalysis

The enzymatic synthesis of α-D-xylosylated maltooligosaccharides by the phosphorylase-catalyzed α-xylosylation using α-D-xylose 1-phosphate (Xyl-1-P) was reported (Scheme 4) [10]. Because the structural difference of Xyl-1-P from Glc-1-P was only the absence of a CH_2OH group at a position 6, a highly possibility for recognition of this nonnative substrate by phosphorylase was supposed. When the enzymatic reaction using Xyl-1-P as a glycosyl donor and Glc_4 as a glycosyl acceptor was carried out catalyzed by phosphorylase, xylosylated oligosaccharides were produced, that was confirmed by the MALDI-TOF MS and ^{1}H NMR spectra of the crude products. Furthermore, the MALDI-TOF MS spectrum showed small peaks assignable to the oligosaccharides consisting of two xylose units in addition to the main peaks ascribable to the oligosaccharides having one xylose unit. However, the above analytical data did not provide sufficient evidence to determine the structures of the products, in which the xylose unit was positioned at the non-reducing end of the xylosylated oligosaccharides. Consequently, the treatment of the reaction mixture with glucoamylase (GA, EC 3.2.1.3), which catalyzed an exowise hydrolysis at the non-reducing end of α-1,4-glucans, to reveal whether the xylose unit was positioned at the non-reducing end. The ^{1}H NMR spectrum of the treated products indicated remaining the signals due to the xylosidic bonds, supporting that the xylose unit was positioned at non-reducing end. Furthermore, the main product was isolated from the crude mixture using high performance liquid chromatography (HPLC) and the structure was confirmed by the ^{1}H NMR spectrum to be α-D-xylosyl-1,4-Glc_4. For further analysis, the formation of the xylosylated maltooligosaccharides vs. reaction time in the phosphorylase-catalyzed α-xylosylation was evaluated under the conditions as the feed molar ratios of Glc_4 to Xyl-1-P = 1 : 10, 1 : 5, and 1 : 1. The total yields of the xylosylated products were calculated by the integrated ratios of the signal due to the anomeric protons of the α-xyloside units to the signals due to those of the reducing end in the ^{1}H NMR spectra of the crude products. Highest yields of 44% (1 : 10), 25% (1 : 5), and 10% (1 : 1) based on the amounts of Glc_4 used were obtained after 48 h. Schwarz et al. has reported the kinetic analysis using Glc-1-P and Xyl-1-P in the phosphorylase-catalyzed reaction. Phosphorylase displayed a very large kinetic selectivity ratio, i.e., $k_{catGlc1P}/k_{catXyl1P}$ = 20000 [11].

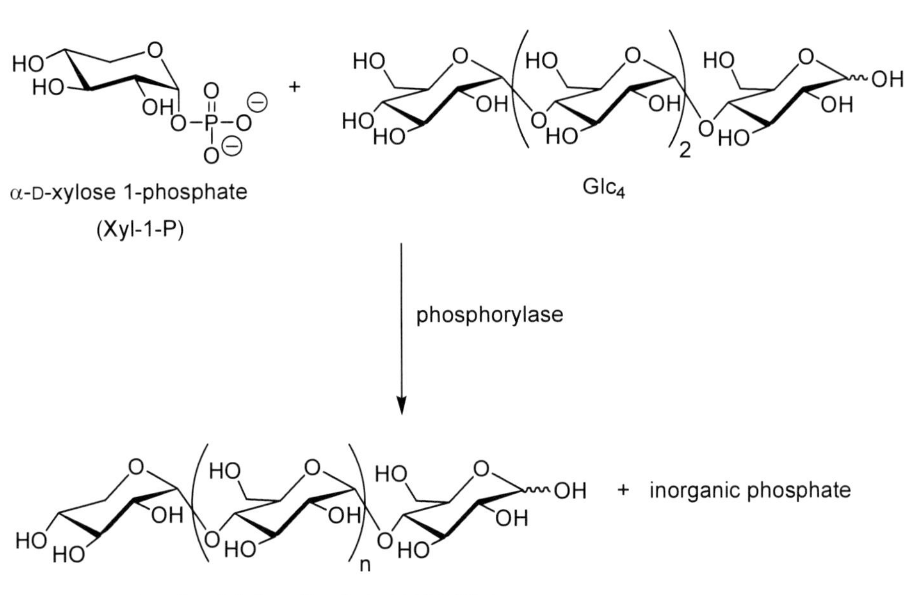

Scheme 4. Phosphorylase-catalyzed α-xylosylation of Glc_4 using Xyl-1-P.

The phosphorylase-catalyzed α-mannosylation using α-D-mannose 1-phosphate (Man-1-P) was also examined (Scheme 5) [12]. Although the equatorial hydroxy group is not essential for the enzyme, the 2-epimer of Glc-1-P, i.e., Man-1-P is expected to have a low affinity to the enzyme due to steric hindrance by the axial hydroxy group. In the crude products obtained by the phosphorylase catalyzed reaction of Man-1-P with Glc_4, mannosylated compounds were found. A pentasaccharide fraction was isolated by size exclusion chromatography and the structures were confirmed by the ^{1}H NMR spectrum to be a mixture of Glc_5 and α-D-mannosyl-1,4-Glc_4. This mixture could not be separated completely. Particularly, the position of the mannose unit was assigned due to the lower intensity of the terminal H-4 signal in comparison with the integral of the α- and β-H-1 at the reducing end.

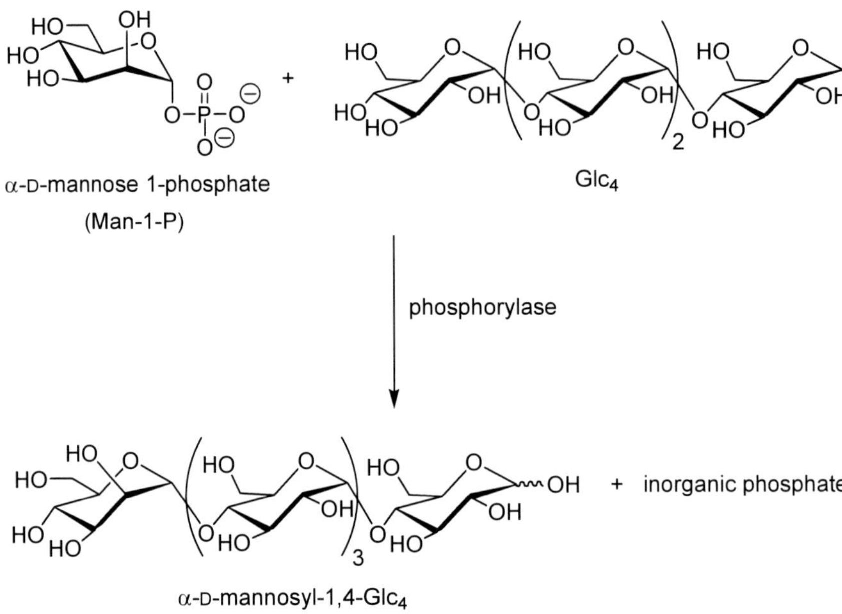

Scheme 5. Phosphorylase-catalyzed α-mannosylation of Glc_4 using Man-1-P.

Withers et al. enzymatically prepared 2-deoxy-α-D-glucose 1-phosphate (dGlc-1-P) in a two step process according to Scheme 6 [13]. First, a 2-deoxy-D-glucose unit is transferred to the α-glucan primer that is catalyzed by inorganic phosphate. In the second step, 2-deoxy-D-glucose is released by phosphorolysis to yield dGlc-1-P and in the overall reaction the primer remains unchanged. On the basis of this study, 1,2-dideoxy-D-glucose (D-glucal) was applied as a glycosyl donor in considerable excess in order to shift the equilibrium to the chain-elongation with a glucal : primer ratio of 15 : 1 for occurrence of 2-deoxy-α-glucosylation in the presence of inorganic phosphate (Scheme 7) [14]. After the phosphorylase-catalyzed reaction in the presence of D-glucal, Glc_4, and only 0.05 equiv of phosphate for 6 h, 2-deoxy-α-D-glucosylated penta-, hexa-, and heptasaccharides were separated by size exclusion chromatography in 17, 12, and 8% yields, respectively. Additionally, a fraction of higher molecular weight with an average DP of 12 was obtained in 33% yield.

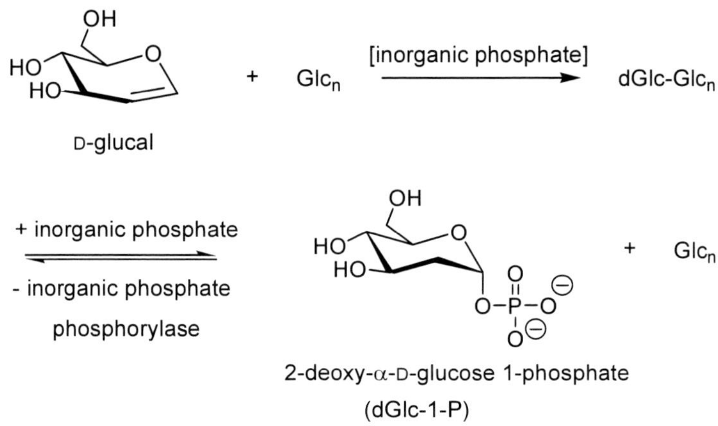

Scheme 6. Two-step synthesis of dGlc-1-P in the presence of inorganic phosphate and phosphorylase.

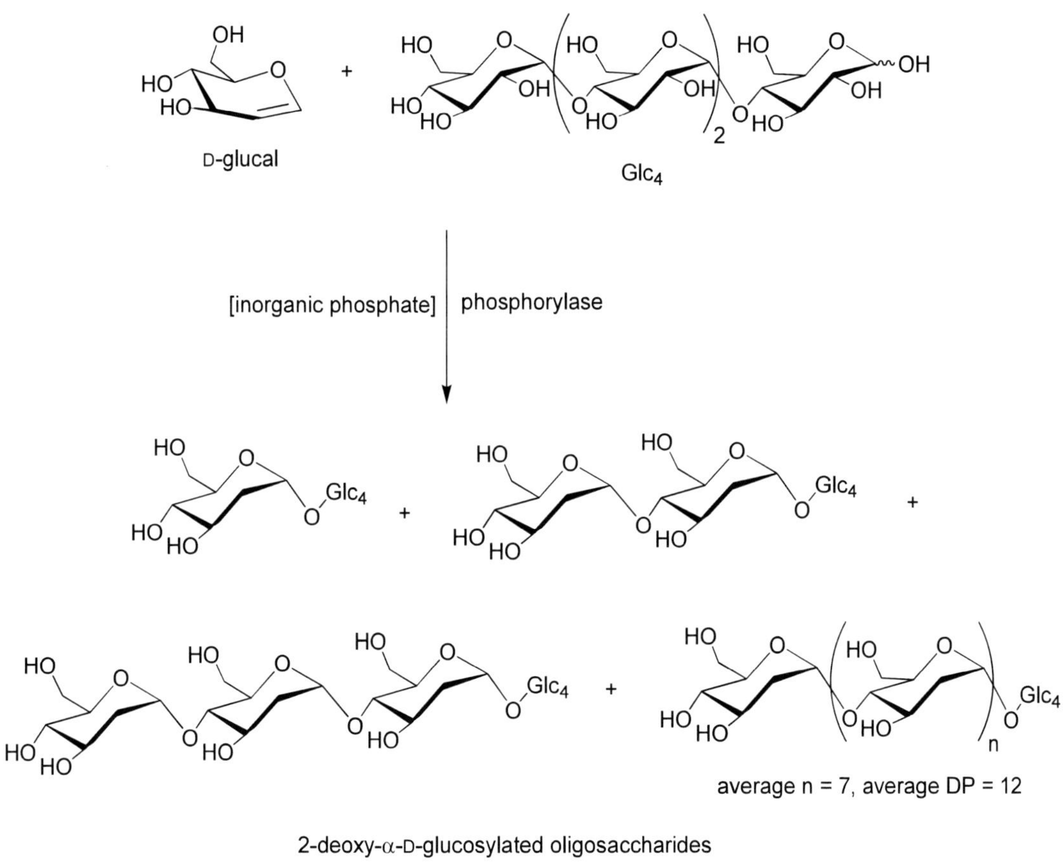

Scheme 7. Phosphorylase-catalyzed synthesis of 2-deoxy-α-D-glucosylated oligosaccharides.

Withers also reported the phosphorylase-catalyzed chain-elongation of glycogen with 3- or 4-deoxy-α-D-glucose 1-phosphates. However, only averages of up to 1.5 units were transferred [15].

4. Synthesis of α- D-Glucosaminylated Maltooligosaccharides by Phosphorylase Catalysis

Oligosaccharides containing 2-amino-2-deoxy-D-glucopyranose (D-glucosamine, GlcN) units and its derivatives, such as 2-acetamido-2-deoxy-D-glucopyranose (N-acetyl-D-glucosamine, GlcNAc), serve key functions in living organisms such as in cell-cell recognition and immune responses. The preparation of saccharide chains containing 2-amino-2-deoxysugar residues, therefore, has been frequently required for the various studies in glycoscience. Much effort has been focused on glycosylation using glycosyl donors derived from GlcNAc and other N-substituted GlcN residues, such as oxazoline glycosylation [16]. However, the glycosylation of a GlcN donor with a free amino group had hardly been achieved. On the basis of these considerations, it was reported that 2-amino-2-deoxy-α-D-glucopyranose 1-phosphate (GlcN-1-P) was recognized as the glycosyl donor in the phosphorylase-catalyzed glycosylation to form α-1,4-glucosaminyl bond [17]. This was a first example of the enzymatic α-glucosaminylation using the GlcN donor with a free amino group. In contrast, The GlcN unit was enzymatically incorporated into chitooligosaccharides linked via the β-1,4-configration by bovine β-1,4-galactosyltransferase (EC 3.4.1.38) [18].

The phosphorylase-catalyzed α-glucosaminylation was performed using GlcN-1-P as a glycosyl donor and Glc_4 as a glycosyl acceptor (Scheme 8). After the reaction mixture was lyophilized, N-acetylation was carried out using acetic anhydride, and the transfer of a GlcN unit to the primer was evaluated by MALDI-TOF MS measurement. Because the difference in the molecular masses of the anhydroglucose and anhydroglucosamine units was only 1 (162 and 161, respectively), which could be made larger by the N-acetylation of the latter unit, the measurement was performed on the N-acetylated material. In the MALDI-TOF MS spectrum of the N-acetylated crude products, a significant peak corresponding to the mass of a pentasaccharide containing one GlcNAc unit was observed. This data indicated that one GlcN unit was transfer from GlcN-1-P to Glc_4 by the phosphorylase-catalyzed α-glucosaminylation. To confirm the presence of the GlcN unit at a non-reducing end of the produced oligosaccharide, the treatment of the N-acetylated crude products with GA was performed. In the MALDI-TOF MS spectrum of the treated products, the peak assigned to the molecular mass of the glucosaminylated Glc_4 remained intact, supporting that the GlcNAc unit was positioned at the non-reducing end. If the transfer of one GlcN residue from GlcN-1-P to Glc_4 proceeded once, further glucosaminylation was probably suppressed because the glucosaminylated Glc_4 was not recognized as the acceptor by phosphorylase. The main product was isolated from the N-acetylated crude mixture after the treatment with GA using HPLC. The ^{1}H NMR spectrum of the isolated material fully supported the structure of N-acetyl-α-D-glucosaminyl-1,4-Glc_4. In particular, there was no signal assigned to the H-4 position of the glucose residue at the non-reducing end of Glc_4, whereas the signal ascribed to the free H-4 position of GlcNAc was observed. This observation indicated that the GlcNAc unit was positioned at the non-reducing end bound with the α-1,4-linkage.

Although the phosphorylase-catalyzed glycosylation of Glc_4 using 2-acetamido-2-deoxy-α-D-glucopyranose 1-phosphate (GlcNAc-1-P) as a glycosyl donor was also performed under the same conditions as those using GlcN-1-P, the MALDI-TOF MS spectrum of the crude products did not show peaks assignable to the molecular masses of oligosaccharides having a GlcNAc unit. This result indicated that the GlcNAc-1-P was not recognized by phosphorylase, probably because the bulky acetamido group in GlcNAc-1-P blocked approach to the active site.

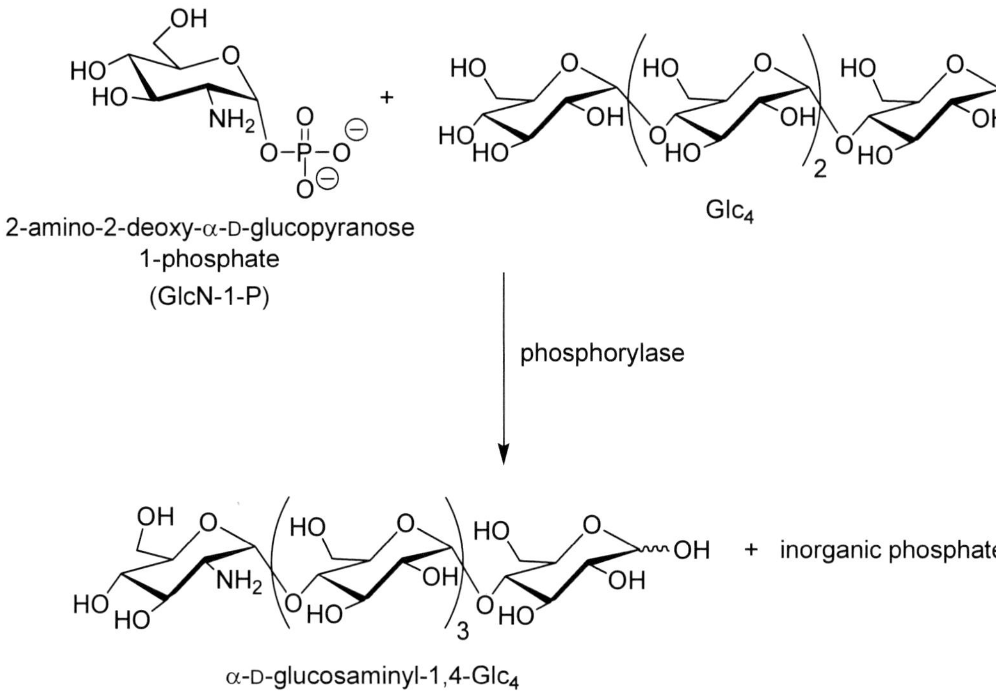

Scheme 8. Phosphorylase-catalyzed α-glucosaminylation of Glc_4 using GlcN-1-P.

On the other hand, the following paper reported that 2-deoxy-2-formamido-α-D-glucopyranose 1-phosphate (GlcNF-1-P) [19], which had a smaller substituent, i.e., a formamide, was recognized as a glycosyl donor by phosphorylase. This allowed the N-formyl-α-glucosaminylation of maltooligosaccharides to give N-formyl-α-D-glucosaminylated maltooligosaccharides (Scheme 9). The MALDI-TOF MS spectrum of the crude products obtained by the phosphorylase-catalyzed N-formyl-α-glucosaminylation using GlcNF-1-P and Glc_4 (5 : 1) showed only a significant peak corresponding to the mass of a pentasaccharide containing one GlcNF unit. This data indicated that the transfer of one GlcNF residue from GlcNF-1-P to Glc_4 occurred. Then, the treatment of the crude products with GA was carried out to reveal whether the GlcNF unit was positioned at the non-reducing end. In the MALDI-TOF MS spectrum of the treated products, the peak assigned to the molecular mass of the produced pentasaccharide remained intact, supporting that the GlcNF unit was positioned at the non-reducing end. Moreover, the pentasaccharide was isolated from the crude products after the treatment with GA. The ^{1}H NMR of the isolated material fully supported the structure of N-formyl-α-D-glucosaminyl-1,4-Glc_4. The two signals due to the formyl group was observed, which were assigned to *E*- and *Z*-isomers owing to hindered rotation of the formyl C-N bond (two rotamers). For further analysis, the formation of N-formyl-α-D-glucosaminylated oligosaccharides vs. reaction time in the phosphorylase-catalyzed glycosylation using GlcNF-1-P and Glc_4 (5 : 1) was evaluated and compared with that using GlcNF-1-P as well as Glc-1-P under the same conditions . The total yields of the glycosylated products were calculated on the basis of the amounts of inorganic phosphate produced from the glycosyl donors by the phosphorylase-catalyzed glycosylations, which were measured by a modified Fiske-Subbarow assay [20]. A reaction time of 4 days in the glycosylation using GlcNF-1-P gave 33% yield of the products based on the amount of Glc_4, whereas the yield was 72% in the glycosylation using GlcN-1-P. These data indicated that phosphorylase recognized GlcN-1-P more efficiently than GlcNF-1-P. However, the reaction using Glc-1-P was much faster than such two reactions and the yield reached almost 100% in 45 min, because Glc-1-P is the native substrate for phosphorylase.

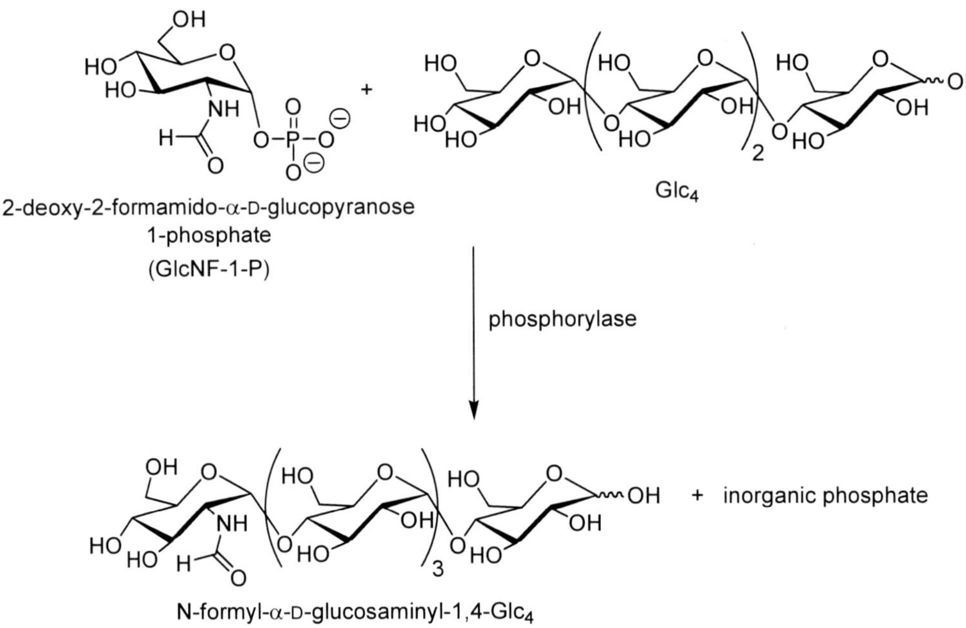

Scheme 9. Phosphorylase-catalyzed N-formyl-α-glucosaminylation of Glc_4 using GlcNF-1-P.

The phosphorylase-catalyzed N-formyl-α-glucosaminylation using GlcNF-1-P was also performed in the presence of maltotriose (Glc_3) or Glc_5. In the MALDI-TOF MS spectrum of the crude products using Glc_3, no peaks assignable to the molecular masses of oligosaccharides having a GlcNF unit were observed, indicating no occurrence of N-formyl-α-glucosaminylation of Glc_3 by GlcNF-1-P. Because the smallest glycosyl acceptor for the phosphorylase-catalyzed glycosylation was Glc_4, Glc_3 was not recognized by phosphorylase. In the MALDI-TOF MS spectrum of the crude products using Glc_5, on the other hand, several peaks separated by $m/z = 162$ were observed, which corresponded to the molecular masses of pentasaccharide – octasaccharide containing a GlcNF unit. This finding indicated the occurrence of both the N-formyl-α-glucosaminylation by GlcNF-1-P and α-glucosylation by Glc-1-P when Glc_5 was used as the glycosyl acceptor (Scheme 10). Because Glc_5 is the smallest substrate for phosphorolysis in the presence of inorganic phosphate, Glc-1-P was possibly produced by phosphorolysis of Glc_5 with simultaneously production of Glc_4 at the early stage of the reaction, where inorganic phosphate was formed by the transfer of a GlcNF unit from GlcNF-1-P to Glc_5. Then, Glc-1-P was recognized more efficiently by phosphorylase than GlcNF-1-P. Thus, the maltooligosaccharides with larger DPs such as Glc_6 and Glc_7 were produced by the transfer of the glucose residue from Glc-1-P to Glc_4 or Glc_5.

Furthermore, if N-formyl-α-glucosaminylation of the produced maltooligosaccharides by GlcNF-1-P proceeded once, subsequent glycosylation was suppressed because the N-formyl-α-D-glucosaminylated oligosaccharides were less efficiently recognized by phosphorylase.

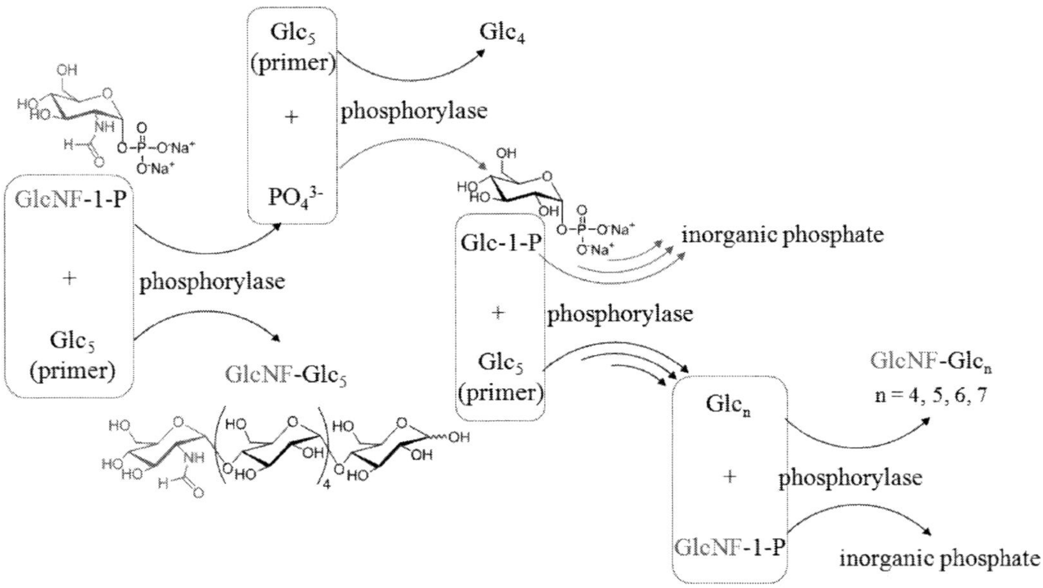

Scheme 10. Plausible pathway in phosphorylase-catalyzed N-formyl-α-glucosaminylation of Glc_5.

Conclusion

This chapter reviewed the synthesis of unnatural oligosaccharides by the phosphorylase-catalyzed glycosylation using different glycosyl donors from Glc-1-P, a native glycosyl donor. Although all the glycosylations described here proceeded with controlling stereo- and regioselectivities to produce unnatural oligosaccharides with well-defied structures, development of new glycosyl donors for the phosphorylase catalysis has not been studied in as much detail as other enzymes. Because it has recently been much attention and importance to provide new unnatural oligosaccharide substrates in the fundamental and application research fields of glycoscience and biotechnology, however, this type of study presented here will necessarily be developed much more. This review concludes with the further expectation that unnatural oligosaccharides as described herein will exhibit important applications in the future.

References

[1] (a) Sharon, N. & Lis, H. (1993). Carbohydrates in cell recognition. *Sci. Am.*, *268*, 82-89. (b) Sharon, N. & Lis, H. (1993). Protein glycosylation - structural and functional-aspects. *Eur. J. Biochem.*, *218*, 1-27. (c) Kobata, A. (1993). Glycobiology - an expanding research area in carbohydrate-chemistry. *Acc. Chem. Res.*, *26*, 319-324. (d) Varki, A. (1993). Biological roles of oligosaccharides - all of the theories are

correct. *Glycobiology*, *3*, 97-130. (e) Dwek, R. (1996). Glycobiology: Toward understanding the function of sugars. *Chem. Rev.*, *96*, 683-720.

[2] (a) Feizi, T. & Childs, R. A. (1985). Carbohydrate structures of glycoproteins and glycolipids as differentiation antigens, tumour-associated antigens and components of receptor systems. *Trends Biochem. Sci.*, *10*, 24-29. (b) Hounsell, E. F. (1987). Tate and Lyle Lecture. structural and conformational characterization of carbohydrate differentiation antigens. *Chem. Soc. Rev.*, *16*, 161-185.

[3] Stick, R. V. Carbohydrates: The sweet molecules of life. London: Academic Press; 2001.

[4] Paulsen, H. (1982). Advances in selective syntheses of complex oligosaccharides. *Angew. Chem. Int. Ed. Engl.*, *21*, 155-173.

[5] Kobayashi, S., Uyama, H., & Kimura, S. (2001). Enzymatic polymerization. *Chem. Rev.*, *101*, 3793-3818. (b) Shoda, S., Izumi, R., & Fujita, M. (2003). Green process in glycotechnology. *Bull. Chem. Soc. Jpn.*, *76*, 1-13. (c) Seibel, J., Jördening, H. –J., & Buchholz, K. (2006). Glycosylation with activated sugars using glycosyltransferases and transglycosidases. *Biocatal. Biotransform.*, *24*, 311-342. (d) Kobayashi, S. & Makino, A. (2009). Enzymatic polymer synthesis: An opportunity for green polymer chemistry. *Chem. Rev.*, *109*, 5288-5253. (e) Kadokawa, J. & Kobayashi, S. (2010). Polymer synthesis by enzymatic catalysis. *Curr. Opi. Chem. Biol.*, *14*, 145-153.

[6] Kitaoka, M. & Hayashi, K. (2002). Carbohydrate-processing phosphorolytic enzymes. *Trends Glycosci. Glycotechnol.*, *14*, 35-50.

[7] (a) Fujii, K., Takata, H., Yanase, M., Terada, Y., Ohdan, K., Takaha, T., Okada, S., & Kuriki, T. (2003). Bioengineering and application of novel glucose polymers. *Biocatal. Biotransform.*, *21*, 167-172. (b) Yanase, M., Takaha, T., & Kuriki, T. (2006). α-Glucan phosphorylase and its use in carbohydrate engineering. *J. Sci. Food Agric.*, *86*, 1631-1635.

[8] Ziegast, G. & Pfannemüller, B. (1987). Phosphorolytic syntheses with di-, oligo- and multi-functional primers. *Carbohydr. Res.*, *160*, 185-204.

[9] Takata, H., Takaha, T., Okada, S., Takagi, M., & Imanaka, T. (1998). Purification and characterization of α-glucan phosphorylase from *Bacillus stearothermophilus*. *J. Fermnet. Bioeng.*, *85*, 156-161.

[10] Nawaji, M., Izawa, H., Kaneko, Y., & Kadokawa, J. (2008). Enzymatic synthesis of α-D-xylosylated maltooligosaccharides by phosphorylase-catalyzed xylosylation. *J. Carbohydr. Chem.*, *27*, 214-222.

[11] Schwarz, A., Pierfederici, F. M., & Nidetzky, B. (2005). Catalytic mechanism of α-retaining glucosyl transfer by *Corynebacterium callunase* starch phosporylase: the role of histidine-334 examined through kinetic characterization of site-directed mutants. *Biochem. J.*, *387*, 437-445.

[12] Evers, B. & Thiem, J. (1997). Further syntheses employing phosphorylase. *Bioorg. Med. Chem.*, *5*, 857-863.

[13] Percival, M. D. & Withers, S. G. (1988). Application of enzymes in the synthesis and hydrolytic study of 2-deoxy-α-D-glucopyranosyl phosphate. *Can. J. Chem.*, *66*, 1970-1972.

[14] Evers, B., Mischnick, P., & Thiem, J. (1994). Synthesis of 2-deoxy-α-D-*arabino*-hexopyranosyl phosphate and 2-deoxy-maltooligosaccharides with phosphorylase. *Carbohydr. Res.*, *262*, 335-341.

[15] Withers, S. G. (1990). The enzymic synthesis and N.M.R. characterization of specifically deoxygenated and fluorinated glycogens. *Carbohydr. Res.*, *197*, 61-73.

[16] (a) Kobayashi, S., Ohmae, M., Fujikawa, S., & Ochiai, H. (2005). Enzymatic precision polymerization for synthesis of glycosaminoglycans and their derivatives. *Macromol. Symp.*, *226*, 147-156. (b) Ohmae, M., Fujikawa, S., Ochiai, H., & Kobayashi S. (2006). Enzyme-catalyzed synthesis of natural and unnatural polysaccharides. *J. Polym. Sci., Part A: Polym. Chem.*, *44*, 5014-5027.

[17] Nawaji, M., Izawa, H., Kaneko, Y., & Kadokawa, J. (2008). Enzymatic α-glucosaminylation of maltooligosaccharides catalyzed by phosphorylase. *Carbohydr. Res.*, *343*, 2692-2696.

[18] Atkinson, E. M., Palcic, M. M., Hindsgaul, O., & Long, S. R. (1994). Biosyntehsis of *Rhizobium meliloti* lipooligosaccharide Nod factors: NodA is required for an N-aceltransferase activity. *Proc. Natl. Acad. Sci. USA*, *91*, 8418-8422.

[19] Kawazoe, S., Izawa, H., Nawaji, M., Kaneko, Y., & Kadokawa, J. (2010). Phosphorylase-catalyzed N-formyl-a-glucosaminylation of maltooligosaccharides. *Carbohydr. Res.*, *345*, 631-636.

[20] (a) Fiske, C. H. & Subbarow, Y. (1925). The colorimetric determination of phosphorus. J. Biol. Chem., 66, 375-400. (b) Saheki, S., Takeda, A., & Shimazu, T. (1985). Assay of inorganic phosphate in the mild pH range, suitable for measurement of glycogen phosphorylase activity. *Anal. Biochem.*, *148*, 277-281.

In: Oligosaccharides: Sources, Properties and Applications ISBN 978-1-61122-193-0
Editor: Nicole S. Gordon

Chapter 11

OLIGOSACCHARIDES: SOURCES, PROPERTIES AND APPLICATIONS

Kaoshan Chen and Yungui Bai
Laboratory of Biochemistry and Molecular Biology. School of Life Science, Shandong University , 27 Shanda Nanlu, Jinan 250100,China

ABSTRACT

It has been demonstrated that oligosaccharides have diverse bioactivities and functions. In plants, oligosaccharides act as early signal molecules and play an important role in the plant growth, development and morphogenesis as well as plant defense responses. Oligosaccharides are characterized as elicitors to trigger plant systemic acquired resistance against a variety of pathogens. Furthermore, oligosaccharides have been used to restore and maintain the intestinal balance by stimulating the growth and reproduction of probiotic strains such as *Lactobacillus* and *Bifidobacterium* and restraining the infection of pathogenic bacteria. In murine and human, the immunemodulatory activities of oligosaccharides which can stimulate the proliferation of splenocytes and the activation of macrophages are confirmed by many studies. Meanwhile, oligosaccharides significantly induce the accumulation of immune regulation factors and inhibit the growth of cancer cell, which act as assistant agents in treating immune system diseases such as hepatitis and cancer. Oligosaccharides also can regulate blood lipid, blood sugar and blood pressure. Therefore, these biological properties of oligosaccharides have made great contribution in the application of these oligosaccharides in agriculture, functional food and pharmaceutical products.

INTRODUCTION

Oligosaccharide is a linear or branched chain saccharide polymer typically containing two to ten units of monosaccharides, linked together by glycosidic bonds. Oligosaccharides are widely found in plants, animals and microorganisms. In addition to natural extraction, oligosaccharides can also be obtained by enzymolysis, chemical degradation and chemical synthesis. Oligosaccharide is classified into homooligosaccharide comprised of a single kind of monosaccharide and hetero-oligosaccharide comprised of different monosaccharides. It can also be classified into reducing oligosaccharide containing hemiacetal hydroxyl and non-reducing oligosaccharide without hemiacetal hydroxyl. Oligosaccharides with biological activity are called functional oligosaccharides. As an important active material in living organisms, oligosaccharide has advantages of small molecular weight, low polymerization degree, simple structure and high solubility. In the recent years, research work of functional oligosaccharides is mainly focused on fructooligosaccharide, lactulose, soybean oligosaccharide, galactooligosaccharide, chitooligosaccharide, xylooligosaccharide, lactosucrose and so on. This review article gives the full details on functional oligosaccharides (called oligosaccharides in short hereinafter). The properties of oligosaccharides and their application are described.

The Role of Oligosaccharides in Microorganisms

Inhibitory Effect on Bacterial Growth

Many studies have examined the effect of oligosaccharides on the growth of bacteria. It was shown that oligosaccharides such as chitooligosaccharides, fructooligosaccharide and galactooligosaccharide were very effective for bacterial inhibition. You-Jin Jeon et al. (2001) investigated the bactericidal effects of chitooligosaccharides against various microorganisms including four Gram-negative, five Gram-positive and four kinds of lactic acid bacteria. The chitooligosaccharides showed an antimicrobial activity at varying degrees for all the microorganisms examined [1]. Christakopoulos et, al (2003) also obtained similar results [2].

Promote the Proliferation of Intestinal Bacteria

Various intestinal bacteria inhabit the intestines of humans and animals, forming an intestinal microflora. Bifidobacteria are the most predominant bacteria in the intestinal flora of infants. Both the number and species of bifidobacteria change in adults. In order to maintain human health, bifidobacterium bacteria should occupy a dominant position in intestinal flora. Studies have shown that ingestion of oligosaccharides may benefit the proliferation of intestinal bacterial species and suppress the growth of harmful bacteria such as Clostridium perfringens. For example, fructooligosaccharide (FOS) promotes the growth of bifidobacteria in vivo [3], and xylooligosaccharide (XOS) could be extensively metabolized by several species of bifidobacteria [4].

Antiviral Activity

Sulfated oligosaccharides have potent anti-human immunodeficiency virus (HIV) activities. Katsuraya et al. (1999) synthesized sulfonated laminara oligosaccharide glycosides and examined the effects of the number of glucose residues and the alkyl chain-length on anti-HIV activity [5]. It was concluded that the anti-HIV activity of sulfated laminara oligosaccharides increased with increasing degree of sulfation.

The Role of Oligosaccharides in Plants

Oligosaccharides derived from plant and microbial cell wall polysaccharides are important regulatory molecules in plants pathogen defense, growth, development and morphogenesis. These oligosaccharides with regulatory activities are referred to as oligosaccharins [6-7].

Plants have the ability to initiate various defense reactions such as the production of phytoalexins, antimicrobial proteins, reactive oxygen species, and reinforcement of cell walls when they are infected by pathogens such as fungi, bacteria and viruses. Antimicrobial phytoalexins could be synthesized and accumulated in response to microbial attack. The term "elicitors" was commonly used for molecules that stimulate any plant defence mechanism. Oligosaccharins are one class of elicitors which can induce defence responses such as activation of defense-related genes, synthesis of phytoalexins and accumulation of pathogenesis-related (PR) proteins at a very low concentration e.g. nM [8]. Application of oligosaccharins provides a feasible strategy for disease control in agriculture.

Oligosaccharides From Fungal Cell Wall

β-glucan Oligosaccharides

Oligosaccharide elicitors derived from the β-glucans of pathogenic Olmycetes, such as Phytophthora sojae, have been characterized. It was shown that a doubly-branched hepta-β-glucoside which originated from P. sojae glucan by partial acid hydrolysis was an active elicitor for glyceollin biosynthesis in soybean cotyledon cells [9]. From detailed studies with a set of naturally obtained and synthetic oligosaccharides, it has been indicated that linear β-1, 3-linked glucooligosaccharides were active elicitors. Yamaguchi et al. (2000) reported that a elicitor-active pentasaccharide purified from the rice blast fungus (Magnaporthe grisea) could induce phytoalexin biosynthesis in rice cells at 10 nM concentrations [10].

Chitin and Chitosan Oligosaccharides

Chitin is a common component of fungal cell walls, and the fragments of chitin, N-acetylchitooligosaccharides, have been shown to function as elicitor signals in several plant systems, for example, induction of lignification in wheat [11], ion flux and protein phosphorylation in cultured tomato cells [12], chitinase activity in melon [13], and gene expression of glucanase in cultured barley cells [14]. In suspension-cultured rice cells, the action of chitin fragments has been extensively studied. It was shown that chitin fragments can induce the biosynthesis of terpenoid phytoalexins [15-16], membrane depolarization [17-18], K^+ and Cl^- efflux, cytoplasmic acidification [19], generation of reactive oxygen species

[20], biosynthesis of jasmonic acid [21] and expression of unique early responsive genes and typical defence related genes [22-24].

Oligosaccharides From Plant Cell Wall

Oligogalacturonides

Oligogalacturonides are also well known as elicitor molecules that derived from pectic polysaccharides of plant cell walls. Oligogalacturonides have been indicated to induce biosynthesis of phytoalexins in the cotyledons of soybean [25] and kidney bean [26], proteinase inhibitors in tomato leaves [27], and lignification of cucumber cotyledons [28] and cultured castor bean cells [29].

Xyloglucan Oligosaccharides

Xyloglucan is a major polysaccharide in cell walls of dicotyledons. Xyloglucan fractions were firstly extracted from cell walls of *Daucus carota* L. cells cultured in suspension [30]. The xyloglucan oligosaccharide could function as an "anti-auxin" to inhibit 2, 4-dichlorophenoxyacetic acid (2, 4-D)-induced elongation of etiolated pea stem segments [31]. This conclusion supports the hypothesis that xyloglucan oligosaccharide play an important role in growth regulation.

Glycopeptides

Glycopeptides generated by cleavage of yeast invertase with α-chymotrypsin were potent elicitors of ethylene biosynthesis and phenylalanini ammonia-lyase in tomato cells cultured in suspension. Active glycopeptide elicitor carries a high mannose N-linked glycan ($Man_{10-12}GlcNAc_2$) and the third α-1, 6-linked mannose residue is necessary for the elicitor activity. However, oligosaccharides released from the elicitor-active glycopeptides by N-glycanase could suppress the activity of the glycopeptide elicitors [32].

The Role of Oligosaccharides in Animal and Human

Difficult to Digest, Low Calorific Value

Oligosaccharides escape hydrolysis by intestinal digestive enzymes due to the configuration of their osidic bonds. Oligosaccharides can be fermented by anaerobic in the colon and have a lower net caloric value. Therefore, oligosaccharides are suitable sweeteners for special populations, such as obesity, diabetes and hypoglycemia patients [33].

Anti-Caries

Oral bacteria such as *Streptococcus mutans* can convert sucrose into acids through fermentation. These acids may cause demineralization in the surface of tooth and form dental plaque. The presence of certain types of dental plaque is considered a major contributory factor to the subsequent development of a number of pathological conditions such as dental caries and periodontal lesions. Oligosaccharides which can not be fermented by *Streptococcus mutans* were shown to be effective as anti-cariogenic agents in preventing bacterial adherence to teeth by inhibiting the formation of the bacterial plaque [34].

INFULENCE THE LIPID AND SERUM CHOLESTEROL

Ingestion of oligosaccharides such as inulin may influence the concentrations of serum and liver lipid, cecal short-chain fatty acid and the fecal excretion. The mechanisms of hypocholesterolemic effect of oligosaccharides may be that oligosaccharides could interfere with the cholesterol absorption in jejunum and ileum [35]. Oligosaccharides are beneficial to healthy population as well as those with cadiovascular diseases.

Improve Immunity

Oligosaccharides could stimulate systemic immune responses of humans and animals. Diets supplemented with nondigestible oligosaccharides (NOD), such as FOS and inulin increase mesenteric lymph node lymphocyte blastogenic activity in response to the T-cell stimulants, concanavalin A (ConA) and phytohaemagglutinin (PHA) [36]. Such diets are combined with shifts in cytokine profiles [37], increase Peyer's patch lymphocyte T-cell responses [36], and lower IgE production [36-37]. These conclusions conjuncted with the higher natural killer cells activities and the increased macrophage response to the intracellular pathogen Listeria for mice fed these diets, consistent with hypothesis that oligosaccharides trigger a shift to a greater dependence on TH1 cell-mediated immunity [36]. Furthermore, the elevated T lymphocyte functions of rodents fed diets with inulin and oligofructose were associated with increased resistance to chemical carcinogens [38] and tumor cells [39].

Anti-Tumor Activity

In order to obtain an oligomeric prodrug of 5-fluorouracil (5FU) with reduced side-effects, affinity for tumor cells and high antitumor activity, 5FU was covalently attached to three chito-oligosaccharides (COS) through hexamethylene spacer groups via carbamoyl bonds. Ouchi T et al. (1990) tested the ability of these compounds to extend the life of lymphocytic leukemia mice (following their intraperitoneal administration) and their anti-tumor effects on Meth-A fibrosarcoma or MH-134 hepatoma mice (following their subcutaneous administration) [40]. These compounds prolonged the survival time of p-388 leukemia mice and inhibited the growth of the solid tumor more effectively than either 5FU,

COS, or blends of 5FU and COS. The conjugate drug did not cause acute toxicity and drastic reduction of body weight.

APPLICATIONS

Oligosaccharides in the Functional Food Market

Oligosaccharides are sweeteners with a low caloric value [33]. Oligosaccharides are recognised as important prebiotic agents benefit the proliferation of bifidobacteria [3]. Many products containing oligosaccharides such as inulin and oligofructose, claiming to have beneficial effects on intestine health and general well-being, are starting to become popular in the European market. Inulin and oligofructose are legally classified as food or food ingredients in all countries in which they are used. They are well accepted for food use without limitations [41]. Thus, consumption of oligosaccharides in the sugar market is expected to continue to increase.

Oligosaccharides Application in Feed Industry

Oligosaccharides are applied as feed ingredients in many countries such as Japan, Canada, European. To reduce stress during weaning period of piglet and decrease the negative effects on the pig's health, subtherapeutic levels of feed grade antibiotics are used extensively. However, the use of antibiotics in animal feed is a cause for concern, because of the risk of selection of resistant strains of microorganisms. To maintain animal health and productivity in such a scenario, alternatives have to be evaluated [42]. Many studies have shown that oligosaccharides such as FOS could be used as feed grade antibiotics. Oligofructose are among the most studied prebiotic agents that benefit the proliferation of bifidobacteria [3] and resist pathogens like E. coli and Salmonella through secretion of volatile fatty acids and competition for nutrients and binding sites [43]. In addition, oligosaccharides could increase systemic immune reponses of animals [36]. Therefore oligosaccharides are accepted as feed grade antibiotics which stimulate fermentation in the intestinal tract and help protect the animal from colonization of threatening strains of bacteria.

Oligosaccharides Application in Agriculture

Humans have utilized pesticides to protect the crops for thousands of years. Before the 15th century, toxic chemical substance like mercury, arsenic and lead were applied to kill pests. Then nicotine sulfate, pyrethrum were extracted as insecticides. Until the 1970s, DDT was replaced by organophosphates and carbamates. Since then pyrethrin compounds became the dominant insecticide. These pesticides are effective but cause a lot of environmental concerns such as water polluion and soil contamination. New pepticides are designed to protect the environment and reduce health risks, including biological and botanical derivatives.

Oligosaccharides are a class of developing alternative bio-pesticides with characteristics of super efficacy, safety and compatibility with environment. Oligosaccharides have been demonstrated to be elicitors that stimulate plant defense mechanism and regulate the growth and development of crops. Oligosaccharides could be degraded easily and generate residues without any side effect on the environment.

Oligosaccharides Application in Pharmaceutics

Oligosaccharides have been demonstrated to resist tumor cells [39], stimulate systemic immune responses [36], inhibit bacterial growth [1] and promote calcium absorption in human gut [44]. A number of acute and subacute toxicity assays of oligosaccharides have been carried out. It is shown that oligosaccharides are products without toxic and side effects. Oligosaccharides can be used alone or together with other drugs for the treatment of cancer, cadiovascular diseases and diabetes.

Over the past decade, functional oligosaccharides have attracted attention of population. They can be extracted or obtained by enzymolysis, chemical degradation and chemical synthesis. Scientists have been working for large-scale production and application of oligosaccharides. The characteristics of oligosaccharides like high security, unique physiological function and non-pollution to the environment make oligosaccharides utilized widespreadly. The market of oligosaccharides is expanding rapidly.

References

[1] Jeon, Y.J., P.J. Park, and S.K. Kim, Antimicrobial effect of chitooligosaccharides produced by bioreactor. Carbohydr Polym, 2001. 44: p. 71-76.

[2] Christakopoulos, P., et al., Antimicrobial activity of acidic xylo-oligosaccharides produced by family 10 and 11 endoxylanases. Int J Biol Macromol, 2003. 31(4-5): p. 171-5.

[3] Hidaka, H., et al., Effects of fructooligosaccharides on intestinal flora and human health. Bifidobact Microflora, 1986. 5: p. 37-50.

[4] Okazaki, M., S. Fujikawa, and N. Matsumoto, Effect of xylooligosaccharide on the growth of bifidobacteria. Bifidobact Microflora, 1990. 9: p. 77-86.

[5] Katsuraya, K., et al., Synthesis of sulfated oligosaccharide glycosides having high anti-HIV activity and the relationship between activity and chemical structure. Carbohydr Res, 1999. 315(3-4): p. 234-42.

[6] Albersheim, P. and A.G. Darvill, Fragments of the plant cell wall have been discovered that serve as regulatory molecules. Sci Amer, 1985. 253(3): p. 58-64.

[7] Cote, F. and M.G. Hahn, Oligosaccharins: structures and signal transduction. Plant Mol Biol, 1994. 26(5): p. 1379-411.

[8] Shibuya, N. and E. Minami, Oligosaccharide signaling for defence responses in plant. . Physiol Mol Plant Pathol, 2001. 59: p. 223-233.

[9] Sharp, J.K., B. Valent, and P. Albersheim, Purification and partial characterization of a beta-glucan fragment that elicits phytoalexin accumulation in soybean. J Biol Chem, 1984. 259(18): p. 11312-20.

[10] Yamaguchi, T., et al., Differences in the recognition of glucan elicitor signals between rice and soybean: beta-glucan fragments from the rice blast disease fungus Pyricularia oryzae that elicit phytoalexin biosynthesis in suspension-cultured rice cells. Plant Cell, 2000. 12(5): p. 817-26.

[11] Barber, M.S., R.E. Bertram, and J.P. Ride, Chitin oligosaccharides elicit lignification in wounded wheat leaves. Physiol Mol Plant Pathol, 1989. 34: p. 3-12.

[12] Felix, G., M. Regenass, and T. Boller, Specific perception of subnanomolar concentrations of chitin fragments by tomato cells: induction of extracellular alkalinization, changes in protein phospholylation, and establishment of a refractory state. Plant J, 1993. 4: p. 307-316.

[13] Roby, D., A. Gadelle, and A. Toppan, Chitin oligosaccharides as elicitors of chitinase activity in melon plants. Biochem Bioph Res Co, 1987. 143: p. 885-892.

[14] Kaku, H., et al., N-acetylchitooligosaccharide elicitor expression of a single (1→3)-β-glucanase gene in suspension-cultured cells from barley (Hordeum vulgare). Physiol Plantarum, 1997. 100: p. 111-118.

[15] Ren, Y.Y. and C.A. West, Elicitation of Diterpene Biosynthesis in Rice (Oryza sativa L.) by Chitin. Plant Physiol, 1992. 99(3): p. 1169-78.

[16] Yamada, A., et al., Induction of phytoalexin formation in suspension cultured rice cells by N-acetylchitooligosaccharides. Biosci Biotechnol Biochem, 1993. 57: p. 405-409.

[17] Kikuyama, M., K. Kuchitsu, and N. Shibuya, Membrane depolarization induced by N-acetylchitooligosaccharide elicitor in suspension-cultured rice cells. . Plant Cell Physiol, 1997. 38: p. 902-909.

[18] Kuchitsu, K., M. Kikuyama, and N. Shibuya, N-acetylchitooligosaccharides, biotic elicitor for phytoalexin production, induce transient membrane depolarization in suspension-cultured rice cells. Protoplasma, 1993. 174: p. 79-81.

[19] Kuchitsu, K., et al., Transient cytoplasmic pH change and ion fluxes through the plasma membrane in suspension cultured rice cells triggered by N-acetylchitooligosaccharide elicitor. Plant Cell Physiol, 1997. 38: p. 1012-1018.

[20] Kuchitsu, K., et al., EPR evidence for generation of hydroxyl radical triggered by N-acetylchitooligosaccharide elicitor and a protein phosphatase inhibitor in suspension-cultured rice cells. . Protoplasma, 1995. 188: p. 138-142.

[21] Nojiri, H., et al., Involvement of Jasmonic Acid in Elicitor-Induced Phytoalexin Production in Suspension-Cultured Rice Cells. Plant Physiol, 1996. 110(2): p. 387-392.

[22] Minami, E., et al., Two novel genes rapidly and transiently activated in suspension-cultured rice cells by treatment with N-acetylchitoheptaose, a biotic elicitor for phytoalexin production. Plant Cell Physiol, 1996. 37(4): p. 563-7.

[23] Nishizawa, Y., et al., Regulation of the chitinase gene expression in suspension-cultured rice cells by N-acetylchitooligosaccharides: differences in the signal transduction pathways leading to the activation of elicitor-responsive genes. Plant Mol Biol, 1999. 39(5): p. 907-14.

[24] Takai, R., et al., Isolation and analysis of expression mechanisms of a rice gene, EL5, which shows structural similarity to ATL family from Arabidopsis, in response to N-acetylchitooligosaccharide elicitor. Plant Sci, 2001. 160(4): p. 577-583.

[25] Nothnagel, E.A., et al., Host-Pathogen Interactions : XXII. A Galacturonic Acid Oligosaccharide from Plant Cell Walls Elicits Phytoalexins. Plant Physiol, 1983. 71(4): p. 916-26.
[26] Dixon, R.A., et al., Elicitor-active components from French bean hypocotyls. Physiol Mol Plant Pathol, 1989. 34: p. 99-115.
[27] Farmer, E.E., et al., Oligosaccharide signaling in plants. Specificity of oligouronide-enhanced plasma membrane protein phosphorylation. J Biol Chem, 1991. 266(5): p. 3140-5.
[28] Robertsen, B., Elicitors of the production of lignin-like compounds in cucumber hypocotyls. Physiol Mol Plant Pathol, 1986. 28: p. 137-148.
[29] Bruce, R.J. and C.A. West, Elicitation of lignin biosynthesis and isoperoxidase activity by pectic fragments in suspension cultures of castor bean. Plant Physiol, 1989. 91(3): p. 889-97.
[30] Michael, E. and U.S. Hanns, Influence of a specific xyloglucan-nonasaccharide derived from cell walls of suspension-cultured cells of Daucus carota L. on regenerating carrot protoplasts. Planta, 1990. 182: p. 174-180.
[31] York, W.S., A.G. Darvill, and P. Albersheim, Inhibition of 2,4-dichlorophenoxyacetic Acid-stimulated elongation of pea stem segments by a xyloglucan oligosaccharide. Plant Physiol, 1984. 75(2): p. 295-7.
[32] Basse, C.W., K. Bock, and T. Boller, Elicitors and suppressors of the defense response in tomato cells. Purification and characterization of glycopeptide elicitors and glycan suppressors generated by enzymatic cleavage of yeast invertase. J Biol Chem, 1992. 267(15): p. 10258-65.
[33] Carabin, I.G. and W.G. Flamm, Evaluation of safety of inulin and oligofructose as dietary fiber. Regulatory Toxicology and Pharmacology, 1999. 30(3): p. 268-282.
[34] Speigel, J.E., et al., Safety and benefits of fructooligosaccharides as food ingredients. Food Technol, 1994. 48: p. 85–89.
[35] Kim, M., The water-soluble extract of chicory reduces cholesterol uptake in gut-perfused rats. Nutrition Research, 2000. 20(7): p. 1017-1026.
[36] Taga, T. and T. Kishimoto, Signaling mechanisms through cytokine receptors that share signal transducing receptors components. Curr Opin Immunol, 1995. 7: p. 17-23.
[37] Field, C.J. and M.I. McBurney, Interaction of fiber fermentation and immunology of the gastrointestinal tract. Iams Nutrition Symposium Proceedings, 1998: p. 523-532.
[38] Buddington, R.K., J. Donahoo, and C.H. Williams, The colonic bacteria and small intestinal nutrient transport of mice fed diets with inulin and oligofructose. Microbial Ecol Health Dis, 2001. 12: p. 233-240.
[39] Buddington, K.K., J.B. Donahoo, and R.K. Buddington, Dietary oligofructose and inulin provide protection against some enteric and systemic pathogens and cancer challenges. J Nutr, 2002. 132: p. 472-477.
[40] Ouchi, T., et al., Synthesis and antitumor activity of conjugates of 5-fluorouracil and chito-oligosaccharides involving a hexamethylene spacer group and carbamoyl bonds. Drug Des Deliv, 1990. 6(4): p. 281-7.
[41] Coussement, P.A., Inulin and oligofructose: safe intakes and legal status. J Nutr, 1999. 129(7 Suppl): p. 1412S-7S.

[42] Gebbink, G.A.R., et al., Effects of addition of fructooligosaccharide (FOS) and sugar beet pulp to weaning pigs diets on performance, microflora and intestinal health. Swine Day, 1999: p. 53-59.

[43] Koopman., J.P. Mroz, and B.A. Williams, Voeding en gezondheid van het maagdarmkanaal. Productschap diervoeder, 1999.

[44] Zafar, T.A., et al., Nondigestible oligosaccharides increase calcium absorption and suppress bone resorption in ovariectomized rats. J Nutr, 2004. 134(2): p. 399-402.

INDEX

C

D

E

F

N

O

P

Q

R

S

X

Y